Analytical Biochemistry

Analytical Biochemistry

Edited by **Artie Weissberg**

SYRAWOOD
PUBLISHING HOUSE

New York

Published by Syrawood Publishing House,
750 Third Avenue, 9th Floor,
New York, NY 10017, USA
www.syrawoodpublishinghouse.com

Analytical Biochemistry
Edited by Artie Weissberg

International Standard Book Number: 978-1-68286-038-0 (Hardback)

The publisher's policy is to use permanent paper from mills that operate a sustainable forestry policy. Furthermore, the publisher ensures that the text paper and cover boards used have met acceptable environmental accreditation standards.

Trademark Notice: Registered trademark of products or corporate names are used only for explanation and identification without intent to infringe.

Printed in the United States of America.

Contents

Preface

This book has been a concerted effort by a group of academicians, researchers and scientists, who have contributed their research works for the realization of the book. This book has materialized in the wake of emerging advancements and innovations in this field. Therefore, the need of the hour was to compile all the required researches and disseminate the knowledge to a broad spectrum of people comprising of students, researchers and specialists of the field.

Analytical biochemistry as a discipline is concerned with understanding the methods for analyzing various structures and processes in biological and biochemical sciences. The chapters included in this book are a compilation of topics ranging from the basic to the most complex advancements in the field of molecular and cell biology, human and plant genetics, etc., and also contains researches contributed by international experts. It will prove to be an asset for students, academicians, professionals, or readers in general interested in analytical chemistry.

At the end of the preface, I would like to thank the authors for their brilliant chapters and the publisher for guiding us all-through the making of the book till its final stage. Also, I would like to thank my family for providing the support and encouragement throughout my academic career and research projects.

Editor

D-3-hydroxybutyrate oxidation in mitochondria by D-3-Hydroxybutyrate dehydrogenase in *Tetrahymena pyriformis*

Omar Akil[2], Zakaria El Kebbaj[1,2], Norbert Latruffe[1]* and M'Hammed Saïd El Kebbaj[2]

[1]INSERM U866; Université de Bourgogne, Laboratoire de Biochimie Métabolique et Nutritionnelle, Faculté des Sciences, 6 Bd Gabriel, 21000 Dijon cedex, France.
[2]Laboratoire de Biochimie et Biologie Moléculaire, Université Hassan II - Aïn Chock, Faculté des Sciences, Casablanca, Morocco.

Tetrahymena pyriformis a ciliated protozoan, is considered as a good indicator of water pollution. However its energy supply is poorly understood. This work was focused on the metabolism of hydroxybutyrate through the study of the membrane bound mitochondrial NAD^+-dependent D-3-hydroxybutyrate dehydrogenase (EC. 1.1.1.30) (BDH), a ketone body catalysing enzyme involved in the interconversion of D-3-hydroxybutyrate to acetoacetate. Due to lack of informations, the physico-chemical properties and kinetic parameters of the enzyme were examined. The results are the following: 1) D-3-hydroxybutyrate is a good substrate for mitochondria. 2) The enzyme catalytic process follows a bi bi-ordered mechanism where the coenzyme binds first, then allowing the substrate linkage to the active site. 3) Two optimal pH values of 8 and 6.5 corresponding to D-3-hydroxybutyrate oxidation and to acetoacetate reduction respectively. On the other hand, pH changes affect the coenzyme binding to the active site. 4) The BDH activity was found strongly linked to submitochondrial vesicles indicating that the protozoan enzyme is membranous and could require lipids for its function as well as it is for the mammalian enzyme. Moreover, an optimal temperature (40°C) and a break appearing in the Arrhenius plot at 19°C were found. The break suggests a membrane lipid fluidity-dependency of BDH conformational change. 5) Several ligands of the active site including methylmalonate and succinate modulate the BDH activity and are competitive inhibitors toward D-3-hydroxybutyrate. 6) Divalent cations, Mg^{2+}, Mn^{2+} and Zn^{2+} protect BDH against thermal inactivation. The protection is the strongest in the presence of Zn^{2+}. Moreover, Ca^{2+} and Mg^{2+} are enzyme activators and modulate the substrate binding to the active site. On the other hand, EDTA, a chelating agent, inhibits the enzyme but prevents inhibition by substrate excess. This work provides new insights on the energy metabolism of *T. pyriformis* wild strain where D-3-hydroxybutyrate is a choice substrate where the properties of BDH have been established especially the activating role of non heavy divalent cations.

Key words: D-3-hydroxybutyrate dehydrogenase, ketone body, mitochondria, *Tetrahymena pyriformis*.

INTRODUCTION

In mammalian, the NAD^+-dependent D-3-hydroxybutyrate dehydrogenase (EC. 1.1.1.30) (BDH) is the ketone body converting enzyme located in the inner face of the mitochondrial membrane. Its physiological role appears to be dual, that is, in liver, to convert, acetoacetate into D-3-hydroxybutyrate, a reducing equivalent and a potential energetic fuel to be exported to extrahepatic tissues, especially in brain, heart, kidney and in muscle. In these latter tissues, BDH operates in the opposite way; secondly, to prevent acidification of high amount of acetoacetate, a strong acid form of ketone bodies which are overproduced during ketosis appearing during fasting, diabete mellitus, hyperlipidic diet. Concerning BDH properties, until now BDH was largely studied in several organisms and tissues, especially in rat liver (Latruffe and Gaudemer, 1974), beef heart (Nielsen et al., 1973), or in Jerboa

*Corresponding author. E-mail: latruffe@u-bourgogne.fr.

(Mountassif et al., 2008). BDH catalyses the reversible oxidation of BOH to AcAc in presence of NAD(H) as cofactor, according to an ordered bi bi mechanism where the coenzyme binds first to the enzyme catalytic site (Latruffe and Gaudemer, 1974; Nielsen et al., 1973). While the enzyme is soluble in bacteria (Bergmeyer et al., 1967), in contrast, in eucaryotes it is anchored in inner mitochondrial membrane (Latruffe and Gaudemer, 1974; Nielsen et al., 1973). The mitochondrial BDH requires lecithin for its activity where phosphatidylcholine have been demonstrated to induce a discrete but significant conformational change of the active site (El Kebbaj et al., 1986). On the other hand, in bacteria, the role of BDH in the energetic metabolism is to oxidize D-3-hydroxybutyrate to acetoacetate leading to the availability of NADH, the main electron donor molecule, to the membrane bound oxidative phosphorylation machinery producing ATP. In such procaryotic organism, D-3-hydroxyburate is produced from the hydrolysis of poly-β-hydroxybutyrate (PHB), an energy storage polymer.

In the recent past researchers paid attention to *Tetrahymena*, a unicellular ciliated lower eukaryotic of microorganism either in the field of phylogeny (Mukai and Endoh, 2004), mitochondrial dependent-apoptosis (Kobayashi and Endoh, 2003; Kobayashi and Endoh, 2005; Endoh and Kobayashi, 2006), heavy metal pollution (Rico et al., 2009) or mitochondrial linked-phospholipids metabolism (Tellis et al., 2003). Interestingly, BDH is also present in *Tetrahymena pyriformis* and considered as a good indicator of water pollution (Mountassif et al., 2006a). In this protozoan, the enzyme is located in the mitochondria (Conger and Eichel, 1965). However, its metabolic role is unknown. We postulate that acetyl-CoA produced from fatty acid oxidation cycle would be converted into acetoacetyl-CoA. Following this, the carbon chain of acetoacetyl-CoA would be incorporated into poly-β-hydroxybutyrate (PHB). Then when needed, the D-3-hydroxybutyrate produced from PHB hydrolysis would be oxidized by BDH as enzyme coupled to the mitochondrial electron transport chain.

Our group has shown that *T. pyriformis* BDH was inhibited by peroxisome proliferators as it is the case for rat liver BDH and that inhibition process occurs in the active site of the enzyme (El Kebbaj et al., 1995). However its energy supply is poorly understood. This work was focused on the metabolism of hydroxybutyrate through the study of the membrane bound mitochondrial NAD+ - dependent D-3-hydroxybutyrate dehydrogenase (EC. 1.1.1.30) (BDH), a ketone body catalyzing enzyme involved in the interconverting D-3-hydroxybutyrate to acetoacetate. Due to the lack of information concerning the use of BOH as mitochondrial energy substrate and the biochemical properties of the converting enzyme in protozoan, we isolated mitochondria of *T. pyriformis* in order to study its respiratory properties during the BDH oxidation process. Then, we determined BDH catalytic parameters, that is, reaction mechanism, themodynamic characteristics and sensitivity towards cations and cofac-

tors.

The aim of this study was to provide a better understanding in the biology and the metabolism of *T. pyriformis*, a water living unicellular organism which can be exposed to environmental changes, that is, substrate availability, salts, toxic, temperature and pH changes.

MATERIALS AND METHODS

Biological preparations

T. pyriformis, wild strain, was grown aerobically without shaking, in a broth medium containing 1.5% proteose-peptone, 0.25% yeast extract at 27 °C during one week (Latruffe et al., 1982). Preparations of coupled mitochondria were performed as described in (Weinbach, 1961) after washing the cells with the dry medium (0.58 g/l Na- citrate, 0.15 g/l NaH_2PO_4, 0.14 g/l Na_2HPO_4 and 0.2 g/l $CaCl_2$).

Submitochondrial vesicles (SMV) which are inside out inner mitochondrial membrane particles are prepared from mitochondria (Kielley and Bronk, 1958) by swelling mitochondria in 20 mM KH_2PO_4 hypotonic buffer and sonication in order to break mitochondrial membrane and release the matrix. Protein estimation was done using the Bio-Rad assay according to the method of Bradford (Bradford, 1976) with bovine serum albumin as a standard.

Oxygen consumption

Freshly isolated and intact mitochondria (10 mg per essay) were incubated and oxygen consumption was measured using a Clarke Electrode (Oxygraph Gilson). Respiratory control ratio and respiratory chain oxidative activities, in state 4 or in state 3, were measured at 37 °C in the Ernster medium containing 62 mM sucrose, 8 mM $MgCl_2$, 50 mM KCl, 20 mM Hepes, pH 7.3 as described in (Estabrook, 1967) using the following substrates either, glutamate + malate, D-3-hydroxybutyrate, succinate, α-ketoglutarate, citrate or pyruvate at 5 mM (state 3 was obtained with the presence of 0.2 mM ADP and 4 mM inorganic phosphate...).

Enzymatic activities measurements

BDH activity was measured as usual (Mountassif et al., 2006b) at 37 °C by following NADH production at 340 nm ($\varepsilon = 6.22 \times 10^3$ M^{-1}.cm^{-1}) using either SMV or frozen-thawed mitochondria (as indicated in the legends of figures) in a medium containing: 6 mM potassium phosphate, pH 8, 0.5 mM EDTA, 1.27% (v/v) redistilled ethanol, 0.3 mM dithiothreitol, in the presence of 2 mM NAD+ and 2.5 µg rotenone (final addition) to prevent NADH reoxidation by the respiratory chain. The activity was started by the addition of DL-3-hydroxybutyrate to 10 mM final concentration. Kinetic parameters of BDH were determined by measuring the initial rate at 37 °C (excepted if indicated conditions cf Figure 2) in a standard medium as above described for the oxidation of D-3-hydroxybutyrate using the following coenzyme and substrate concentrations:

[NAD$^+$] = 0.2, 0.4, 0.6 or 0.8 mM ; [D-3-hydroxybutyrate] = 1.25, 2.5, 4.5 or 10 mM; or in the same medium at pH 7 without NAD$^+$ and rotenone and in presence of varying acetoacetate concentrations (0.2, 0.4, 0.6 or 0.8 mM) and NADH concentrations (0.2, 0.4, 0.6 or 0.8 mM) (using 0.8 mM NADH, the absorbance was mesued at 366 nm ($\varepsilon = 3.0 .10^3$ M^{-1}.cm^{-1}) in order to measure optical density). In all cases, different NaCl concentrations were ajusted to the medium in order to keep constant salt concentration. Graphical determination of parameters was made from mathematical analysis according to

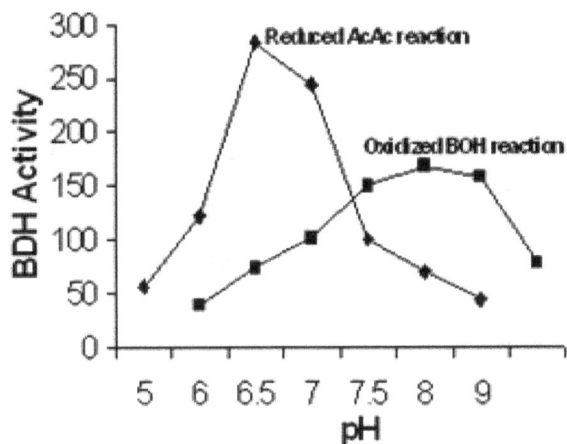

Figure 1. pH dependency of BDH activity on forward –black squares■- or reverse –black diamonds♦- reaction. Aliquots of submitochondrial vesicles (1.7 mg protein/ml) were used to measure BDH activity in function of pH with a buffer coktail containing 50 mM of Tris-HCl, Mes, Hepes, and potassium phosphate ; pH was adjusted at 37°C either with KOH or HCl. BDH activity was expressed in nmol NAD reduced or NADH oxidized. mn^{-1}. mg protein $^{-1}$. Data correspond to one of three representative experiments.

was measured as reported (Bruni et al., 1965) by estimation of inorganic phosphate using the "Fiske and Subbarow" method.

Chemicals

DL-3-hydroxybutyrate (sodium salt) was purchased from Fluka (Buchs Switzerland); NAD+ (free acid) and NADH were from Boehringer (Mannheim, Germany); acetoacetate, methylmalonate and succinate were from Sigma (St Louis, USA) and all other chemicals were of analytical grade.

Statistical data analysis

In each assay, the experimental data represent the mean of three independant assays. Student test was used as value lower than 0.05 was considered significant.

Remark

Despite that *T. pyriformis* grows experimentally at 27°C or possibly ambiante varying temperatures in its natural biotope, we measured enzymatic activities at the usual temperature of 37°C since Figure 2 shows that BDH activity inceases almost regularly from 15 to 37°C (optimum temperature activity).

RESULTS

Mitochondrial oxidative characteristics of *T. pyriformis*

Coupled mitochondria were purified from *T. pyriformis* wild strain. Mitochondrial oxidative activities were measured using several substrates, D-3-hydroxybutyrate, succinate, 2-ketoglutarate, citrate, pyruvate and a mixture

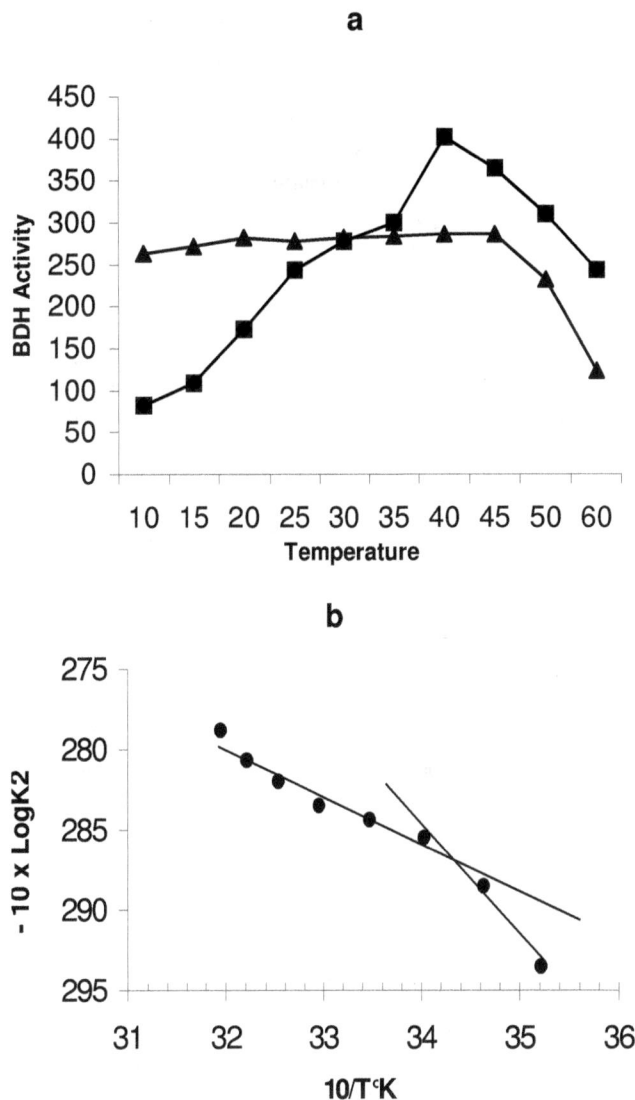

Figure 2. Effect of temperature on BDH activity. For thermal activation –black squares■ of the enzyme, aliquots of submitchondrial vesicles (1.9 mg protein/ml) were used to measure BDH activity at different temperatures (from 10 to 60°C). For thermal inactivation –black diamonds♦ of the enzyme, aliquots of submitochondrial vesicles (1.9 mg protein/ml) were incubated for three minutes at each fixed temperature (from 10°C to 60°C) before performing BDH activity assay at 37°C. BDH activity was expressed in nmol NAD reduced. min^{-1}. mg protein^{-1}. (Figure 2a). Data are from one of five representative experiments. For Arrhenius plot of BDH activity (Figure 2b), the logarithm of the catalytic activity (Ln k 2) is plotted as a function of the reciprocal of temperature (in Kelvin degrees, T°K). The lines represent the best linear regression of data points. The slopes of the line correspond to two activation energy values (-Ea/R). The arrow indicates the break in the activity. The adjacent numbers correspond to apparent Ea values (in Kcal/mol). Data are from two representative experiments.

of glutamate + malate.

As shown in Table 1, the oxygen consumption is the highest with D-3-hydroxybutyrate as substrate. The stimulation of oxidative activities in presence of ADP is si-

Table 1. Mitochondrial oxygen consumption and respiratory control ratio of *T. pyriformis*.

Mean of oxygen consumption (n. atoms O_2 / min / mg prot) at 37°C	Respiratory control ratio		
Substrates	State 4	State 3	State 3 / State 4
Glutamate + Malate	1.22	2.34	1.91
3-hydroxybutyrate	1.70	3.34	1.96
Succinate	1.29	2.20	1.70
2-ketoglutarate	1.27	3.22	2.53
Citrate	1.26	2.16	1.71
Pyruvate	0.95	3.02	3.16

For experimental conditions, see materials and methods. Freshly isolated and intact mitochondria were used in this study. Assays have been done in duplicate and experiments have been repeated twice. No significant variations were observed in these conditions.

Table 2. Specific activities of mitochondrial BDH, SDH and ATPase from *T. pyriformis*.

Enzyme	Specific activity at 37°C (n.mol. product / min / mg prot.)	
	Control	+CAP
BDH	110.15 ± 16.41	107.81 ± 3.12 (x 0.98)
SDH	39.84 ± 3.91	64.84 ± 5.86 (x 1.63)
ATPase	208.59 ± 23.43	100 ± 33.2 (x 0.48)

For experimental conditions, see materials and methods. The specific activities correspond to NADH oxidation, DCIPH2 oxidation or inorganic phosphate production for BDH, SDH and ATPase respectively. The measurements using frozen-thawed mitochondria (5 mg protein) have been done in triplicate for BDH and SDH, and in duplicate for ATPase. Chloramphenicol (CAP), a mitochondrial protein synthesis inhibitor was added in the *T. pyriformis* medium during growth at 50 µg/ml final concentration. Experiments were repeated twice.

milar for BOH and all other substrates. The specific activities of inner mitochondrial bound enzymes were reported in Table 2. As shown, BDH activity is higher than SDH activity but lower than ATPase activity. The presence of chloramphenicol (CAP), a mitochondrial protein synthesis inhibitor (Turner and Lloyd, 1971) in the culture medium, stimulates SDH activity and decreases ATPase activity but does not significantly modify BDH activity.

Catalytic mechanism of BDH

Following the Lineweaver and Burk analysis (Lineweaver and Burk, 1934), the obtained kinetic parameters are summarized in Table 3. It appears, in results not reported, using Cleland analysis (Cleland, 1963), that BDH from *T. pyriformis* follows an ordered bi bi catalytic mechanism where NADH is a competitive inhibitor towards NAD+ (Ki = 0.047 mM) and non competitive towards D-3-hydroxybutyrate (Ki = 0.11 mM), while acetoacetate is non competitive towards both BOH (Ki = 2.65 mM) and NAD+ (Ki = 1.8 mM).

Effect of pH on BDH activity

The variation of BDH activity versus medium pH is shown in Figure 1. The highest activity was obtained at pH 8 for the conversion of D-3-hydroxybutyrate to acetoacetate, while an optimum pH of 6.5 was observed for the conversion of acetoacetate to D-3-hydroxybutyrate.

Effect of temperature on BDH activity

The study of enzyme stability towards temperature is reported in Figure 2. It can be seen that BDH denaturetion process occurs at 40°C. On the other hand, the Arrhenius plot reveals a break at 19°C with two activation energy values, that is, Ea = 6.77 Kcal.mol^{-1} at temperatures higher than the break and Ea = 13.53 Kcal .mol^{-1} at temperatures lower than the break.

Effect of ligands and cations on BDH thermal stability and activity

The effect of temperature on enzyme stability was studied in the presence of cations, coenzyme or substrate. It appeared (Table 4) that Zn^{2+} strongly protects BDH activity against denaturation (93% after 20 min of incubation). A low but significant protection was also observed with Mg^{2+}, Mn^{2+}, Ca^{2+} and with NAD(H). Thermodynamic parameters of the NAD$^+$ binding to BDH were determined at 37°C. The calculated values were the following: ΔG = -3.91 Kcal.mol^{-1}, ΔH = 5.97 Kcal.mol^{-1}, ΔS = 32.18 cal

Table 3. Apparent kinetic parameters of mitochondrial BDH from *T. pyriformis* in forward (oxidation) and in reverse (reduction) reactions.

Forward reaction (Vmax = 130 nmol NADH produced / mg of protein / min)			Reverse reaction (Vmax = 280 nmol NADH oxidized / mg of protein / min)		
K_M NAD$^+$ (mM)	K_D NAD$^+$ (mM)	K_M BOH (mM)	K_MNADH (mM)	K_D NADH (mM)	K_MAcAc (mM)
0.33 ± 0.04	2.5 ± 0.5	2.22 ± 0.13	0.11 ± 0.03	1.11 ± 0.05	1.15 ± 0.07

Aliquots of submitochondrial vesicles (SMV) (3 - 5 mg/ml) were used to measure BDH activity at 37°C at various substrates concentrations as indicated in materials and methods section. The mean values are given ± SEM as the result of three independent experiments.

Table 4. Effect of coenzymes, substrates and cations on BDH activity and against thermal inactivation.

	Remaining BDH activity (%) after preincubation at 37°C			
	0 min (instant measurement)	5 min	10 min	20 min
Control	100	60 ± 10	40 ± 10	20 ± 3.0
NAD$^+$ (1.5 mM)	-----	73.7 ± 10	52.6 ± 15	25 ± 7.0
NADH (1.5 mM)	-----	100 ± 00	75 ± 3.0	30 ± 12
NADP$^+$ (2 mM)	-----	57 ± 25	35.7 ± 11	14.3 ± 10
BOH (1.3 mM)	-----	60 ± 7.0	50 ± 14	20 ± 7.0
AcAc (1 mM)	-----	66 ± 8.0	41 ± 8.0	20 ± 10
MgCl$_2$ (3 mM)	180 ± 10 (+ 80%)	63 ± 7.0	63 ± 9.0	47.3 ± 6.0
Control + EDTA 1.66 mM	50	----------	----------	----------
Control + EDTA 1.66 mM + MgCl$_2$ (1 mM)	90	----------	----------	----------
CaCl$_2$ (3 mM)	150 ± 10 (+ 50%)	84.2 ± 4.0	63 ± 17	26 ± 4.0
MnCl$_2$ (3 mM)	100 (0)	89 ± 8.0	58 ± 8.0	52.6 ± 3.0
ZnCl$_2$ (3 mM)	98.5 (- 1.5%)	93 ± 8.0	93 ± 4.0	93 ± 10
NaCl (3 mM)	100 ± 6 (0%)	62 ± 16	45 ± 6.0	22 ± 9.0

Submitochondrial vesicles (SMV) (3 mg protein / ml) were preincubated in 50 mM Hepes, pH 7.4 at 37°C in presence or in absence of effectors. Aliquots were removed at different times to measure BDH activity, as described in materials and methods. Results are the means ± SEM of three separate experiments.

Table 5. Effect of divalent cations (Mg^{2+} and Ca^{2+}) on BDH kinetic parameters and on BDH competitive inhibitor-inhibition constant.

Kinetic constants	K_M NAD$^+$ (mM)	K_D NAD$^+$ (mM)	K_M BOH (mM)	Ki methyl malonate (mM)
Control	0.33 ± 0.04	2.22 ± 0.02	2.5 ± 0.1	0.12 ± 0.02
+ Mg^{2+} (2.5 mM)	0.22 ± 0.03[(NS)]	1.25 ± 0.36[(NS)]	1.29 ± 0.38[(NS)]	3.33 ± 0.15[(*)]
+ Ca^{2+} (1.5 mM)	0.34 ± 0.10[(NS)]	1.33 ± 0.4[(NS)]	1.16 ± 0.08[(NS)]	1.50 ± 0.2[(*)]

Aliquots of submitochondrial vesicles (3 mg protein/ ml) were used to determine kinetic parameters in the presence or in the absence of cations, at 37°C and at pH 8 (see Materials and Methods, and legend of table I). The mean values ± SEM represent three independent experiments. (NS): non significant differences (*): Statistically significant differences

cal.mol^{-1} and Ea = 6.77 Kcal.mol^{-1}.

Two D-3-hydroxybutyrate structural analogs (methylmalonate and succinate) were tested as competitive inhibitors towards BOH; methylmalonate is the strongest inhibitor as shown by its inhibition constant value (Ki = 0.12 mM) as compared to the one for succinate (Ki 1.8 mM). This difference could be due to the presence of the branched methyl group beared by methylmalonate.

The effects of some monovalent and divalent cations on BDH activity were also studied. One can see in Table 4 that Ca^{2+} and Mg^{2+} are the only activators. These two cations increase the affinity binding of NAD$^+$ (decrease of KD) and decrease the affinity binding of BDH (strong increase of Ki methylmalonate binding) (Table 5). In the order to confirm the activating role of Mg^{2+}, the effect of EDTA, a chelating compound, was tested. It was found that EDTA inhibits BDH activity (Table 4) and this inhibition was reversed by the addition of Mg^{2+}. The effect of Mg^{2+} or EDTA is a concentration dependent process where high concentrations of Mg^{2+} lead to a decrease of BDH activity (not shown). Moreover, the presence of EDTA leads to a lack of catalytic process inhibition in excess of D-3-hydroxybutyrate or NAD+ (not shown). The maximum velocity is reduced at an average of 20%. In this case, two molecules of each substrate bind to the enzyme and their dissociation constants were estimated following the Kuhn's method (Kuhn, 1974) and corresponded to KD = 3.1 mM and KD'= 1.07 mM for NAD^+, and KD = 2.08 mM and KD'= 19.95 mM for BOH.

DISCUSSION

Mitochondrial oxidative characteristics of *T. pyriformis* were investigated following the oxidation of several substrates and by measuring the activities of BDH, SDH and ATPase in absence or in presence of chloramphenicol (CAP) since it is known that the presence of CAP in the growth medium modifies the phopholipid metabolism (Turner and Lloyd, 1971) but also inhibits mitochondrial protein synthesis. BDH and SDH are encoded by nuclear genes while several ATPase (F1) subunits are encoded by mitochondrial DNA. Thus, as expected the presence of CAP gives only to a decrease of ATPase activity. It must be pointed out that frozen-thawed mitochondrial leads to the formation of mixed particles when most of them are inside-out mitochondrial inner membrane. In this case, swelling interference on substrate permeability and accessibility can be considered negligible since BDH, SDH and ATPase activities are all located on the inner face of the inner mitochondrial membrane.

Concerning BDH, we found (not shown) that the activity was linked to submitochondrial vesicles (SMV) and that the freezing and thawing of SMV leads to the progressive loss of the activity. The activity could be recovered by the addition of phospholipid vesicles in accordance with the need of lipid requirement of this enzyme (not shown). Related to this, the lack of CAP effect on BDH activity was due to the fact that the possible phospholipid changes surrounding BDH in mitochondria remain unchanged after CAP treatment while it does for SDH since an activation is observed. Alternatively, the stimulation of SDH activity after CAP treatment may reveal a compensation of the decrease of oxidative phosphorrylation by an increase of succinate dehydrogenase units. The kinetic study of BDH revealed that the enzyme catalytic mechanism follows an ordered bi bi. The same mechanism was previously shown for BDH from other species, that is, rat liver (Latruffe and Gaudemer, 1974), beef heart (Nielsen et al., 1973) and bacterial *Rhodopseudomonas spheroides* (Preuveneers et al., 1973).

The highest activity obtained at pH 8 for the D-3-hydroxybutyrate oxidation and at pH 6.5 for the acetoacetate reduction was similar with BDH of other species like bacterial *Pseudomonas lemoignei* (Delafield et al., 1965) and

rat liver (Latruffe and Gaudemer, 1974). In results not shown, we found that the NAD+ binding increased at optimum pH while the affinity for D-3-hydroxybutyrate is not significantly affected. This was as a result of a pH induced - change in the coenzyme binding site. Such change has been previously reported by our group in rat liver BDH active site concomitant with a charge transfer (El Kebbaj and Latruffe, 1997).

The BDH denaturation process occurs at 40 °C and the break in Arrhenius plot at 19 °C with two activation energy values. These values are lower than those obtained for rat liver BDH (Latruffe and Gaudemer, 1974). This break suggests either a change of tertiary structure of the enzyme, as proposed for rat liver BDH (El Kebbaj and Latruffe, 1986) or a phase transition dependency of phospholipid surrounding the enzyme, as reported in other systems (Lenaz et al., 1972). The thermodynamic parameters values (Ross et al., 1981) are in agreement with the existence of hydrophobic interactions between coenzyme and BDH. The protection by NAD+ against thermal denaturation is in accordance with results for rat liver BDH (Wise and Lehninger, 1962). In contrast NADP+ (an inactive coenzyme for BDH) and substrate alone (BOH or AcAc) do not exhibit thermal protection. This is in agreement with the lack of substrate binding to the active site without the present of the specific coenzyme.

The activating and protecting effects of divalent cations, especially Ca^{2+} and Mg^{++} can be compared with the effect of such cations on protein conformation previously reported for several mitochondrial enzymes : L-glycerol-3-phosphate dehydrogenase from different mam-malian tissues (Beleznai et al., 1988) and pig heart 2-ketoglutarate dehydrogenase complex (Panov and Scarpa, 1996). The inhibition of BDH activity by EDTA and the removal of this inhibition restored by the addition of Mg^{2+} show that the EDTA exhibits its effect through chelating action. Similar results were reported for BDH from bacterial Rhodopseudomonas spheroides (Bergmeyer et al., 1967). All these results are in favor of the implication of divalent cations on the structure-function relationships of BDH. This is supported by the fact that the NAD^+ dissociation constant and the methylmalonate inhibition constant were modified in the presence of the two studied cations (Ca^{2+} and Mg^{2+}) where the highest change occurs with Ca^{2+} (Table 5). These two cations may exhibit their effect on BDH binding site either directly or indirectly by modifying the membrane structure as previously reported for beef heart BDH (Mc Intyre et al., 1988) where calcium or magnesium cations induces enzyme conformational changes. Interestingly, the positive and protecting effects of non heavy metal cations contrast with the inhibitory effect of the water pollutant heavy metal cations (Mountassif et al., 2006a; Rico et al., 2009).

In conclusion, the present work provides new insights on the energy metabolism of *T. pyriformis* wild strain where D-3-hydroxybutyrate is a choice substrate. Moreover, the properties of BDH reveal the following features: 1. the BDH kinetic study shows that the enzyme follows a bi bi order catalytic mechanism; 2. the enzyme is sensitive to several ligands. Indeed, Ca^{2+} and Mg^{2+} are activators, while Zn^{2+} strongly stabilizes the BDH against thermal inactivation; 3. EDTA inhibits the enzyme while leads to the lack of inhibition process by excess of substrate or of coenzyme; 4. succinate and methylmalonate are competitive inhibitors where methylmalonate gives the strongest effect; 5. BDH is sensitive to pH and temperature, where pH affects the binding of the coenzyme while the temperature dependency shows a break near 19°C. Such break suggests a conformational change of BDH in relationship with change of phospholipid physical state surrounding BDH.

ACKOWLEDGMENTS

This work has been supported in part by the programme d'appui à la Recherche PARS Biologie n° 134 and PROTARS III Morocco and the Action Intégrée Franco-Marocaine (n° MA/01/22 and MA/05/134). We thank Dr Roland Perasso (Orsay) and Jean-Jacques Curgy (Lille) for their help in the initial experiments on the mitochondria from different Tetrahymenas pyriformis strains.

ABBREVIATIONS

AcAc, Acetoacetate; **BDH**, D-3-Hydroxybutyrate dehydrogenase (EC 1.1.1.30); **BOH**, DL-3-hydroxybutyrate; **CAP**, Chloramphenicol; **DCIP (H2)**, Dichloroindolphenol oxidized and reduced form; **EDTA**, Ethylenediamine tetraacetic acid; **Hepes**, 4-(2-hydroxyethyl)-1-piperazine ethane sulfonic acid; **Mes**, 4-(N-morpholino) ethanesulfonic acid; **NAD + (H)**, Nicotinamide adenine dinucleotide oxidized/reduced forms; **SDH**, Succinate dehydrogenase; **SMV**, Submitochondrial vesicles; **T. pyriformis**, Tetrahymena pyriformis; **Tris**, Trihydro-xymethyl-aminomethane.

REFERENCES

Beleznai Z, Szalay L, Jancsik V (1988). Ca^{2+} and Mg^{2+} as modulators of mitochondrial L-glycerol-3-phosphate dehydrogenase. Eur. J. Biochem . 170: 631-639,.

Bergmeyer HU, Gawehn K, Klozsch H, Krebs HA, Williamson DH (1967). Purification and properties of cristalline 3-hydroxybutyrate dehydrogenase from Rhodopseudomonasspheroides. Biochem. J .102: 423-431.

Bradford M (1976). A rapid and sensitive method for the quantitation of microgram quantities of protein utilizing the principle of protein dye binding. Anal Biochem 73: 248-254,.

Bruni , Luciani S, Bortignon (1965). Comparative reversal by adenine nucleotides of atractyloside effect on mitochondrial energy transfert. Biochim Biophys. Acta. 97: 434 – 441,.

Cleland WW (1963). The kinetics of enzymes catalysed reaction with two or more substrates or products. I. nomenclature and rate equations. Biochim. Biophys. Acta . 67: 104-137.

Conger NE, Eichel HJ (1965). Solubilisation and some properties of beta-hydroxybutyrate dehydrogenase from Tetrahymena pyriformis. Fed Proc 24: 370.

Delafield FD, Cooksey KE, Doudoroff M (1965). D-3-hydroxybutyric dehydrogenase and dimer hydrolase of Pseudomonas lemoignei. J.

Biol. Chem . 240: 4023-4027.

El Kebbaj MS, Latruffe N, Monsigny M, Obrenovitch A. (1986.) Interactions between apo-(D-3-hydroxybutyrate dehydrogenase) and phospholipids studied by intrinsic fluorescence. Biochem. J. 237: 359-364,

El Kebbaj MS, Cherkaoui Malki M, Latruffe N. (1995) Effect of peroxisome proliferators and hypolipemic agents on mitochondrial membrane linked D-3-hydroxybutyrate dehydrogenase (BDH). Biochem. Mol. Biol. Inter. 35 : 65-77,

El Kebbaj M S, Latruffe N. (1997). Alkylation at the active site of the D-3-hydroxybutyrate dehydrogenase (BDH), a membrane phospholipid-dependent enzyme, by 3-chloroacetyl pyridine dinucleotide (3-CAPAD). Biochimie 79: 37-42,

Endoh H, Kobayashi T(2006). Death harmony played by nucleus and mitochondria: nuclear apoptosis during conjugation of tetrahymena. Autophagy 2:129-31,

Estabrook R.W (1967). Mitochondrial respiratory control of the polarographic measurement of ADP : O ratios. In : R. W. Estabrook and M. E. Pullman (eds). Methods in Enzymology. Acad Press, New York. 10: 41-47.

Kielley WW, Bronk JR. (1958). Oxidative phosphorylation in mitochondrial fragment obtained by sonic vibration. J. Biol .Chem . 230: 521-533,

Kobayashi T, Endoh H (2003) Caspase-like activity in programmed nuclear death during conjugation of Tetrahymena thermophila. Cell Death Differ.10: 634-40,.

Kobayashi T, Endoh H (2005). A possible role of mitochondria in the apoptotic-like programmed nuclear death of Tetrahymena thermophila. FEBS J. 272: 5378-87.

Kole G, Roessner A, Themann H (1976). Effect of clofibrate (-p-chlorophenoxy-isobutyryl-ethyl-ester) on male rat liver. Virchows Arch Abst B 22: 73-87,.

Kuhn W, (1974). Cinétique et mécanisme d'action. In : Masson et Cie (eds.). L. Penasse "les enzymes". Paris, p. 34.

Latruffe N, Gaudemer Y. (1974). Propriétés et mécanisme cinétique de la D-3-hydroxybutyrate déshydrogénase de particules sous-mitochondriales de foie de rat ; effet comparés de différents agents thiols. Biochimie 56: 435-444,.

Latruffe N, Perasso R, Curgy JJ (1982). Variations de l'activité d'enzymes respiratoires chez une souche sauvage de Tetrahymenas pyriformis et chez son mutant mitochondrial resistant au chloramphénicol (CAP). Biol. Cell. 44: 73a,.

Lenaz G, Sechi AM, Parenti-castelli G, Ladi L, Bertoli E (1972). Activation energies of different mitochondrial enzymes breaks in Arrhenius plots of membrane-bound enzymes occur at different temperature. Biochem Biophys Res Commun 49: 536-542,.

Lineweaver M, Burk D (1934). The determination of enzyme dissociation constants. J. Am .Chem .Soc. 56: 658-666.

Mc Intyre JO, Latruffe N, Brenner SC, Fleischer S (1988). Comparison of D-3-hydroxybutyrate dehydrogenase from bovine heart and rat liver. Arch Biochem Biophys 262: 85-98.

Mountassif, D, Kabine, M, Manar R, Bourhim N, Zaroual Z, Latruffe N, El Kebbaj MSE (2006). Physiological, morphological and metabolic changes in Tetrahymena pyriformis for the in vivo cytotoxicity assessement of metallic pollution: Impact on D-beta-hydroxybutyrate dehydrogenase. Ecological Indicators 7: 882-94.

Mountassif D., Kabine E, M.. Latruffe N. & El Kebbaj M.S.E (2006). Characterization of two D-3-hydroxybutyrate dehydrogenase populations in heavy and light mitochondria from jerboa (Jaculus orientalis) liver. Comp Biochem. Physiol .B 143: 285-293.

Mountassif, D, Andreoletti P, El Kebbaj, Z, Moutaouakkil A, Cherkaoui Malki M, Latruffe N, El Kebbaj MSE (2008), A new purification method for characterization of the mitochondrial membrane-bound D-3-hydroxybutyrate dehydrogenase from Jaculus orientalis BMC .Biochem. sept 30 ; 9(1): 26.

Mukai A, Endoh H (2004). Presence of a bacterial-like citrate synthase gene in Tetrahymena thermophila: recent lateral gene transfers (LGT) or multiple gene losses subsequent to a single ancient LGT? J .Mol .Evol. 58: 540-549.

Nielsen MC, Zahler WL, Fleischer S. (1973). Mitochondrial D-3-hydroxybutyrate dehydrogenase.IV. Kinetic analysis of reaction mechanism. J. Biol .Chem . 248: 2556-2562.

Panov A, Scarpa A (1996). Independent modulation of activity of 2-ketoglutarate dehydrogenase complex by Ca 2+ and Mg 2+. Biochemistry 35 : 427-432,.

Preuveneers MJ, Peacock D, Crook EM, Clark JB, Brocklehurst K (1973). D-3-hydroxybutyrate dehydrogenase from Rhodopseudo-monas spheroides . Kinetic mechanism from steady-state kinetics of the reaction catalysed by the enzyme in solution and covalently attached to diethylaminomethyl-cellulose. Biochem. J. 133: 133-157,

Rico D, Martín-González A, Díaz S, de Lucas P, Gutiérrez JC (2009). Heavy metals generate reactive oxygen species in terrestrial and aquatic ciliated protozoa. Comp. Biochem. Physiol. Toxicol. Pharmacol.149: 90-6.

Ross PD, Subramaniam S (1981) Thermodynamics of protein association reactions: forces contributing to stability. Biochemistry 20: 3096-3102.

Sekuzu I, Jurtshuk P, Green DE (1961). On the isolation and properties of D-3-hydroxybutyrate dehydrogenase of beef heart mitochondria. Biochem. Biophys. Res. Commun. 6: 71-75,.

Tellis C, Pantazi D, Ioachim E, Galani V, Lekka ME. (2003). Locali-zation of an alkyl-acetyl-glycerol-CDP-choline: cholinephosphotrans-ferase activity in submitochondrial fractions of Tetrahymena pyriformis. Eur. J .Cell Biol. 82: 573-8.

Turner G, Lloyd D. (1971) The effect of chloramphenicol on growth and mitochondrial function of ciliate protozoan, Tetrahymena pyriformis strain ST. J. Microbiol. 67: 175 –188,.

Weinbach EC. (1961). A procedure for isolating stable mitochondria from rat liver and kidney. Ann. Biochem. 2: 335-343,.

Wise JB, Lehninger AL. (1962). The stability of mitochondrial D-3-hydroxybutyrate dehydrogenase and its relationship to respiratory chain. J. Biol. Chem. 237: 1363-1370,.

Comparative effects of the leaves of *Vernonia amygdalina* and *Telfairia occidentalis* incorporated diets on the lipid profile of rats

C. E. Ugwu[1]*, J. E. Olajide[1], E. O. Alumana[2] and L. U. S. Ezeanyika[2]

[1]Department of Biochemistry, Kogi State University, Anyigba, Nigeria.
[2]Department of Biochemistry, University of Nigeria, Nsukka, Enugu State, Nigeria.

The study compared the effects of the leaves of *Vernonia amygdalina* (VA) and *Telfairia occidentalis* (TO) incorporated diet on the lipid profiles of rats. The rats were fed for 28 days on diets specially formulated to contain 5, 15 and 30% by weight respectively of the leaves of each plant while the control group was fed standard rat diet. The serum total cholesterol (TC), triacylglycerol (TG), high-density lipoprotein cholesterol (HDL-C) and low- density lipoprotein cholesterol (LDL-C) were determined on blood samples collected on the 28th day using standard methods. The results showed that at the 5 and 15% treatments, TO significantly lowered the serum TC level relative to the effect of VA ($P<0.05$). The results also showed that the effects of VA and TO on TG and LDL-C were similar. For HDL-C, the TO diet preparation induced a significantly higher serum HDL-C level relative to the effect of the VA diet at the 15% treatment. Overall, the incorporation of *V. amygdalina* and *T. occidentalis* in diet preparations were anti-lipidaemic where TO showed a greater effect compared to the VA diet preparation.

Key words: *Vernonia amygdalina, Telfairia occidentalis*, lipid profile.

INTRODUCTION

Hypercholesterolaemia is recognized as an important risk factor in the development of coronary artery disease (Wald and Law, 1995; Nagra et al., 2003). The clinical manifestation of cholesterol build-up in arteries servicing the heart muscle cause more death and disability than all types of cancer combined (Lloyed-Jones et al., 2009; Daniels et al., 2009). This is an important outcome for a common polycyclic lipid with the humble function of maintaining the permeability and fluidity of cell membranes. Other lipid related risk factors, which have also been implicated in the development of chronic artery disease (CAD), include increased serum triacylglycerols, low density lipoprotein-cholesterol (LDL-C) and low level of high density lipoprotein-cholesterol (HDL-C) (Durrington et al., 1988; Lamendala, 2000). The treatment of hypertension

has failed to show definitive effect on the incidence of coronary heart disease, which has aroused interest in lipid metabolism in hypertensive therapy (Karim et al., 2004). A number of African indigenous plants have been credited with a lot of chemotherapeutic potentials (Farombi, 2003).

Vernonia amygdalina (Compositae) also called bitter leaf in Nigeria because of its bitter taste, is a shrub that grows predominantly in Tropical Africa. The leaves have found relevance in traditional folk medicine as an antihelminth, a laxative herb and an antimalarial as they are known as quinine substitute (Farombi, 2003). It is also used in the treatment of cough and hypertension (Amira and Okubadejo, 2002).

Telfairia occidentalis (Cucurbitacea) leaves and young shoots are frequently eaten as a potherb (Tindal, 1968; Okigbo, 1977; Okoli and Mgbeogu, 1983). The root and leaves have been shown to contain highly toxic alkaloids and saponins (Alada, 2000). In Nigeria, the herbal preparation of the plant has been employed in the treatment

*Corresponding author. E-mail: ugwuchidiksu@yahoo.com.

of sudden attack of convulsion, malaria and anaemia (Gbile, 1986).

Following the recorded pharmacological outcomes of the leaves of these plants, the present study was designed to compare the effect of the two diet preparations on the lipid profile of rats.

MATERIALS AND METHODS

Animals

20 albino rats (Wistar strain) that weighed between 93-120 g were used as the experimental animals. The rats were kept in cages for two weeks to acclimatize, and were allowed free access to food and water *ad libitum*. The protocol was in line with the guidelines of the National Institute of Health (NIH) (NIH Publication 85-23, 1985) for laboratory animal care and use. The experimental animals were randomly distributed into four groups of five animals each. Group 1 rats were fed standard rat diet (Vital Feeds, Nigeria), and served as the control, while groups 2, 3 and 4 were fed on diets that contained 5, 15 and 30% by weight of *V. amygdalina* leaves respectively for 28 days. The protocol was repeated with *T. occidentalis* leaves for another set of rats.

Feed formulation

The leaves of *V. amygdalina* and *T. occidentalis* were purchased from a local market in Anyigba, Kogi State, Nigeria. The botanical identification and authentication were confirmed at the Department of Biological Sciences, Kogi State University, Anyigba. The leaves were dried at room temperature for 2 weeks to a constant weight and then powdered in a mortar. The standard rat diet was similarly milled. The feed for each leaf type was mixed with the standard rat diet to contain 5, 15 and 30% by weight of the leaves for groups 2, 3 and 4 respectively.

Sample collection

Overnight, prior to treatment, the animals were starved of food. Blood was collected from the ocular median-cantus vein of the rats with the aid of capillary tubes, transferred to test tubes, allowed to clot and subsequently centrifuged to obtain the serum component used for lipid analysis.

Lipid analysis

The lipid profiles were determined using kits manufactured by TECO Diagnostics Lakeview Ave, Anaheim, CA, USA. Serum total cholesterol (TC) was determined by the method of Aliain et al. (1974), while triacylglycerols was determined by the method of Burstein et al. (1980). The lipoproteins, very low-density lipoprotein (VLDL) and HDL were precipitated using phosphotungstic acid and magnesium chloride. After centrifugation, the supernatant contained the high-density lipoprotein cholesterol (HDL-C) fraction which was assayed for cholesterol by the method of Grove (1979). The low-density lipoprotein cholesterol (LDL-C) was calculated using the method of Friedewald et al. (1972).

Statistical analyses

Data collected were subjected to analysis of variance (ANOVA). In order to test whether or not significant differences existed between groups, the mean values with the paired t- test was analyzed. The mean ± SD of each parameter was taken for each group. Test probability value of $P < 0.05$ was considered significant. The analyses were carried out on SPSS for Windows version 10.

RESULTS

The effect of *V. amygdalina* and *T. occidentalis* diet preparations on the serum total cholesterol and triacylglycerols concentrations is shown in Table 1. There was a significant decrease in the effect of the vegetable leaves on the serum cholesterol and triacylglycerols levels relative to the control ($P < 0.05$). The results showed that there was a significant difference in the effect of the two vegetable leaves even at the same concentrations ($P < 0.05$) on serum cholesterol of the animals. For serum triacylglycerols, there was no significant difference between the effect of *V. amygdalina* and *T. occidentalis* diet preparations ($P > 0.05$).

Table 2 depicts the effects of *V. amygdalina* and *T. occidentalis* diet preparations on the serum LDL-C and HDL-C (mg/dl). From the results, there was a significant decrease on the serum LDL-C concentration relative to the control ($P < 0.05$) while there were no significant differences ($P > 0.05$) between the effect of *V. amygdalina* and *T. occidentalis* diet preparations at the various levels of treatment. The results also showed that the *T. occidentalis* diet induced a relatively lower serum LDL-C at the 5 and 15% treatments while *V. amygdalina* produced a lower LDL-C at the 30% treatment. The results also showed that there was a significant increase in the serum HDL-C levels of both *V. amygdalina* and *T. occidentalis* diet preparations ($P < 0.05$) relative to the control. From the results, the *T. occidentalis* diet preparation induced a significantly higher serum HDL-C concentration relative to *V. amygdalina* diet preparation.

DISCUSSION

Cholesterol is an essential substance involved in many cellular functions, including the maintenance of membrane fluidity, production of vitamin D on the surface of the skin, production of hormones and possibly helping cell connections in the brain (Champe and Harvey, 1994; Adaramoye et al., 2005; Daniels et al., 2009). It is of vital necessity that the body cells should have adequate supply of cholesterol. However, when cholesterol levels rise in the blood, they can have deleterious consequences. In particular, cholesterol has generated considerable notoriety for its causative role in artherosclerosis, the leading cause of death in developed countries around the world (Stamler et al., 2000; Daniels et al., 2009). Great efforts have been put into reducing the risk of cardiovascular diseases through the regulation

Table 1. Effects of *Vernonia amygdalina and Telfairia occidentalis* diet preparations on the serum total cholesterol and triacylglycerols (mg/dl).

Treatment group	Total cholesterol (mg/dl)		Triacylglycerol (mg/dl)	
	V. amygdalina	*T. occidentalis*	*V. amygdalina*	*T. occidentalis*
Control	133.80 ± 5.95	133.80 ± 5.95	150.73 ± 6.8	150.73 ± 6.8
5%	118.35 ± 5.95[*a]	115.55 ± 4.89[*b]	144.00 ± 2.39[*a]	146.37 ± 4.05[*a]
15%	112.55 ± 4.42[*a]	107.31 ± 4.69[*b]	121.17 ± 2.40[*a]	121.18 ± 2.40[*a]
30%	101.92 ± 3.24[*a]	103.99 ± 2.81[*b]	111.62 ± 4.89[*a]	111.62 ± 4.89[*a]

Results are mean ±SD. Values with different alphabetical superscripts for the same parameter in a row are significant with respect to each other ($P<0.05$). Values with the superscript * in a column are significant with respect to the control ($P<0.05$).

Table 2. Effects of *Vernonia amygdalina* and *Telfairia occidentalis* diet preparations on the serum LDL- C and HDL-C (mg/dl).

Treatment group	LDL-C (mg/dl)		HDL-C (mg/dl)	
	V. amygdalina	*T. occidentalis*	*V. amygdalina*	*T. occidentalis*
Control	127.33 ± 5.64	127.33 ± 5.64	22.13 ± 2.49	22.13 ± 2.49
5%	124.44 ± 3.81[*b]	120.49 ± 2.68[*b]	28.69 ± 3.71[*a]	29.40 ± 3.00[*a]
15%	116.72 ± 2.03[*b]	115.30 ± 3.68[*b]	30.59 ± 3.21[*a]	35.85 ± 3.48[*b]
30%	104.66 ± 3.84[*b]	110.65 ± 5.34[*b]	38.72 ± 2.21[*a]	37.50 ± 4.87[*a]

Results are mean ±SD. Values with different alphabetical superscripts for the same parameter in a row are significant with respect to each other ($P<0.05$). Values with the superscript * in a column are significant with respect to the control ($P<0.05$).

of cholesterol, thus the therapeutic benefits of plant foods have been the focus of many extensive dietary studies (Yokozawa et al., 2006; Zhang et al., 2007). For centuries, traditional plant remedies have been used in the treatment of illnesses (Akhtar and Ali, 1984) but only a few have been evaluated scientifically. Therefore, the effects of the diet preparations of *V. amygdalina* and *T. occidentalis* leaves on the lipid profile of rats were studied and compared.

From the results obtained in the study, it is apparent that the two diet preparations lowered the serum cholesterol levels of rats though the *T. occidentalis* diet induced a higher effect at lower concentrations. Adaramoye et al. (2007) had shown that *T. occidentalis* has hypolipidaemic effect in rats fed cholesterol rich diet while Nwanjo (2005) showed that the administration of aqueous leaf extract of *V. amygdalina* induced hypoglycaemic, hypolipidaemic and antioxidant effects in rats. Ahmed-Raus et al. (2001) suggested that the mechanism of this hypocholesterolaemic action may be due to the inhibition of the absorption of dietary cholesterol in the intestine or stimulation of the biliary secretion of cholesterol and cholesterol excretion in faeces.

Triacylglycerols are partly taken up with the diet and partly synthesized in the liver (Anderson et al., 1991). Dietary cholesterol has been shown to reduce fatty acid oxidation, which in turn, increases the levels of hepatic and plasma triacylglycerols (Fungwe et al., 1993). There is evidence that shows that high triacylglycerols are associated with coronary atherosclerosis (Bainton et al., 1994; Cullen, 2003). The results from this study showed that there was no significant difference on the triacylglycerols lowering effects of both diet preparations.

The link between increased levels of low density lipoprotein-cholesterol (LDL-C) in the blood and artherosclerosis is incontrovertible (Cullen, 2003). There is also very strong evidence that this link is causal; lowering of LDL-C by drugs, in particular statins, reduces the morbidity and mortality from cardiovascular disease in general and chronic heart disease in particular (Blakenhorn et al., 1993; Hodis et al., 1994). The mechanisms that manage and utilize LDL are tightly controlled systems evolved to distribute cholesterol through the circulatory system and its cells that require extracellular cholesterol. Unfortunately, LDL-C does not always reach its most appropriate destination, but rather accumulates in artery walls causing atherosclerosis, the leading cause of death and disability in the developed world (Yusuf et al., 2001). For this reason, the quantity of circulating LDL-C is a well-known risk factor for heart disease, and is the primary focus of most lipid lowering therapies (NCEP, 2001). The pathogenicity of LDL and likelihood of artherosclerotic development are heavily influenced by genetic composition of gene products involved with LDL metabolism. Patients with genetic defects that cause severely elevated LDL have familial hypercholesterolaemia, which affects approximately 1:500 people (Koivisto et al., 1993), and is the consequence of mutations in the low density lipoprotein receptor (LDLR) and other genes. The results showed that both preparations significantly lowered the serum LDL-C concentrations relative to the control. However, both diet preparations could not induce a significant effect when compared at equal concentrations.

Beyond the role of LDL-C in the development of arthrosclerosis, growing evidence suggests that high density lipoprotein cholesterol (HDL-C) is a powerful predictor of cardiovascular disease (CVD). Indeed, epidemiological, mechanistic and intervention studies suggest that low HDL-C is a major CVD risk factor and that increasing HDL-C plasma levels may be beneficial, particularly in patients with low HDL-C levels (Philips, 2007). The results from the study showed that the treatment with *V. amygdalina* and *T. occidentalis* diets led to a significant increase in serum HDL-C, showing their protective role in CVD. The comparison of their effects showed that *T. occidentalis* diet preparation induced a significant increase in serum HDL-C compared to *V. amygdalina* diet preparation at the 15% concentration. The protective role of HDL-C against CVD has been suggested to occur in several ways (Nofer et al., 2002). Particles of HDL prevent coronary artery disease by serving as transport particles for excess cholesterol to the liver, where it is converted into bile acids and excreted. In humans, HDL levels are a very well known measure of cardiac health due to their strong inverse relationship with coronary artery disease (Wilson et al., 1988; Stamler et al., 2000). The principal HDL pathway, termed reverse cholesterol transport (RCT) is a major component of lipid homeostasis. Genetic variation in the RCT pathway contributes greatly to phenotypic variations in humans (O'Connell et al., 1988).

Results from this study confirm that *V. amygdalina* and *T. occidentalis* have lipid lowering effects which may be beneficial to people at risk of CVD. *V. amygdalina* and *T. occidentalis* were found to be effective in lowering the levels of serum cholesterol, triacylglycerols and LDL-C, thereby, showing their hypocholesterolaemic property. In conclusion, the results from the study showed that the *T. occidentalis* diet preparation had more anti- lipidaemic property than the *V. amygdalina* diet preparation. These leaves could be beneficial to people at high risk of cardiovascular disease.

REFERENCES

Adaramoye OA, Nwaneri VO, Anyanwu KC, Farombi EO, Emerole GO (2005). Possible antiatherogenic effect of Kolaviron (a *Garciania kola* seed extract) in hypercholesterolemic rats. Clin. Exp. Pharmacol. Physiol., 32: 40-46.

Adaramoye OA, Achem J, Akintayo OO, Fafunso MA (2007). Hypolipidemic effect of *Telfairia occidentalis* (Fluted pumpkin) in rats fed a cholesterol rich diet. J. Med. Food, 10(2): 330-336.

Ahmed-Raus RR, Abdul-Latif EA, Mohammed JI (2001). Lowering of lipid composition in aorta of Guinea Pigs by *Curcuma domestica*.

BMC Compl. Altern. Med., 1: 6.

Alada ARA (2000). The haematological effect of *Telfaira occidentalis* diet preparation. Afr. J. Biomed. Res., 3(20): 185-186.

Akhtar MS, Ali MR (1984). Study of antidiabetic effect of a compound medicinal plant prescription in normal and diabetic rabbits. J. Pak. Med. Assoc., 34: 239-244.

Aliain CC, Pon LS, Chan CSG, Richmond W, Wu PC (1974). Enzymatic determination of total cholesterol. Clin. Chem., 20: 470-475.

Amira CA, Okubadejo NU (2007). Frequency of complementary and alternative medicine utilization in hypertensive patients attending an urban tertiary care in Nigeria. BMC Compl. Altern. Med., 7: 30-48.

Anderson KM, Odelt PM, Wilson PW, Kannel WB (1991). Cardiovascular disease risk profile. Am. Heart J., 21: 293-298.

Bainton D, Sweetnam P, Baker I, Elwood P (1994). Peripherial vascular disease: consequence for survival and association with risk factors in the Speedwell prospective heart disease study. Br. Heart J. 72: 128-132.

Blankenhorn DH, Azen SP, Kramsch DM, Mack W, Cashinhenemphill L, Hodis HN, Deboer LWV, Masteller MJ, Vailas LI, Alaupovic P, Hirsch LJ (1993). The Mars Research Group Coronary angiographic changes with lovastin therapy: The monitored atherosclerosis regression study (MARS). Ann. Intrn. Med., 119: 969-976.

Burstein M, Schnolnic HR, Marlin R (1980). Rapid method for the isolation of lipoprotein from human serum by precipitation with polyanions. Scan. J. Clin. Lab. Invest., 40: 583-595.

Champe PC, Havey RA (1994). Cholesterol metabolism. In: Lippicot's illustrated reviews biochemistry. 2nd edition. Champe PC, Havey RA.(eds). JB Lippincott, Kendallville, pp. 210-225.

Cullen P (2003). Triacylglycerol-rich lipoproteins and atherosclerosis-where is the link? Biol. Soc. Trans., 315: 1080-1083.

Daniels TF, Killinger KM, Michal JJ, Wright RW, Jang Z (2009). Lipoproteins, cholesterol homeostasis and cardiac hearth. Int. J. Biol. Sci., 5: 474-488.

Durrington PM, Hunt I, Ishola M, Arrol S, Bhatnagar D (1988). Apolipoproteins (a), AL and B and parental history in men with early onset ischaemic heart disease. Lancet, 1: 1070-1073.

Farombi EO (2003). African indigenous plants with chemotherapeutic potentials and biotechnological approach to the production of bioactive prophylactic agents. Afr. J. Biotech., 2(12): 602-671.

Friedewald WT, Levy RI, Fredrickson DS (1972). Estimation of the concentration of low density lipoprotein cholesterol in plasma, without use of the preparative ultracentrifuge. Clin. Chem., 18: 499-502.

Fungwe TV, Cagen LM, Cook GA, Wilcox HG, Heimberg M (1993). Dietary cholesterol stimulate biosynthesis of triglyceride and reduces oxidation of fatty acids in rats. J. lipid Res., 34: 933-941.

Gbile ZO (1986). Ethnobotany, taxonomy and conservation of medicinal plant. In: the state of medicinal plant research in Nigeria. Sofoworo,A.O.(ed.), Ife University press, p. 19.

Grove TH (1979). Effect of reagent pH on determinationof high density lipoprotein cholesterol by precipitation with sodium phosphotungstate-magnesium. Clin. Chem., 25: 560-564.

Hodis HN, Mack WJ, Azen SP, Alaupovic P, Pogoda JM, Blankenhorn DH, Hemphill LCN, Kramsch DM, Blankenhorn DH (1994). Circulation, 90: 42-49.

Karim A, Ansari NH, Ahmad M, Iqbal SD, Mehmood A (2004). Effect of amlodpine on lipid profile. The professional, 11(4): 437-442.

Koivisto UM, Hamalainen L, Taskinen MR, Kettunen K (1993). Prevalence of familial hypercholesterolemia among young North Karelian patients with coronary heart disease: A study based on diagnosis by polymerase chain reaction. J. lipid. Res., 34: 269-277.

Lamendala C (2000). Hypertriglycerideaemia and low high-density lipoprotein; risks for coronary artery disease. J. Cardiovsc. Nurs., 21 (2): 79-90.

Lloyd-Jones D, Adams R, Carnethon M, Desimone G, Ferguson BI, Flegal K, Ford E, Furie K, Go A, Greenland K, Haase N, Hailpern S, Ho M, Howard V, Kissela B, Kittner S, Lackland D, Lisabert L, Marelli A, McDermott M, Meigs J, Mozffarian D, Nichol G, O'Donnell C, Roger V, Rosamond W, Sacco R, Sorlie P, Stafford R, Steinberger R, Steinberger J, Thom T, Sylvia-Wassertheiel S, Wong N, Wylie-Rosett J, Hong Y (2009). Heart disease and stroke statistics- 2009 update: A report from the American heart Association statistics committee and stroke statistics subcommittee. Circulation, 119: 480-486.

Nagra MH, Hussain I, Alam Z, Amin K, Javed M (2003). Relationship of coronary artery disease (CAD) with total cholesterol and LDL-cholesterol/HDL-cholesterol ratio. The professional, 10(3): 1-4.

National Cholesterol Education Program (NCEP) (2001). Executive summary of the third report of the NCEP expert panel on detection, evaluation, and treatment of high blood cholesterol in adults (Adults treatment panel III). JAMA, 285: 2486-2497.

Nofer JR, Kehrel B, Fobker M, Levkau B, Assmann G, Von Eckard-stein A (2002). HDL and artheriosclerosis: beyond reverse cholesterol transport. Atherosclerosis, 161: 1-16.

Nwanjo HU (2005). Efficacy of aqueous leaf extract of *Vernonia amygdalina* on plasma lipoprotein and oxidative status in diabetic rat models. Nigerian J. Physiol. Sci., 20(1-2): 39-42.

O'Connell DL, Heller RF, Roberts DC, Allen JR, Knapp JC, Steele PL, Slove D (1988). Twin study of genetic and environmental effects on lipid levels. Genet. Epidemiol., 5: 323-341.

Okigbo BN (1977). Neglected plants of horticultural and nutritional importance in traditional farming systems of tropical Africa. Acta. Hortic., 55: 131.

Okoli BE, Mgbeogu CM (1983). Fluted pumkin (*Telfaira occidentalis*). West Afr. Veg. crop. Eco. Bot., 37(2): 145.

Philip BMD (2007). HDL cholesterol, very low levels of LDL cholesterol, and cardiovascular event. New Engl. J. Med., 357: 1301-1310.

Stamler J, Daviglus ML, Garsideetal DB, Dyer AR, Greenland P, Neaton JD (2000). Relationship of baseline serum cholesterol levels in 3 large chorts of younger men to long-term coronary, cardiovascular, and all-cause mortality and longevity. JAMA, 284: 311-318.

Tindal HD (1968). Commercial vegetable growing. Oxford Press, London, p. 69. Trease GE, Evans WC (1983). Textbook of Pharmacognosy. 12th edition. Balliese Tindall and Company publishers, London, pp. 343-368.

Wald WJ, Law M A (1995). Serum cholesterol and ischaemic heart disease. Atherosclerosis, 118: 1-5.

Wilson PW, Abbott RD, Castelli WP (1988). High density lipoprotein cholesterol and mortality. The Framingham heart study. Arteriosclerosis, pp. 737-741.

Yokozawa T, Cho EJ, Sasaki S, Satoh A, Okamato T, Sei Y (2006). The protective role of Chinese prescription Kangen-karyu extract on diet-induced hypercholesterolemia in rats. Biol. Pharm. Bull., pp. 29: 270-765.

Yusuf S, Reddy S, Ounpuu S, Anand S (2001). Global burden of cardiovascular diseases part 1: general considerations, epidemiologic transition, risk factors, and impact of urbanization. Circulation, 104: 2746-2753.

Zhang HW, Zhang YH, Lu MJ, Tongwei-Jun CAO (2007). Comparison of hypertension, dyslipidaemia and hyperglycaemia between buckwheat seed- consuming and non- consuming Mongolian-Chinese population. Clin. Expt. Pharmacol. Physiol., 34: 838-844.

3

Hepatoprotective role of Garcinia kola (Heckel) nut extract on methamphetamine: Induced neurotoxicity in mice

Gabriel Oze*, Iheanyi Okoro, Austin Obi and Polycarp Nwoha

Institute of Neuroscience and Biomedical Research, College of Medicine, Imo State University, Owerri, Nigeria.

The hepatoprotective effect of aqueous extract of Garcinia kola (AEGK) was studied in 60 mice of mixed sexes. The animals were divided into 6 groups of 10 mice each. Group I received normal saline, groups II and III got 100 and 200 mg/kg AEGK (orally), respectively. Group IV received 10 mg/kg methamphetamine (MAM) (s.c.) only. Groups V and VI got 100 and 200 mg/kg of AEGK respectively, before 10 mg/kg methamphetamine which was used to induce neurotoxicity. The serum levels of AST, ALT, ALP, total bilirubin and its conjugated metabolite were used to assess liver damage. Fifty percent of the animals in group IV died. 30% died in group V and none in group VI after 10 - 30 min interval of MAM administration. The serum levels of some of the marker enzymes and bilirubin were decreased significantly in groups VI at 200 mg/kg of AEGK (P < 0.05). The Blood glucose level increased transiently in the MAM treated groups. There was a slight rise in serum WBC after an initial fall at 100 mg/kg AEGK. The results suggest a possible hepatoprotective potential of AEGK. This may justify their local use in the management of some hepatic dysfunction and stress-related conditions.

Key words: Garcinia kola extract, methamphetamine, neuroprotection, mice.

INTRODUCTION

Garcinia kola (Gultiferae) is an evergreen plant found in the equatorial forest of Sub-Saharan Africa. The plant grows wild and is also domesticated because of the wide medicinal values of the extract of its various components in folk medicine. The G. kola nut (GKN) is culturally and socially significant in some parts of South Eastern Nigeria (West Africa) where the yellow nut is served for traditional hospitality in private, social and cultural functions. It is commonly called bitter kola because of the bitter taste of the nut.

As a result of its wide spread consumption, especially among the Ibos of South Eastern Nigeria, some studies have been carried out on the extract of various components of the plant. The phytochemical studies shows that GKN contains phenolic compounds, steroids, xanthines, benzophenones (Etkin, 1981; Iwu, 1982), tannins, guttiferins and saponins (Etkin, 1981). Animal and human studies revealed that the extracts of GKN

exhibit aphrodisiac effects on male subjects (Iwu, 1993; Orie and Okon, 1993) for which reason they are sometimes called "male kola in some parts of Nigeria. It is reported to suppress ovulation and delay fertility in female subjects (Iwu and Igboko, 1982). GKN extracts have been shown to possess antipyretic, anti-inflammatory, analgesic (Olaleye et al., 2000), antiviral, hepatoprotective (Iwu, 1985; Akintonwa and Essien, 1990), CNS stimulant (Orie and Okon 1993), antidepressant, antioxidant (Adaramoye et al., 2005), antidiabetic (Braide et al., 2003; Akpanta et al., 2005) activities. The registration of GKN formulation as a hop substitute in the brewing of beer and wine was under consideration by the Food and Drug Administration (FDA, 1999).

The serum levels of alanine aminotransferase (ALT), aspartate aminotransferase (AST), alkaline phosphatase (ALP), total bilirubin (TB) and conjugated bilirubin (CB) were used as hepatic markers (Mercer and Tainuno, 1982; Muragesh et al., 2005). The levels of these non - functional enzymes correlate hepatic damage. However, ALT is more specific for hepacellular injury than the

---

*Corresponding author. E-mail: gabrieloze@yahoo.com.

other enzymes (Alagbonna and Onyeyilli, 2003).

Methamphetamine (MAM) is the d - isomer of the parent drug, amphetamine. It is a potent sympathomimetic agent with greater pressor effect when compared with the parent compound on equimolar basis, especially on the CNS. The induction of neurotoxicity is achieved at higher doses of about 10 - 40 mg/kg s.c. (Innis and Nickerson, 1975; Bloom, 1985; Imam and Ali, 2000). MAM is useful in the induction of experimental neurotoxicity in animal models, especially in mice (Imam et al., 2001). The ability of GKN extract to attenuate the raised serum levels of the liver marker enzymes is an indication of its hepatoprotective effect.

However, there is paucity of data on the neurological effect of AEGK. This study evaluates the possible neuroprotective effects of the aqueous extract of Garcinia kola (AEGK) using liver and kidney functions and some blood parameters as indices. The study aims at providing some information on the possible neuroprotective role of AEGK. It also aims at providing some scientific basis for the use of the GKN in folk medicine to treat hepatitis and diabetes - related conditions.

MATERIALS AND METHODS

Plants materials

The nuts of Garcinia kola were purchased from Owerri Municipal Markets, Owerri, Imo State, Nigeria. It was authenticated by a plant taxonomist, Dr. C. Okeke, Department of Plant Science and Biotechnology, Imo State University, Owerri. A voucher specimen is deposited in the University Herbarium.

Extraction

The nuts were dehusked and chopped into bits, sun-dried to constant weight and pulverized using a mechanical grinder (Thomas Contact Mill, Pye Unicam, Cambridge, England). 200 g of the power was obtained. This was soaked in distilled water in a soxhlet apparatus and extracted after 24 h. The solvent was evaporated using oven (Acumex, India) (50°C) and rotatory evaporator (Laborato 400, China). The dry residue (1.24 g) was constituted in normal saline (NS) (100 mg/ml) and used for the experiment.

Animals

The mice (BALD strain), weighing 20 - 35 g of mixed sexes were obtained from the Animal House of the Department of Pharmacology, University of Port Harcourt. They were housed in stainless steel cages and allowed to acclimatize for two weeks in the Animal House of the College of Medicine, Imo State University, Owerri, under 12 h light/ dark cycle before commencement of the experiment. The animals had access to standard feed (Guinea Feeds, Ltd, Ewu, Edo State, Nigeria) and water ad libitium.

Drugs and chemicals

The methamphetamine (Desoxgn) used was the product of Glaxosmith Kline, England.

Experimental design

The Animals were divided into 6 groups of 10 mice each. Group 1 received normal saline plus feed and water for two weeks and served as the negative control. Groups ii and iii received 100 and 200 mg/kg of AEGK, respectively for 2 weeks only. Group iv received normal saline for 6 weeks and also served as negative control, while Groups v and vi received 100 and 200 mg/kg AEGK for 6 weeks. All the 6 weeks segment groups additionally received 10 mg/kg methamphetamine (s.c.) on the 6th week. The AEGK were administered to the mice by oral intubation between 9 - 10:00 am daily. They also had feed and water ad libitium.

Biochemical studies

The animals were starved 24 h prior to the collection of blood samples. The samples were drawn by cardiac puncture into marked sample bottles and allowed to clot for 45 min at room temperature. The serum was obtained by centrifugation at 2500 rpm at 30°C for 5 min using Wisperfuge centrifuge (model 1384, Samson, Holland).

The serum was separated using Pasteur pipette into sterile serum sample tubes from where they were drawn for the biochemical assay. The method of Reithman and Frankel (1957) was adopted for the ALT and AST assay. ALP was estimated using the method of King and King (1954) as adapted by Cheesbrough (2000). Bilirubin and glucose were determined using the methods of Malloy et al. (1937) as modified by Tietz (1996). Serum sodium and potassium were estimated using reagent titrimetric method. Serum chloride was determined by the method of Schales and Schales (1941).

Histological studies

At the end of the experiment, the mice were sacrificed under chloroform anaesthesia .The livers were dissected out and immediately fixed in 10% formal saline. Using a standard tissue processor, the tissues were then dehydrated in ascending grades of alcohol: 70, 95% and absolute alcohol in 2 changes each. After which clearing was done with xylene/absolute alcohol [50:50 v/v ratio]. This was followed by infiltration in molten paraffin wax at 60°C in 2 changes. They were further processed for staining with haematoxyline and eosin (H&A) as described by John et al. (1990). Photomicrographs of the slides were taken for histological examination.

Statistics

The results were analyzed using Duncan multiple range test. Data were expressed as mean ± standard deviations. Differences between means were considered at 95% confidence limit and probability level of 0.05 was taken as significant.

RESULTS

The results of the biochemical parameters are presented in Table 1. In the absence of the neurotoxin (MAM), the serum levels of ALT, ALP and TB were significantly reduced ($p < 0.05$) relative to group I control. Although the AST level decreased, it was non - significant ($p > 0.05$). The CB concentration was slightly raised against the control. This reduction did not follow a dose dependent pattern.

Table 1. The effect of AEGK on methamphetamine-induced hepatotoxicity in mice (mean ± SD) (n = 6 - 10).

Group	TB (mg/l)	CB (mg/l)	ALT mg/l	AST (mg/l)	ALP, (mg/l)	Na$^+$ (meq/l)	K$^+$ (meq/l)	Cl$^-$ (meq/l)	Glucose (mg/l)	WBC Counts/mm^2
I (Control)	2.7 ± 0.8	0.2 ± 0.01	4.75 ± 0.3	5.67 ± 0.22	48.3 ± 0.5	120.7 ± 1.1	5.2 ± 0.7	121.3 ± 1.25	84.7 ± 1.64	2433.3 ± 115.4
II (100 mg/kg)	2.0 ± 0.26	0.38 ± 0.25	3.5 ± 0.13	5.17 ± 0.24	48.5 ± 0.79	121.7 ± 1.0	5.6 ± 0.4	107.3 ± 11.0	83.3 ± 0.53	2366.7 ± 59.7
III (200 mg/kg)	2.0 ± 0.57	0.4* ± 0.01	3.5 ± 0.5	4.3 ± 0.10	43.3 ± 0.76	121.0 ± 1.0	5.1 ± 0.4	104.7 ± 4.17	69.3 ± 1.53	2500 ± 0.00
IV Control	4.9 ± 0.7*	1.0 ± 0.015*	7.25 ± 0.84*	5.5 ± 0.16	66.0 ± 0.17*	121.7 ± 2.08	4.78 ± 0.06	122.4 ± 11.0	98.0 ± 1.14	1366.7 ± 115.4*
V (100 mg/kg)	4.16 ± 0.28	0.67 ± 0.56	7.10 ± 0.85	5.43 ± 0.54	48.27 ± 0.69	122.4 ± 2.5	5.03 ± 0.59	107.0 ± 0.57	106.8 ± 11.0	1500 ± 200
VI (200 mg/kg)	3.97 ± 0.59	0.62 ± 0.09	3.89 ± 0.84*	4.73 ± 0.28	44.33 ± 0.51*	121.8 ± 1.75	59.0 ± 0.40	107.0 ± 0.57	106.8 ± 11.0	1750 ± 150

Key: TB = Total bilirubin, CB = conjugated bilirubin, ALT = alanine aminotransferase, AST = aspartate aminotransferase, ALP = alkaline phosphatase, WBC = white blood cell, * = significantly different from control (p < 0.05).

The serum electrolyte levels were relatively unchanged, except for chloride ion which fell transiently. The blood glucose was reduced significantly (p < 0.05) in the absence of MAM at 200 mg/kg but the WBC count was slightly raised at 200 mg/kg of AEGK, relative to group I animals. In the presence of the hepatotoxin, the marker enzymes and metabolites were elevated significantly (p < 0.05) (Group I and IV) suggesting the induction of hepato-neurotoxicity, except for AST whose level remained relatively constant. The serum electrolytes were relatively unchanged. There was also significant decrease in the plasma WBC and transient rise in blood glucose when compared with the control. When 200 mg/kg AEGK was challenged with 10 mg/kg (s.c.) MAM, the serum AST, ALT, TB and CB decreased relative to their respective controls in group IV. The decease was significant for ALT and ALP (p < 0.05). However the decrease at 100 mg/kg AEGK was transient and non - significant (p > 0.05). The histological studies (Figures 1 - 4) support this observation. The sodium level was not altered at both doses in the presence of the hepatotoxin, but the K+ concentration increased slightly (p > 0.05). Although there was a fall in the serum level of chloride ion, it was not significant (p > 0.05) at the dose levels of AEGK tested.

Clinical observation

About 10 - 30 min after the administration of the neurotoxin, 50% of the animals in control Group IV died. 30% of those in Group V also died. No death was recorded for the animals in Group VI; though the animals were slightly debilitated, but finally recovered fully. The remaining animals in the neurotoxin treated groups subsequently recovered but the recovery in Group VI was faster and sustained.

Histological observation

Figure 1 show a section of the liver of the mice in Group 1 used as negative controls, having received only normal saline. Normal hepatocytes are seen radiating from the central vein.

However, in Figure 2 which shows a liver section from Group 4 mice (positive controls); there was gross distortion of the cell cords. There are vacoulations in the cytoplasm of the cell. This is evidence of cellular damage.

Figure 3 is a liver section from Group 5 mice which received 100 mg/kg of AEGK and methamphetamine. There was minimal distortion of the cell cords.

Figure 4 shows the section from Group 6 mice which received 200 mg/kg of AEGK and methamphetamine. The cell cords are fairly well preserved like those of the Group 1 mice. This is evidence of protection.

DISCUSSION

Methamphetamine (MAM) is a potent N-methyl d-analog of phenylethylamine. It belongs to the class of sympathomimetic CNS stimulants but acts as a neurotoxin at higher doses (10 - 40 mg/kg s.c.) with more profound action on the CNS, relative to the periphery (Shockley, 1991; Hoffman et al., 1996). Like other amphetamines, it blocks neuronal re-uptake of the catecholamine neurotransmitters, especially dopamine, at the nerve endings of sympathetic neuron (Hoffmann et al., 1996; Imam et al., 2001). This event

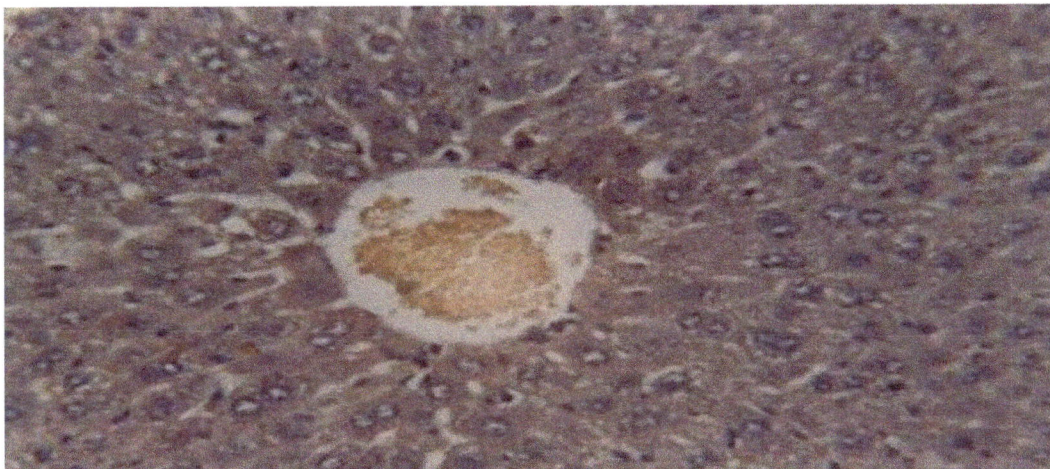

Figure 1. Section from the liver of mice which received normal saline and no methamphetamine. It shows normal hepatocytes and cell cords.

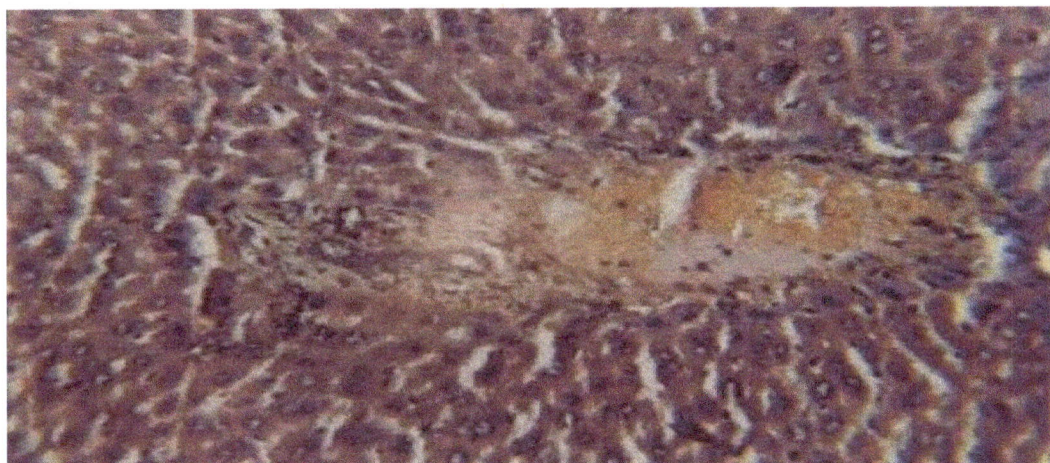

Figure 2. Section of the liver from the mice which received normal saline and methamphetamine (Positive control). It shows much vacuolation and distortion of the hepatic cell cords.

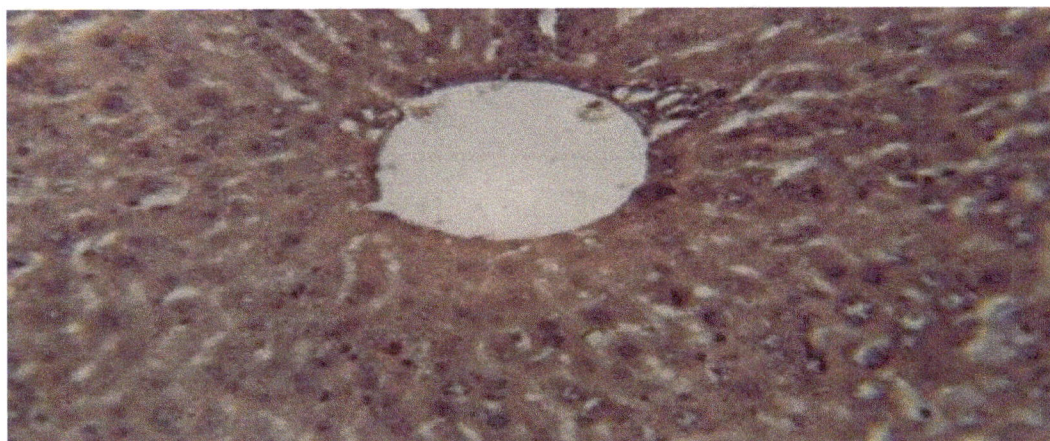

Figure 3. Section of liver from mice which received 100 mg/kg and methamphetamine. It shows slight vacuolation with minimal distortion of the cell cords.

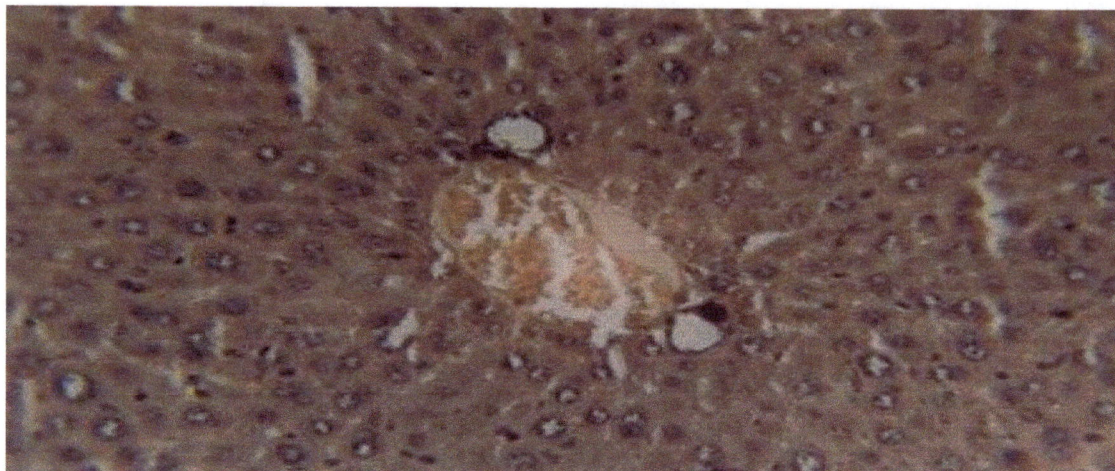

Figure 4. Section of liver from mice which received 200 mg/kg AEGK and the methamphetamine. It shows no vacuolation and has reasonable presentation of the cell cords. Looks almost like the tissue from normal saline control which received no toxin.

exacerbates the pharmacological actions of the affected neurotransmitters, exposing them to extensive metabolic degradation. Under this condition, there may be hypertensive crisis, cerebral hemorrhage, increased cardiac output and increased renal blood flow. Subsequently, the rate of enzymic degradation exceedsthe rate of neuronal biosynthesis, leading to depletion of the neurotransmitter, neuronal failure and possibly death.

Imam and Ali (2000) and Imam et al. (2001) have demonstrated that MAM neurotoxicity may be associated with the neuronal generation of free radicals such as peroxynitrites, super oxides and hydroxyl from dopamine metabolism, leading to oxidative damage and possibly death. They also showed that antioxidants such as selenium or glutathione could offer neuro-protection in these cases. Kampen et al. (2003) made similar observations for methylphenyltetrahydropyridine (MPTP), a potent neurotoxin.

In the assessment of liver damage, the serum concentrations of the hepatic marker enzymes, AST, ALT and ALP, are important as well as serum total bilirubin and its metabolites. The hepatocellular marker enzymes are non-functional, with negligible systemic concentrations in normal situations, but leak into general circulation when there is necrosis or damage to the hepatic cells; neuronal or skeleton muscle cells; consequently their serum concentrations rise above normal values (Murray et al., 2000). However, ALT is more specific for the liver tissue because the other enzymes and metabolites could be released as a result of damage to other organs.

Damage to hepatocytes and blood cells exacerbate therelease and metabolism of haem and its metabolites including bilirubin and its excretory products. Thus, the serum level of bilirubin and its conjugate metabolites would be expected to rise with a corresponding decrease in blood cell count due to mild haemolysis. This

observation was made by Uko et al. (2001) in rats and agrees with the outcome of the present study (Table 1).

The renal functional status could be assessed using the serum levels of electrolytes and proteins. The sodium and chloride levels are complementary and may be altered in nephrotoxicity (Murtey et al., 2000; Nwanjo et al., 2006). The plasma potassium concentration inversely relates to the sodium concentration.

The liver is innervated by the xth cranial nerve and the ciliac ganglia and is therefore subject to neuronal control. The sympathetic stimulation of the adrenal medulla causes the release of adrenaline into systemic circulation. As a hormone, adrenatine regulates blood glucose via the glycogen metabolic pathway in the liver (Glew et al., 1989). In neuronal damage involving the liver, the glycogen metabolic pathway would be affected adversely. A complementary immuno-response to cellular damage may lead to transient compensatory increase in blood glucose concentration.

In this study, 10 mg/kg (s.c.) MAM was used to induce hepato-neurotoxicity in the mice (Table 1, Groups I and IV). This is shown by the significant ($p < 0.05$) elevation of serum levels of the liver marker enzymes and metabolites. The induction of nephrotoxicity could not be ascertained from the current observations due to the marginal alterations in the level of the electrolytes in the presence of the neurotoxin.

The study evaluated the hepato-protective role of AEGK in MAM - induced neurotoxicity in mice. During the course of treatment with MAM, it observed that 50% of the mice in Group IV died 10 - 30 min after the administration of the neurotoxin. Thirty percent of the mice in Group V also died, while none died in Group VI. The animals may have died either due to neuronal exhaustion or oxidative damage or both, while the animals in Group VI survived possibly due to a neuroprotective action offered by the 200 mg/kg AEGK.

The 100 mg/kg AEGK caused a decline in the serum levels of the marker enzymes, especially in the cases of conjugated bilirubin and ALT, where the fall were significant ($p < 0.05$), indicating a possible hepatocellular protective effect as reported by Muragesh et al. (2005). A similar pattern of attenuation for ALT, TB, CB ALP and AST were seen in the presence of the neurotoxin at 200 mg/kg of the AEGK. The results tend to suggest a possible hepatoprotection by AEGK in mice. This line of reasoning is also supported by attenuated serum levels of TB and CB which followed similar trend as the hepatocellular marker-enzymes in the presence of the neurotoxin. Muragesh et al. (2005) noted that the return of the marker enzymes towards normal levels in the presence of the extract is an indication of membrane stabilization as well as repair of tissue damage via rapid regeneration of parenchyma cells. They also noted that decrease in ALP and TB serum levels may be indications of restoration of secretary mechanisms of the hepatocytes. The above observations were made in the current study (Table 1).

The mechanism by which the AEGK could cause hepatoprotectof in the presence of the neurotoxin is not known. It may be related to the ability of the AEGK to cause membrane stabilization via reduced neuronal firing and arrest of free radical reaction. It is important to note that Adaramoye et al. (2005) had demonstrated the antioxidant potential of AEGK. The histopathological studies (Figures 1 - 4) showed that the AEGK reversed the damage of hepatocytes associated with the hepatotoxin especially at the dose of 200 mg/kg in 6th weeks.

It is not possible to draw similar conclusion for the renal function because there seemed to be no induction of nephrotoxicity. The serum levels of the electrolytes remained relatively unchanged in the absence and in the presence of the neurotoxin. Although there was a decrease in the chloride level in both phases, and at 100 and 200 mg/kg, this was marginal and may not make meaningful impact on the functional status of the kidney. Gidado and Umar (2001) obtained similar results on renal parameters for Dutura strumonium (Jimson weed) in rats.

The blood glucose was reduced in the absence and presence of MAM. This agrees with the work of Braide and Grill (1990) who reported the antidiabetic activity of the aqueous extract of GKN in rats. The blood sugar rose slightly in the presence of MAM relative to the control ($p > 0.05$). The rise at 100 mg/kg was greater than for 200 mg/kg AEGK; which tends to suggest an initial rise and subsequent decline in the sugar level as the dose of AEGK increased. The initial rise in blood glucose in the presence of the neurotoxin was expected because of the hyperglycemic action of the neuro-endocrine hormone (adrenaline) whose release is stimulated by the neurotoxin.

These outcomes lend more support to the possible hepato-protective effect of AEGK. Iwu (1985) and

Akintonwa and Essien (1990) previously reported the hepatoprotective potentials of AEGK. Moreover, Iwu et al. (1990) and Adaramoye et al. (2005) reported the hypoglycemic effect of AEGK in rats. It is concluded that the AEGK could play a hepatoneuroprotective role in mice. This tends to justify the use of the aqueous extract in folk medicine as antidiabetic, antihepatitis and possible mood stabilizing agent.

REFERENCES

Adaramoye OA, Farombi EO, Adeyemi EO, Emerole GO (2005). Comparative study on the Antioxidant properties of flavonoids of Garcinia kola seeds. Pakistan J. Med. Sci. 21(3): 1-2.

Alagbonna OP, Onyeyili (2003). Effects of Rhaptopetalum coriacerum Stem Bark Extract on Serum Enzyme Activities and Histopathological changes in Rats. J. Expt. Clin. Anat. 2(1): 30 – 33.

Akintonwa A, Essien AR (1990). Protective Effect of Garcinia kola seed extract against Paracetamol – induced hepatoxicity in rats. J. Ethnopharmacol. 29: 207 – 211.

Akpantah AO, Oremosu AA, Ajala MO, Skanlawon AO (2003). The effect of Crude Extract of Garacinia kola seed on the Histology and Hormonal milieu of male spragne – Dawley Rats, Reproductive Organs. Nig. J. Health Biomed. Sci. 2(1): 6.

Akpanta AO, Oremosu AA, Noronha CC, Ekanem TB, Okanlowon AO (2005). Effectof Gareinia kola seed extract on ovulation, estrouscycle and foetal development in female spragne – Dawley rats. Nig. J. physrol. Sci. 20(1-2): 58 – 62.

Braide VB, Agube CA, Essien GE, Udoh, FV (2003). The effect of crude kola seed alkaloiad extract on level of gonadal hormornes and pituitary gonadotrophins in rat serum. Nig. J. Physiol. Sci. 18(12): 59 – 64.

Cheesbrough M (1981). Estimation of Sodium and potassium in plasma. in District Laboratory Practice in Tropical Countries. Cambridge University Press, Cambridge pp. 389 – 394.

Cheesbrough M (2000). Clinical chemistry in: District and Laboratory Practice in Tropical Countries, Part 2, Cambridge University Press, Cambridge p. 318.

Ebomiyi MI, Iyawe VI (2000). Peak Expiratory Flow Rate (PEFR) in Young Adult Nigerians following ingestion of Garcinia kola (Heckel) seeds. Afr. J. Biomed. Res. 3(3): 187- 189.

Esimore CO, Nwafor SV, Okolie CO, Chah KF, Uzuegbu DB, Chibundu CI, Eche MA, Adikwu MU (2002). In-vivo Evaluation of Interaction between aqueous seed extract of Garcinia kola Heckel and ciprofloxacin – Hydrochloride. Am. J. Therapeutics 9(4): 275 – 280.

Esomonu GUI, El-Taalu AB, Aunka JA, Ndodo ND, Salim AM, Kabiru M (2005). Effect of ingestion of extract of Garcinia kola seed on erythrocytes in Wistar Rats. Nig. J. Physiol. Sci. 20(1 – 2): 30 – 32.

FDA (1999). Dept of Health and Human Services. Washington, DC 20204. GRN 000025.

Gidado A, Umar AI (2001). Effect of Aqueous Preparations of Datura stramorium (Jmson weed) level and seeds on some indices of liver and kidney function in rats. Nig. J. Bioch. Molecular. Boil 16: 17 – 18.

Hoffman BB, Leftwowit R (1996). Catecholamine, sympathommetics, Adrenergic receptor antagonists, in, Goodman and Gilman's Phamacol. Basis of Therapeutics. 9th Ed. Hardman JG, Molinof PB, AG (ed). M. Graw Hill. New York pp. 119 – 125.

Imam SZ, Ali SF (2000). Selenium an antioxidant, attenuates methamphetamine-induced dopaminergic toxicity and peroxynitrite generation. Brain Res. 855: 186-191.

Imam SZ, Newport GD, Hzak Y, Cadel FI, Slikker W, Ali SF (2001). Peroxyntrite play a role in Methanphetamine – induced dopaminergic neurotoxicity. Evidence from mice lacking neuronal nitric – oxide synthase gene or overexpressing copper – zinc superoxide dismutase. J. Neurosci. 26: 745 – 749.

John DM, Alan S, David RT (1990).Sandard haematoxyline and eosin stain for paraffin sections. Theory and practice of histological technique,Churchhill Livingston,London,3rd Ed. p .112.

Kampen JV, Robertson H, Hagg T,Drobitch R (2003). Neuroprotective action of the ginseng extract G115 in two rodent models of parkinson's disease. J. Neurochem. 76: 745 – 749.

Innes IR, Nickerson M (1975). Noradrenaline, Adrenaline, and the sympatominetic amines, in Pharmacol. Basis of Ther. Goodman & Gilmanis (5th ed.) McMillan, N.Y. pp. 447 – 512.

Iwu MM, Igboko O (1982). Flavonoid of Garcinia kola seeds. J. Nat. Prod. 45: 650 – 651.

Iwu MM (1985). Antihepatotoxic constituents of G. kola seeds Experimentia 41: 699 – 700.

Iwu MM, Igboko OA, Okunji CO, Tempesta MS (1990). Antidiabbetic and Aldose Reductase Activity of Biflavonones of Garcinia kola. .J. Pharmac. Pharamacol. 42: 290 - 292.

King EJ, King PR (1954). Estimation of serum phosphatase by determination of hydrolysed phenol with amino – antipyrene. J. Clin. Path. 7: 322 – 327.

Mercer DW, Talamo TS (1987). The role of biochemical markers in the management of cancer, In, Clinical Studies in Medical Biochemistry: Glew RH, Peters SP (ed). Oxford University Press. 25 – 34.

Murrey RK, Granner DK, Mayer PA, Rodwell VN (2000-). Harpers' Biochemistry (25th ed.) McGraw Hill. New York pp. 242 – 245.

Muragesh KS, Yeliogor VC, Maiti BC, Maity TK (2005). Hepatoprotective and Antioxidenat role of Berberis tinetoria Lesch leaves on Paracetamol-induced hepatic damage in rats. Iranian J. Pharmacol. 22: 107 – 109.

Nwafor A, Ogheneaga IE (1992). Influence of Garcinia kola on in-vivo secretion of gastric acid. Afr. J. Pharmacol. 22: 107-109.

Nwanjo HU, Okafor MC, Oze GO (2005). Changes in biochemical parameters of kidney function in rats co-administered with chloroquine and aspirin. J. Clin. Sci. 23: 10-12.

Nwaoha PU (2007). Garcinia kola, Health related properties. J. ENV. Neurosci. Biomedicine 1(1): 26 – 32.

Ojiako OA, Nwanjo HU (2006). Effect of Co-administration of chloroguine with paracetamol or ibuprofen on renal function of rabbits. Afr. J. Biotech. 5(8): 668 – 670.

Olaleye SB, Farombi EO, Adewoye EA, Owoyele VB, Onuasanwo, SA (2000). Analgasic and anti inflammatory effect of kolaviron (A G. kola seed extract). Afr. J. Biomed. Sci. 3(3): 171 – 174.

Orie NN, Okon Eu (1993). The Brochodilatory effect of Garcinia kola seeds. East Afr. Med. J. 70(3): 143 – 145.

Pasternak CA (1975). Introduction to Human Biochemistry Oxford. Medical Pub. New York pp. 199 – 205.

Rreithman, Frankel (1957). A colorimetric method for the determination of serum glutamic oxaloacelate and glutamic pyruvic transaminase. Am J. Clin Path. 28: 56.

Schaless O, Schales SS (1941). Estimation of serum chloride using mercuric citrate method. J. Boil. Chem. 140: 879 – 884.

Stockley IH (1991). Drug interaction (2nd ed). Blackwell Sci. Pub. Lon. p 540.

Tietz N (1996). Lever Function Tests, Nitrogen Metabolites and Renal Function In: Fundamentals of Clinical Chemistry. 3rd ed. W.B. Saunders, Philadelphia pp. 476 – 576.

Uko OJ, Usma A, Ataja MA (2001). Some biological activities of Garcinia kola in growing rats. Vaternarski Archive 7(5): 287 – 297.

Yakubu MT, Akanji MA, Balu TO (2005). Protective effect of ascorbic acid in some selected tissue of ranitidine treated rats. Nig. J. Biochem. Molecular. Boil 16: 17 – 18.

4

Betaine reduction of hyperhomocysteinemia and enhancement of 5-hydroxyindoleacetic acid in ethanol-induced hyperhomocysteinemia in rabbits

Masoud Alirezaei[1]*, Mehdi Saeb[1], Katayoun Javidnia[2], Saeed Nazifi[3], Najmeh Khalighyan[2], and Saeedeh Saeb[1]

[1]Department of Biochemistry, School of Veterinary Medicine, Shiraz University, Shiraz, 71345, Iran.
[2]Medicinal and Natural Products Chemistry Research Center, Shiraz University of Medical Sciences, Shiraz, Iran.
[3]Department of Clinical Pathology, School of Veterinary Medicine, Shiraz University, Shiraz, 71345, Iran.

Hyperhomocysteinemia is a hypothesis for the association of homocysteine with cerebrovascular diseases, neurodegenerative diseases and depression of mood. Thus, we examined whether oral betaine can act as a preventive agent in ethanol-induced hyperhomocysteinemia on the monoaminergic system. A total of 32 New Zealand White rabbits were divided into four groups (n=8) among which is the control group (C). The ethanol group (E) was administered ethanol at a dosage of 4 g/kg daily. The betaine group (B) received betaine at a dosage 1.5% (w/w) of the diet daily, and the betaine and ethanol group (B and E) was administered with the betaine group diet; after one hour the rabbits received ethanol at a dosage of 4 g/kg daily. Blood samples were taken in the morning of the day before beginning treatment (0.0 day) and on the 30th, 60th and 90th day of the treatment. Serum folate and vitamin B12 levels were determined using a radioimmunoassay, total plasma homocysteine (tHcy) level was determined by homocysteine EIA kit, and 5- hydroxyindoleacetic acid (5-HIAA) of plasma was measured with HPLC-ECD. There was a significant negative correlation between 5-HIAA and tHcy in the E group (r=-0.473, P=0.02), and compared to the E group the concentrations of 5-HIAA in the B and E group increased considerably (p<0.05). In contrast to the E group, significantly high concentrations of 5-HIAA were observed in the B and C groups. While the serum concentrations of vitamin B12 showed no significant difference in the B and E group on the 90th day compared to the control group, the serum concentrations of folate on the 90th day differed significantly (p<0.05). However, no significant difference was observed between tHcy and gender. Overall, oral pretreatment with betaine significantly prevented ethanol-induced hyperhomocysteinemia, subsequently increasing 5-HIAA in the plasma as well as vitamin B12 and folate in the serum. Thus, betaine may be recommended as a pretreatment method for depressive patients with alcoholism.

Key words: Betaine, hyperhomocysteinemia, 5-HIAA, ethanol, vitamin B12, folate.

INTRODUCTION

Chronic alcoholism leads to elevated plasma homocysteine levels, as shown by clinical investigations and animal experiments (Bleich et al., 2004). Homocysteine, a metabolite of the essential amino acid methionine, can be either remethylated to methionine by enzymes that require folate or cobalamin or catabolized by cystathionine β-synthase, a pyridoxine-dependent enzyme, to form cysteine (Figure 1) (Kruman et al.,

2000). The formation of methionine from homocysteine can occur either via betaine or via 5-methyl-tetrahydrofolate (MTHF). Animal studies have shown that both pathways are equally important and betaine is a vital methylating agent (Craig, 2004). Liver betaine homocysteine methyl transferase (BHMT) concentrations increase when rats are fed diets supplemented with betaine or choline, showing an adaptive change in the catabolism of betaine (Finkelstein et al., 1983). Elevated total homocysteine (tHcy) level has been observed as a result of chronic alcohol consumption in rats and ethanol has altered sulfur amino acid metabolism, including decreased conversion of methionine to S-adenosylmethionine (SAMe) and homocysteine to methionine (Cravo et al., 1996).

Folate and homocysteine are related through the one-carbon cycle, which involves the production of S-adenosyl methionine from adenosine triphosphate and methionine. SAMe, which is uniformly distributed in the brain, serves as the major donor of methyl groups required in the synthesis of neuronal messengers and membranes (Papakostas et al., 2005). Folate and vitamin B12 deficiency, hyperhomocysteinemia and the T677 allele of the methylenetetrahydrofolate reductase (MTHFR) gene, which cause impaired methylation reactions in the central nervous system, have been associated with depressive disorders (Kim et al., 2008). In addition, methyl folate has been proved to have an antidepressant effect and correlates with cerebrospinal fluid 5-hydroxindoleacetic acid (5-HIAA) (Atmaca et al., 2005). Patients with severe hyperhomocysteinemia exhibit a wide range of clinical manifestations including neurological abnormalities such as mental retardation, cerebral atrophy, and depression (Kruman et al., 2000). Monoaminergic abnormalities have been implicated in the pathophysiology of depression and alcoholism. For example, lower cerebrospinal fluid (CSF) 5-HIAA levels with alcoholism are associated with higher lethality of suicide attempts in major depression (Sher et al., 2007).

It is well known that alcoholism is associated with altered CSF monoamine metabolite levels. The reduction of serotonin metabolite, 5-HIAA, has been observed in the serum samples of depressive patients (Bose et al., 2004). Taking the above into consideration, we hypothesized that the oral administration of betaine prior to ethanol can act as a methylating agent to increase the level of 5-HIAA in ethanol-induced hyperhomocysteinemia in rabbits. We also investigated how plasma tHcy varied with concentrations of vitamin B12 and folate.

---

*Corresponding author. E-mail: Alirezaei_m54@yahoo.com.

Abbreviations: 5-HIAA, 5- hydroxyindoleacetic acid; **tHcy**, total homocysteine; **HPLC-ECD**, high performance liquid chromatography-electrochemical detector; **MTHF**, 5-methyl tetrahydrofolate; **BH4**, tetrahydrobiopterin; **BHMT**, betaine-homocysteine methyltransferase; **DHFR**, dihydrofolate reductase; **HVA**, homovanillic acid; **MTHFR**, methyltetrahy-drofolate reductase; **NMDA**, N-methyl-D-aspartate; **SAMe**, S-adenosyl methionine; **SAH**, S-adenosyl homocysteine; **SAHH**, S-adenosyl homocysteine hydrolase; **CSF**, cerebrospinal fluid; **5HT**, serotonin; **CBS**, cystathionine beta-synthase; **GSH**, glutathione; **DMG**, dimethyl glycine; **MS**, methionine synthase; **MAT**, methionine adenosyl transferase.

MATERIALS AND METHODS

Alcohol (Ethanol 95%) and 1-octanesulfonic acid sodium salt were from Merck Chemical Company (Merck, Darmstadt, Germany). Betaine (Betafin® 96%) was obtained from Biochem Company (Lohne, Germany). 5-Hydroxindoleacetic acid was purchased from Sigma (St Louis, MO, USA). SimulTRAC-SNB Radioassay kit vitamin B12 [57Co]/Folate [125I] was prepared by MP, Biomedical, LLC (CNI Pharmaceutical, USA) and the homocysteine kit was prepared by Axis® Homocysteine EIA (Axis-Shield AS, Germany). All other chemicals used were of analytical grade.

Animals and experimental design

All animal experimentation procedures were approved by the Institutional Animal Care and Use Committee of Shiraz University of Medical Science. A total of 32 adult New Zealand White rabbits (2.0 - 2.5 kg) obtained from the animal house of Shiraz University of Medical Science were housed under standard conditions of temperature (23 ± 2°C) and illumination (12-h light–dark cycle). They were provided with standard chow diet (average 50 g/kg), and water *ad libitum* in 2-week of acclimation period. Then, animals were divided into four groups; the first group (Control) received standard chow diet. Second group (Ethanol) were administered with ethanol with a dosage of 4 g/kg per 500 ml water daily plus standard chow diet. Third group (Betaine), received the standard chow diet, plus betaine with a dosage 1.5% (w/w) of the diet soluble in water daily by using gavage and fourth group (Betaine and Ethanol) administered with the Betaine group diet, after one hour rabbits received ethanol with a dosage of 4 g/kg per 500 ml water daily (pretreatment method) (Ji and Kaplowitz, 2003; Song et al., 2003). Each group consisted of 8 (male and female) animals. The total period of study was 90 days. Weight gains and food consumption was determined at weekly intervals.

Blood samples and biochemical analyses

Blood samples were collected from the rabbits in a fasting state and were taken from the marginal ear vein in the morning of day before beginning treatment (0.0 day) and the 30th, 60th and 90th days of the treatment. 1.0 ml of whole blood were drawn into tubes of ethylenediamine tetra-acetic acid (EDTA), centrifuged, separated into plasma aliquots and the remaining whole blood was placed in another tubes and collection of serum assessed in micro tubes. Serum and plasma aliquots stored at -70°C until analysis. Serum folate and vitamin B12 levels were determined using a radioimmunoassay; (SimulTRAC-SNB Radioassay kit vitamin B12/Folate). In short, the sample volume used was 200 μl diluted serum (1/8). All experiments were performed in duplicate and the linearity of dilution that was used was 100 and 110% for vitamin B12 and folate analyses respectively. The intra- and inter-assay coefficients of variations for the determination of folate were less than 5.8 and 8.9%; and for vitamin B12 they were less than 6.2 and 7.5%, respectively (Ferrucci et al., 2007; Golbahar et al., 2005). Total plasma homocysteine level was determined by Axis® Homocysteine EIA kit. In brief, the sample volume used was 25 μl. Absorbance was measured at a wavelength of 450 nm using ELISA reader (STAT FAX 2100, USA). All estimations were performed in duplicate and the intra-assay coefficient of variation was <10% and the detection limit of the tHcy assay was 2.0 μM/l (Golbahar et al., 2005; Karthikeyan et al., 2007).

Determination of 5-HIAA

Plasma 5-hydroxyindoleacetic acid concentrations were measured

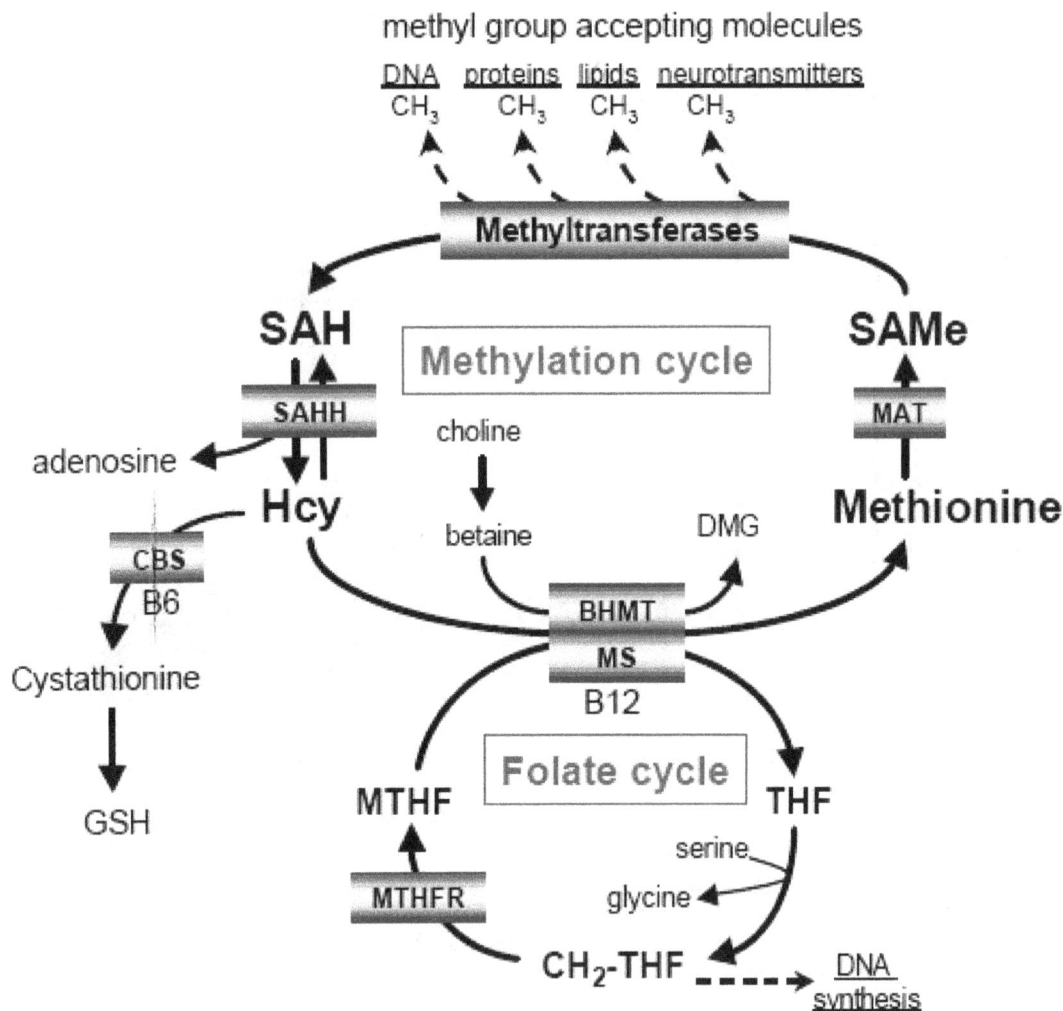

Figure 1. Homocysteine metabolism- Homocysteine has three main metabolic fates: to be remethylated to methionine, to enter the cysteine biosynthetic pathway, and to be released into the extracellular medium. Hcy, homocysteine; CBS, cystathionine beta-synthase; GSH, glutathione; DMG, dimethyl glycine; MS, methionine synthase; BHMT, betaine-homocysteine methyltransferase; MTHF,5-methyl tetrahydrofolate; MTHFR, methyltetrahydrofolate reductase; THF, tetrahydrofolate; MAT, methionine adenosyl transferase; SAMe, S-adenosyl methionine; SAH, S-adenosyl homocysteine; SAHH, S-adenosyl homocysteine hydrolase. (Bottiglieri, 2005). Bottiglieri T. Homocysteine and folate metabolism in depression. Progress in Neuropsychopharmacology and Biological Psychiatry (2005); 7:1103-1112. Bottiglieri T. Homocysteine and folate metabolism in depression. Progress in Neuropsychopharmacology and Biological Psychiatry 2005; 7:1103-1112

by high-performance liquid chromatography with electrochemical detection (Chi et al., 1999). In short, The HPLC system consisted of a Constametric1000 pump (Knauer, Germany), a manual Rheodyne7725 injection valve equipped with a 20-μl loop, a 3 mm particle size (250 × 4.6 mm, I.D.) with C18 analytical column (Knauer, Germany). End-point detection was achieved with an Introamperometric detector (EC3000, GmbH, Germany). The operating potential was 0.75 V. The mobile phase consisted of 0.1 M KH_2PO_4 acetonitrile (84:16, v/v) and 1-octane sulphonic acid (100 mg/l) adjusted to pH 4.75 (with 0.5 M K2HPO4). The flow-rate was 1.0 ml/min. Peak height rather than area in the chromatography was normally measured. Concentration of 5-HIAA was calculated by interpolation of its standard curve. Working standards for the assay were prepared using the mobile phase as the diluent and consisted of six concentration points over the range 2 - 32 ng/ml.

The plasma extraction procedure used was combination of a protein precipitation step via acetonitrile and centrifugation at 14500 g for 5.0 min at 4℃. Normally 20 μl of the supernatant was injected into the HPLC system.

Statistical analysis

Statistical analysis was performed using a computer statistical package SPSS 11.0 for windows (SPSS, Inc., Chicago, I L., U S A). The significance of the differences between the groups was assessed with One-Way ANOVA. Tukey's test was used after One-Way ANOVA to determine statistical differences among all of the groups. The significant differences within the groups at monthly intervals were assessed with repeated measures ANOVA. The

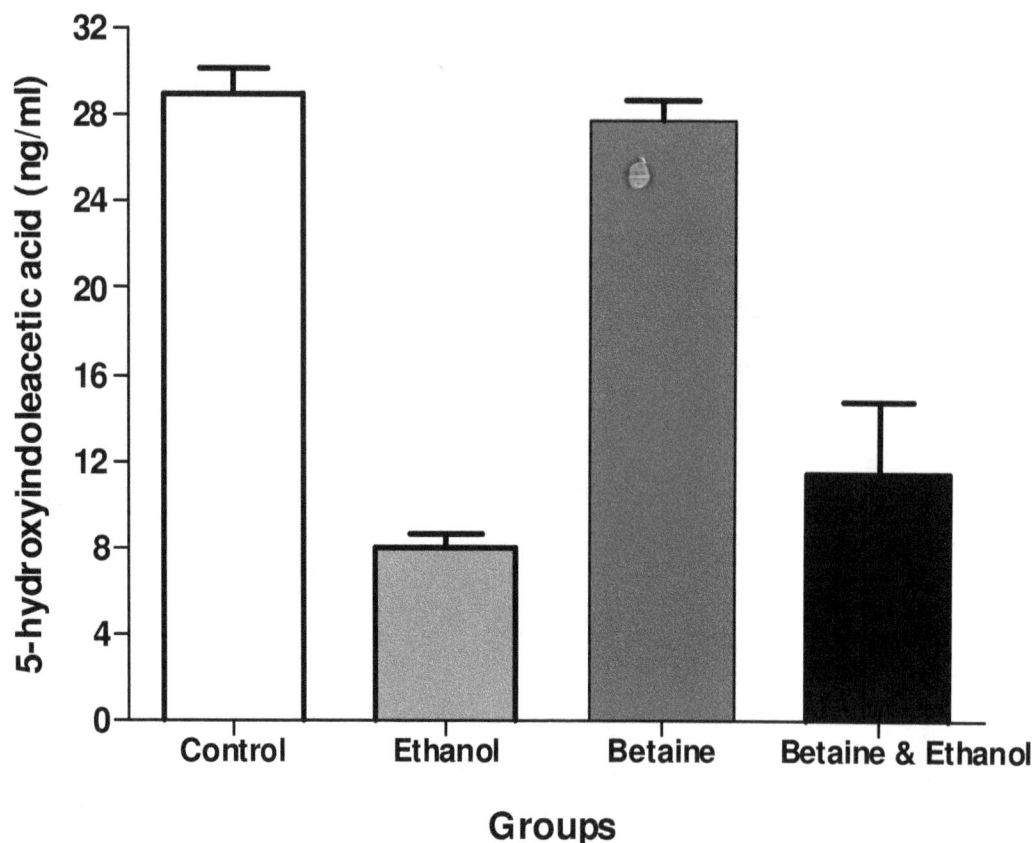

Figure 2. 5- HIAA levels of plasma between the control and treatment groups; on the 90th day of the treatment. Values represent mean ± SD of 5-HIAA; (*, **, ***) indicate statistical difference (p < 0.05) between the groups. Tukey's test was used after One-Way ANOVA to determine statistical differences among all of the groups.

relationship between tHcy and 5-HIAA in the plasma of the E group on the 90th day was calculated by Pearson's correlation test. Independent sample t-Test was used for tHcy in both male and female rabbits from the Ethanol group. Data were expressed as mean ± SD and p-values of <0.05 were regarded as statistically significant.

RESULTS

5-HIAA and tHcy were compared in the treatment groups and control having significant differences between the groups only on the 90th day. Therefore, the differences between the groups on the 90th day of the treatment for 5-hydroxyindoleacetic acid and tHcy have been illustrated in Figures 2 and 3 respectively. There was a significant negative correlation between 5-HIAA and tHcy in the E group (r = -0.473, P = 0.021), and a significant increase in the concentration of 5-HIAA in the B AND E group compared to the E group (p<0.05). Significantly high concentrations of 5-HIAA were also observed in the B and C groups, in contrast to the E group (p<0.05). Figures 4 and 5 show the folate and vitamin B12 concentrations of the control and the treatment groups in

one-month intervals. While vitamin B12 showed no significant difference in the B AND E group on the 90th day compared to the control group, the serum concentrations of folate on the 90th day differed significantly.

Significant differences were observed for folate on the 30th, 60th and 90th days of the treatment with regard to the 0.0 day for the all treatment groups (Figure 4). Also, significant differences for vitamin B12 on the 30th, 60th and 90th day of the treatment in the B and E groups and on the 30th and 60th day in the B AND E group were observed with respect to the 0.0 day (Figure 5). The plasma total homocysteine levels in the E group are not significantly higher in male (16.32 µM/l) versus female rabbits (14.75 µM/l). Therefore, no significant difference was observed between tHcy and gender.

DISCUSSION

Our results support the hypothesis that betaine reduces hyperhomocysteinemia (Finkelstein, 2007) and enhances 5-hydroxyindoleacetic in ethanol-induced hyperhomocysteinemia in rabbits. To the best of our

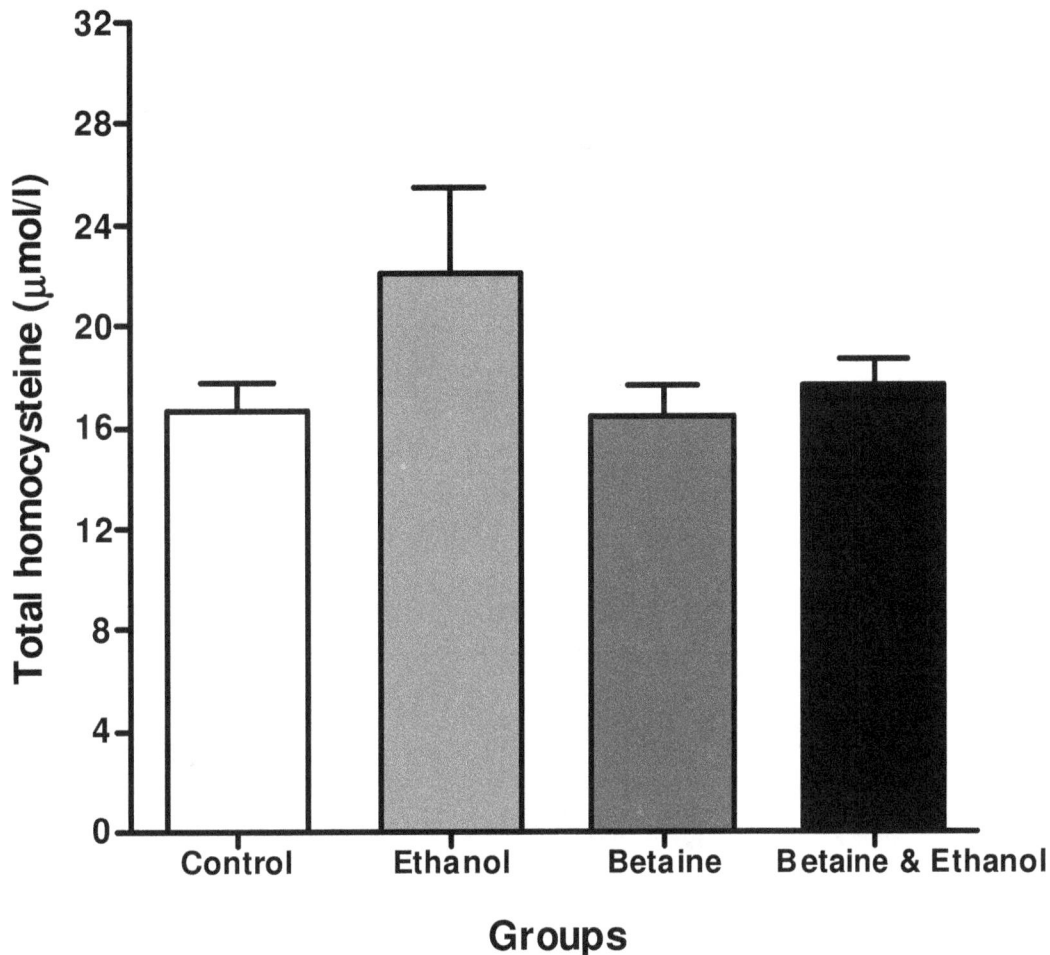

Figure 3. Total homocysteine levels of plasma between the control and treatment groups; on the 90th day of the treatment. Values represent mean ± SD of tHcy; (*, **) indicate statistical difference (p < 0.05) between the groups. Tukey's test was used after One-Way ANOVA to determine statistical differences among all of the groups.

knowledge this study is the first ethanol-induced hyperhomocysteinemia investigation of associations between folate, vitamin B12, total homocysteine and serotonergic system to determine therapeutic effects of betaine. The protective effects of betaine were confirmed by higher folate and vitamin B12 in serum and lower total homocysteine in plasma. The data we found provides new evidence that the tHcy of plasma decline, while 5-HIAA elevate with oral betaine. Also, there is a negative association between the 5-HIAA (metabolite of serotonergic system) and ethanol treatment. While vitamin B12 showed no decrease in E group on the 90th day with regard to the 30th day of the treatment, the concentration of folate differed and folate seems to be associated with increased risk for hyperhomocysteinemia. Furthermore, vitamin B12 showed no significant difference in Betaine and Ethanol group on the 90th day compared to the control group while, the concentration of folate differed significantly. Thus, the important role of folate rather than vitamin B12

is established and confirms by previous studies (Bottiglieri, 2005; Kim et al., 2008; Papakostas et al., 2005). Although, plasma total homocysteine levels were higher in the male versus the female rabbits, it was not significant. It is clear that this difference is related to sex steroids (Giltay et al., 1998).

Homocysteine is a non-essential, thiol containing and potentially cytotoxic 4-carbon α-amino acid formed during methionine metabolism through the demethylation of methionine (Nasir et al., 2007). In recent years increasingly more evidences support the hypothesis that elevated total homocysteine is an independent risk factor for coronary vascular and neurodegenerative diseases (Bidulescu et al., 2009; Chandra et al., 2006; Folstein et al., 2007; Kim et al., 2008; Zieminska and Lazarewicz, 2006). High levels of homocysteine are associated with cerebrovascular disease, abnormality of monoamine neurotransmitters, and depression of mod. A plausible hypothesis for these associations is that of high homocysteine levels (Folstein et al., 2007). Traditional

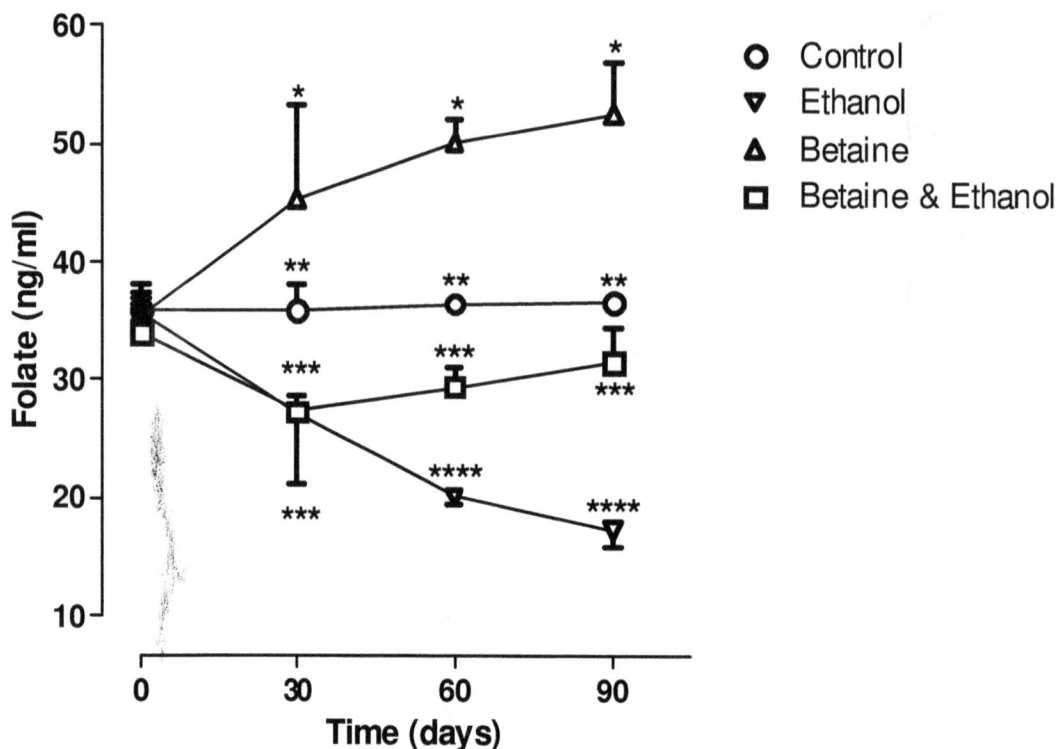

Figure 4. Effect of administration of betaine and ethanol on the serum folate. Values represent mean ± SD of Folate; (*, **, ***, ****) indicate statistical difference (p < 0.05) between the groups on the same days. Tukey's test was used after One-Way ANOVA to determine statistical differences among all of the groups.

explanations of the mechanism of Hcy neurotoxicity point to key role of disturbance in methylation and remethylation process. SAMe accumulated in cells in hyperhomocysteinemia is a very strong competitive antagonist of many transferase (Bottiglieri et al., 2000; Zieminska and Lazarewicz, 2006). Hcy toxicity and impaired methylation may be potential mechanisms involved in the clinical spectrum of MTHFR and cerebral folate deficiency. However, folate may also be directly involved in the regulation of neurotransmitter metabolism. Low concentrations of CSF 5-hydroxyindole acetic acid, a metabolite of serotonin (5-HT) that reflects global CNS tissue levels, have been reported in folate-deficient patients with various neuropsychiatric illnesses and severe depression (Bottiglieri, 2005). Our observation for vitamins, tHcy and 5-HIAA in the treatment groups of the present study supports the idea that vitamin B12 and folate are associated with 5-HIAA through their involvement in homocysteine remethylation.

A possible mechanism linking folate deficiency and perturbed monoamine neurotransmitter function may involve tetrahydrobiopterin (BH4) metabolism, a cofactor required in the synthesis of monoamine neurotransmitters. Due to the structural similarities between folate and BH4, the folate enzymes MTHFR and dihydrofolate reductase (DHFR) have been postulated to be involved in BH4 metabolism. However, in depressed patients, significant correlations between red cell folate and CSF BH4, and also between CSF monoamine metabolites have been reported (Bottiglieri, 2005; Bottiglieri et al., 2000). With regard to the fact that folate appears to influence the rate of synthesis of tetrahydrobiopterin, a cofactor in the hydroxylation of phenylalanine and tryptophan, rate-limiting steps in the biosynthesis of dopamine, norepinephrine, and serotonin, neurotransmitters postulated to play a role in the monoamine hypothesis of affective disorders. In addition when folate has been administered in parenteral and certain oral forms, both SAMe and methyl folate have been shown to have antidepressant efficacy. (Atmaca et al., 2005; Hofmann et al., 1996). In contrast, in some cases causative treatment of hyperhomocysteinemia and depression to reduce the levels of Hcy in human body fluid, include; supplementation of the diet with folic acid and//or Vitamin B12 and B6 it was unsuccessful (Zieminska and Lazarewicz, 2006). Therefore, we designed the present investigation to determine the preventive effect of betaine on ethanol-induced hyperhomocysteinemia. The results of the present study demonstrated that betaine supplementation to alcohol-fed rabbits promotes the generation of hepatic S-adenosyl methionine due to stimulation of methionine synthesis by the alternate BHMT pathway (Bottiglieri, 2005; Finkelstein, 2007). Betaine, a methyl-donor that

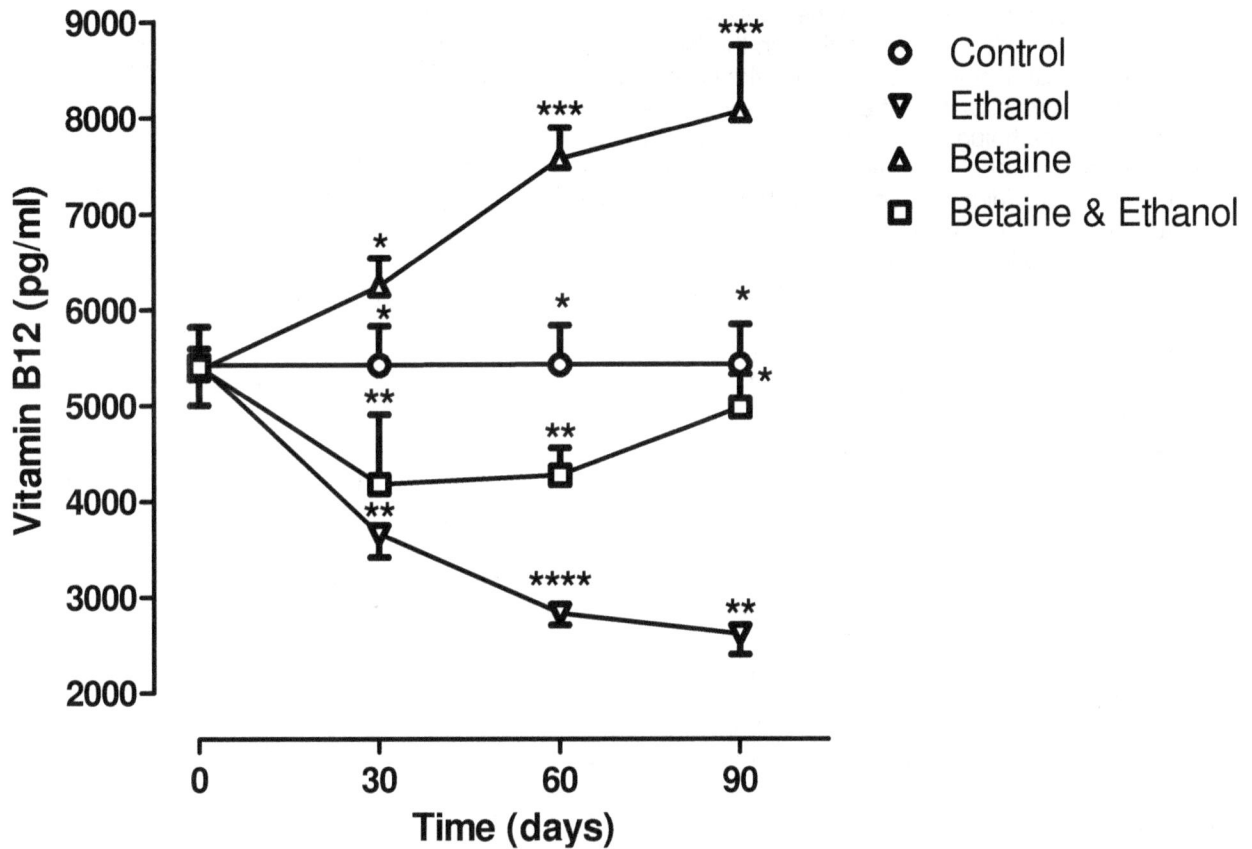

Figure 5. Effect of administration of Betaine and Ethanol on the serum vitamin B12. Values represents mean ± SD of vitamin B12; (*, **, ***, ****) indicate statistical difference ($p < 0.05$) between the groups on the same days. Tukey's test was used after One-Way ANOVA to determine statistical differences among all of the groups.

continuously generates S-denosyl methionine, is shown to lead to long term lowering of plasma homocysteine during supplementation in the dietary intake range of 1.5% (w/w) (Ji and Kaplowitz, 2003).

It seems the intracellular content of S-adenosyl methionine (SAMe) is the likely gauge for the availability of methionine. Changes in the concentrations of SAMe and S-adenosyl homocysteine (SAH) affect the methionine conserving enzymes of both of the methionine cycles. SAMe inhibits the activity of methionine adenosyl transferase (MAT) and also down-regulates the expression of the MAT gene in hepatocytes. Conversely, low concentrations of SAMe allow the expression of this gene in liver cells (Finkelstein, 2007). Thus, SAMe which issuees from BHMT pathway in our study via betaine supplementation can probably inhibit MTHFR, subsequently MS and ultimately increase vitamins (B12, folate) by saving its consumption through the classical pathway involving MS (Figure 1). With regard to some limitations, measurement of depression and the molecular mechanism of homocysteine neurotoxicity are not evaluated from the present investigation. One limitation of the present study is that there was no instrument or program to define

depression of rabbits. Another limitation is the insufficiently sensitive index to measure the changes in depression and the timing of clinical improvement. In contrast, previous studies have reported that homocysteine, being an excitatory neurotoxic by excessive accumulation of cytosolic calcium, N-methyl D-aspartate (NMDA) receptor overstimulation, generating oxidative stress and activation of apoptotic pathway (Chandra et al., 2006).

According to the monoamine hypothesis of affective disorders, depression is due to a deficiency of 5-HT, or norepinephrine, or both these monoamines (Atmaca et al., 2005; Folstein et al., 2007). In the present study, the results of the ethanol group of rabbits showed an abnormality of tHcy and 5-HIAA that is related to deficiency of folate and vitamin B12. In animal studies, low dietary intake of choline and betaine results in aberrant DNA methylation and possible increased atherogenesis and independently of folate; the dietary intake of choline and betaine are inversely associated with plasma homocysteine (Bidulescu et al., 2009). In the present study, it was concluded that betaine is a preventable metabolic agent and that the ingestion of lipotropic agents is part of a preventative strategy.

Betaine prevented a decrease in the content of vitamin B12 and folate and decreased the tHcy in the betaine as well as betaine and ethanol groups of rabbits. It is well known that, depression may be the consequence of a developmental pathology affecting serotonergic and/or dopaminergic neuronal (Atmaca et al., 2005). The present study provides evidences for the serotonergic neurotoxicity of homocysteine (Bleich et al., 2004); it also indicates that the elevated levels of this excitatory amino acid in the ethanol group have been declined in the betaine and ethanol group of rabbits with long term betaine therapy. Therefore, these relationships probably have resulted in elevation of serotonin that plays a major role in the configuration of mood and depression (Bleich et al., 2004; Bottiglieri, 2005; Kim et al., 2008).

Conclusions

The results of the present study may have important implications in understanding the possible role of betaine on the treatment of human depression and betaine may be recommended in the pretreatment method of depressive patients with alcoholism. It appears that a limited betaine-dependent remethylation of homocysteine to methionine (BHMT pathway) also exists within mammalian brains (Finkelstein, 2007). Interestingly, the existence of brain BHMT pathway would enable explains details of our results. Nevertheless, future investigations with controlled human studies are required to confirm whether the betaine exactly increases 5-HT output or not.

ACKNOWLEDGMENTS

This research was financially supported by grant 87-GR-VT-1 from the Shiraz University Research Council. The authors wish to thank Dr. R. Miri (Manager of Medicinal and Natural Products Chemistry Research Center, Shiraz University of Medical Science, Shiraz, Iran), for his kind cooperation. to Dr. N. Tanideh and Dr. A. Tamadon (The members of Stem cell and Transgenic Technology Research Center, Shiraz University of Medical Sciences, Shiraz, Iran) for their scientific cooperation.

REFERENCES

Atmaca M, Tezcan E, Kuloglu M, Kirtas O, Ustundag B (2005). Serum folate and homocysteine levels in patients with obsessive-compulsive disorder. Psychiatr. Clin. Neurosci., 5: 616-620.

Bidulescu A, Chambless LE, Siega-Riz AM, Zeisel SH, Heiss G (2009). Repeatability and measurement error in the assessment of choline and betaine dietary intake: The Atherosclerosis Risk in Communities (ARIC) Study Nutr. J., 1: 14-20.

Bleich S, Degner D, Sperling W, Bönsch D, Thürauf N, Kornhuber J (2004). Homocysteine as a neurotoxin in chronic alcoholism. Progress Neuropsychopharmacol. Biol. Psychiatr. 3: 453-464.

Bose D, Durgbanshi A, Capella-Peiró ME, Gil-Agustí M, Esteve-Romero J, Carda-Broch S (2004). Micellar liquid chromatography determination of some biogenic amines with electrochemical detection. J. Pharm. Biomed. Anal., 2: 357-363.

Bottiglieri T (2005). Homocysteine and folate metabolism in depression. Progress Neuropsychopharmacol. Biol. Psych., 7: 1103-1112.

Bottiglieri T, Laundy M, Crellin R, Toone BK, Carney MWP, Reynolds EH (2000). Homocysteine, folate, methylation, and monoamine metabolism in depression. Br. Med. J., 2: 228-232.

Chandra G, Gangopadhyay PK, Senthil KKS, Mohanakumar KP (2006). Acute intranigral homocysteine administration produces stereotypic behavioral changes and striatal dopamine depletion in Sprague–Dawley rats. Brain Res., 1: 81-92.

Chi JD, Odontiadis J, Franklin M (1999). Simultaneous determination of catecholamines in rat brain tissue by high-performance liquid chromatography. J. Chromatogr. B: Biomed. Sci. Appl., 2: 361-367.

Craig SA (2004). Betaine in human nutrition. Am. J. Clin. Nutr., 3: 539-549.

Cravo ML, Gloria LM, Selhub J, Nadeau MR, Camilo ME, Resende MP, Cardoso JN, Leitao CN, Mira FC (1996). Hyperhomocysteinemia in chronic alcoholism: Correlation with folate, vitamin B-12, and vitamin B-6 status. Am. Soc. Nutr., 63: 220-224.

Ferrucci L, Guralnik JM, Bandinelli S, Semba RD, Lauretani F, Corsi A, Ruggiero C, Ershler WB, Longo DL (2007). Unexplained anaemia in older persons characterised by low erythropoietin and low levels of pro-inflammatory markers. Br. J. Haematol., 6: 849-855.

Finkelstein JD (2007). Metabolic regulatory properties of S-adenosylmethionine and S-adenosylhomocysteine. Clin. Chem. Lab. Med., 12: 1694-1699.

Finkelstein JD, Martin JJ, Harris BJ, Kyle WE (1983). Regulation of hepatic betaine-homocysteine methyl transferase by dietary betaine. J.Nutr., 3: 519.

Folstein M, Liu T, Peter I, Buel J, Arsenault L, Scott T, Qiu WW (2007). The homocysteine hypothesis of depression. Am. J. Psychiatr., 6: 861-867.

Giltay EJ, Hoogeveen EK, Elbers JMH, Gooren LJG, Asscheman H, Stehouwer CDA (1998). Effects of sex steroids on plasma total homocysteine levels: A study in transsexual males and females. Endocrine Soc., 83: 550-553.

Golbahar J, Aminzadeh MA, Hamidi SA, Omrani GR (2005). Association of red blood cell 5-methyltetrahydroafoate with bone mineral density in postmenopausal Iranian women. Osteoporos. Int., 12: 1894-1898.

Hofmann P, Loimer N, Chaudhry HR, Pfersmann D, Schmid R, Wieselmann G (1996). 5-Hydroxy-indolacetic-acid (5-HIAA) serum levels in depressive patients and ECT. J. Psychiatr. Res., 3: 209-216.

Ji C, Kaplowitz N (2003). Betaine decreases hyperhomocysteinemia, endoplasmic reticulum stress, and liver injury in alcohol-fed mice. Gastroenterology, 5: 1488-1499.

Karthikeyan G, Thachil A, Sharma S, Kalaivani M, Ramakrishnan L (2007). Elevated high sensitivity CRP levels in patients with mitral stenosis and left atrial thrombus. Int. J. Cardiol., 3: 252-254.

Kim JM, Stewart R, Kim SW, Yang SJ, Shin IS, Yoon JS (2008). Predictive value of folate, vitamin B12 and homocysteine levels in late-life depression. Br. J. Psychiatr., 4: 268-274.

Kruman II, Culmsee C, Chan SL, Kruman Y, GuoZ, Penix LR, Mattson MP (2000). Homocysteine elicits a DNA damage response in neurons that promotes apoptosis and hypersensitivity to excitotoxicity. J. Neurosci., 18: 6920-6926.

Nasir K, Tsai M, Rosen BD, Fernandes V, Bluemke DA, Folsom AR, Lima JAC (2007). Elevated homocysteine is associated with reduced regional left ventricular function: The Multi-Ethnic Study of Atherosclerosis. Circulation, 2: 180-187.

Papakostas GI, Petersen T, Lebowitz BD, Mischoulon D, Ryan JL, Nierenberg AA, Bottiglieri T,Alpert JE, Rosenbaum JF, Fava M (2005). The relationship between serum folate, vitamin B12, and homocysteine levels in major depressive disorder and the timing of improvement with fluoxetine. Int. J. Neuropsychopharmacol., 4: 523-528.

Sher L, Oquendo MA, Grunebaum MF, Burke AK, Huang Y, Mann JJ (2007). CSF monoamine metabolites and lethality of suicide attempts in depressed patients with alcohol dependence. Eur. Neuropsychopharmacol., 1: 12-15.

Song Z, Zhou Z, Chen T, HillD, Kang J, Barve S, McClain C (2003). S-adenosylmethionine (SAMe) protects against acute alcohol induced hepatotoxicity in mice. J. Nutr. Biochem., 10:591-597.

Zieminska E, Lazarewicz JW (2006). Excitotoxic neuronal injury in chronic homocysteine neurotoxicity studied *in vitro*: The role of NMDA and Group I metabotropic glutamate receptors. Acta Neurobiol. Exp., 301-309.

5

Biochemical profile of sodium selenite on chemically induced hepatocarcinogenesis in male Sprague-Dawley rats

Nasar Yousuf Alwahaibi[1]*, Siti Belkis Budin[2] and Jamaludin Mohamed[2]

[1]Department of Pathology, College of Medicine and Health Sciences, Sultan Qaboos University, Muscat – Oman
[2]Faculty of Allied Health Sciences, Department of Biomedical Sciences, University of Kebangsaan, Kuala Lumpur Malaysia.

Despite the success of experimental, clinical and epidemiological studies on selenium as an anti-cancer agent, basic studies on the effects of selenium are still scanty. This study was designed to investigate the biochemical effects of sodium selenite using preventive and therapeutic approaches on chemically induced hepatocarcinogenesis in rats. Rats were divided randomly into 6 groups: negative control, positive control [Diethyl nitrosamine (DEN) + 2-Acetylaminofluorene (2-AAF)], preventive group, preventive control group (respective control for preventive group), therapeutic group and therapeutic control group (respective control for therapeutic group). The activities of plasma alanine transaminase (ALT), aspartate aminotransferase (AST), gamma-glutamyl transferase (GGT), alkaline phosphatase ALP and concentrations of total protein, albumin and globulin were determined by an auto-analyzer. GGT and ALT activities were significantly higher in the positive control, preventive and therapeutic groups when compared with the negative control. Globulin concentration was significantly lower in the positive and therapeutic group controls and higher in the therapeutic group and its respective control when compared with the negative and positive controls, respectively. Plasma GGT enzyme marker could be used as an early marker for liver neoplasm in rats. The effect of selenium on globulin, as an indicator of immunity status, needs to be clarified.

Key words: Alanine transaminase, gamma-glutamyltransferase, liver neoplasm, selenium.

INTRODUCTION

Since the past decades, there has been increasing interest in the role of selenium in the pathogenesis of cancer including liver neoplasm. The poor prognosis and current limited treatment for liver neoplasm are still a major concern as it has been noted that the incidence of liver neoplasm is on the rise worldwide (Parkin et al., 2005). Thus, other preventive approaches such as chemoprevention have been highly emphasized (Kensler et al., 2003).

Since the report of Scharauzer and his colleagues (1977), stating that selenium is a potential human cancer protective agent, worldwide, research on selenium as an anti-cancer agent has escalated. In addition, the publication of Clark's clinical study (Clark et al., 1996) has attracted more researchers in this field. In the Clark's study, selenium was found to dramatically reduce the incidence of cancer, in addition to improvements in the prognosis of prostate cancer, colorectal cancer and lung cancer patients. Furthermore, no signs of selenium toxicity were reported in the Clark's study. In addition, other works have emphasized the anti-cancer activity of selenium (Raymond, 2001).

Selenium is an essential micronutrient mineral, which

*Corresponding author. E-mail: nasarsidab@yahoo.com.

could have a high clinical value in some cancer patients (Valkoo et al., 2006). As a part of the preventive health strategy 'Prevention rather than treatment', dietary selenium intake is an essential element in the protection from many diseases (Raymond, 2001). The dietary reference intakes (DRIs) for selenium has been set at approximately 55 µg per day for adults (IOM, 2000).

It has been reported that the analysis of liver enzymes in plasma reflects cellular damage (Tazi et al., 1980). Liver function enzymes such as alanine transaminase (ALT), aspartate aminotransferase (AST), gamma-glutamyltransferase (GGT), and alkaline phosphatase (ALP) and other biochemical parameters such as total protein, albumin and globulin, are generally used in humans and animals as indicators of liver injury as well as liver response to medicine (Kim and Park, 1994). In addition, basic studies on the effects of selenium are still scanty. Hence, this study aimed at investigating the effects of selenium on some liver enzymes and biochemical markers on chemically induced hepatocarcinogenesis in rats.

MATERIALS AND METHODS

Chemicals

Sodium selenite, DEN and 2-AAF were obtained from Sigma Chemical Co, Germany. All enzymes and biochemical reagents were obtained from Randox Laboratories Ltd, U.K.

Animals and diet

Male Sprague-Dawley rats (6 – 8 weeks old) were obtained from the Laboratory Animal Resource Unit, Faculty of Medicine, University of Kebangsaan Malaysia, Kuala Lumpur, Malaysia. They were housed in plastic cages (3 – 4 rats per cage) with wood chips for bedding. The animals were acclimatized to standard laboratory conditions [temperature (22 – 25°C), humidity (55 ± 10%) with a 12 h light-dark cycle] for one week before the commencement of the experiments. During the entire period of study, the rats had free access to food and water. The rats were maintained on a basal diet (22% crude protein, 5% crude fiber, 3% fat, 13% moisture, 8% ash, 0.85 - 1.2% calcium, 0.6 - 1% phosphorus and 49% nitrogen free extract) (Mouse pellet 702-P from Gold Coin Co, Limited, Malaysia).

According to the manufacturers of basal diet, mouse pellet 702-P contains 0.2 mg/kg of selenium, which is within the recommended reference range. The recommendations of the University of Kebangsaan Malaysia Animal Ethics Committee (UKMAEC) for the care and use of animals were strictly followed throughout the study (UKMAEC No: FSKB/2006/Jamaludin/22- August/170-December-2006).

Experimental design

Fourty-four rats were randomly divided into 6 groups, 6 or 8 rats in each group, as follow: Group 1 (negative control): rats were given normal rat chow and drinking water. Also, a single intraperitoneal (I.P) injection of saline (0.9%) was given. Group 2 (positive control): liver tumors were induced with a single I.P injection of DEN at a dose of 200 mg/kg body weight in saline (Solt and Farber, 1976).

Two weeks after DEN administration, the carcinogen effect was promoted by 2-AAF (0.02%). The promoter was incorporated into the rat chow for 10 weeks. Group 3 (preventive group): 4 weeks before DEN administration, rats were fed with sodium selenite (4 mg/L) through drinking water and stopped at week 4 (the day of commencement of DEN administration). Group 4 (preventive group control): rats in this group served as controls for group 3. Rats were given sodium selenite for 4 weeks only. No DEN or 2-AAF was given instead a single I.P injection of saline (0.9%) was given. Group 5 (therapeutic group): 4 weeks after the start of DEN administration (as in Group 2), the rats were treated with sodium selenite (4 mg/L) through drinking water and this continued until the completion of the experiment (8 weeks). Group 6 (therapeutic group control): rats in this group served as controls for group 5. Rats were given sodium selenite for 8 consecutive weeks. No DEN or 2-AAF was given instead a single I.P injection of saline (0.9%) was given. 16 weeks after the initiation of the experiment, all the rats were fasted overnight and then killed by cervical dislocation under ether anesthesia. Sodium selenite supplementation in drinking water and normal drinking water was renewed every 2–3 days. Diet with 2-AAF was freshly prepared and wood chips for bedding were changed weekly.

Collection of blood and liver tissues

Under ether anesthesia, the blood of all experimental rats was taken by cardiac puncture using 21 G needle and 10 ml syringe. Samples were then collected in EDTA plastic tubes. Plasma was prepared by centrifuging at 3000 g for 15 min. The plasma, in the supernatant, was then pipetted into 2.0 ml Eppendorf cups. In addition, portions of the livers were fixed in 10% neutral buffered formalin for routine histopathological examination.

Enzyme analysis

The activities of ALT, AST, GGT, ALP, total protein and albumin were determined by an auto-analyzer (Selectra E, Vital Scientific N.V, Netherlands). The mean of duplicate reading was taken. Globulin activity was measured as the difference between total protein and albumin.

Statistical analysis

Data were expressed as means ± standard deviation (SD). The data were analyzed using Statistical Package for Social Sciences (SPSS) version 13. Shapiro – Wilk test was used to check the normality of the variable. Accordingly, Student's t and Mann – Whitney's U tests were used to analyze data that follow normal or non–normal behavior of distribution pattern, respectively (Mahajan, 1997). Differences in statistical analysis of data were considered significant at $P < 0.05$.

RESULTS

Histopathological examination of the liver in the positive control showed a completely disrupted architecture. The normal liver cords were displaced with variably-sized neoplastic nodules. The hepatocytes were more than 2 cells thick, paler and showed enlarged vesicular nuclei with prominent nucleoli (Figure 1). However, the preventive and therapeutic groups revealed that the liver was

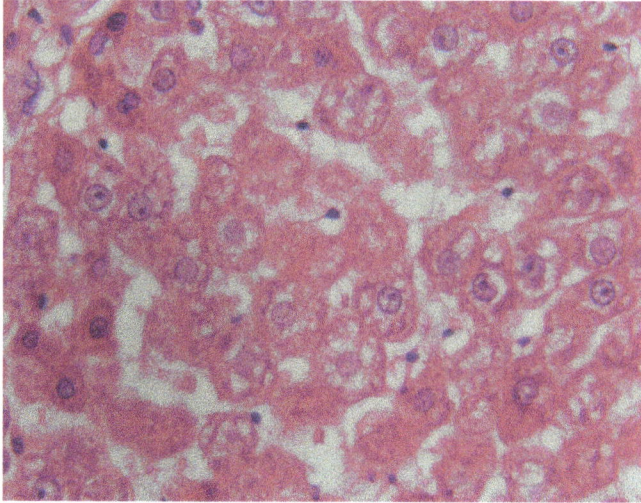

Figure 1. Neoplastic liver, showing enlarged nuclei with prominent nucleoli, as seen in the positive control (received a single I.P injection of DEN at a dose of 200 mg / kg body weight in saline and two weeks later, the carcinogen effect was promoted by 2-AAF (0.02%) and continued for 10 weeks). Hematoxylin and eosin (X 60).

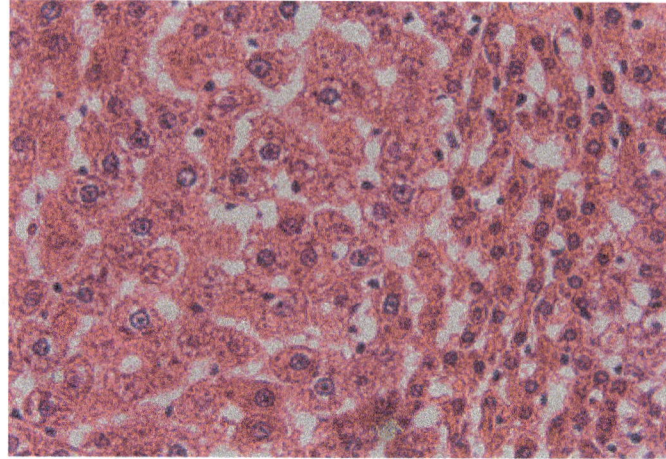

Figure 2. Hyperplastic liver with largely preserved architecture as seen in the preventive group (received sodium selenite and stopped at week 4, the day of commencement of DEN administration as in group 2) and therapeutic group (received a single I.P injection of DEN as in group 2 and 4 weeks later, rats were treated with sodium selenite for 8 weeks). Hematoxylin and eosin (X 40).

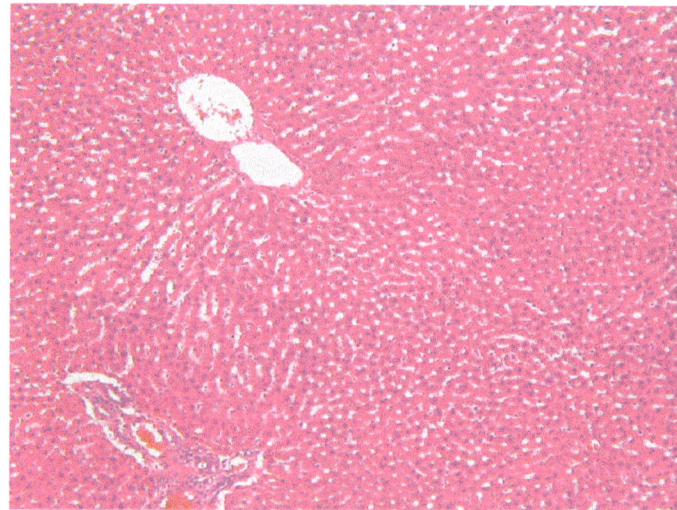

Figure 3. Normal liver architecture as seen in the negative (received normal rat chow and drinking water), preventive (treated with sodium selenite alone for 4 weeks and served as control for group 3) and therapeutic controls (treated with sodium selenite alone for 8 weeks and served as control for group 5). Hematoxylin and eosin (X 10).

nodular but with largely preserved architecture. The majority of varying-sized nodules were hyperplastic (1 cell thick) (Figure 2). The negative, preventive and therapeutic controls were free of any abnormality (Figure 3).

The activities of ALT, AST, ALP and GGT are shown in Table 1. The activity of GGT was significantly higher in the positive, preventive and therapeutic groups when compared with the negative control. On the contrary, GGT activity was significantly lower in the preventive and therapeutic controls (selenium treated groups without carcinogens) when compared with the positive control. Interestingly, the therapeutic group (group 5) showed a significant higher activity when compared with the positive control. In addition, the preventive (group 3) and therapeutic groups showed significantly higher activities of GGT when compared with their respective controls, 4 and 6, respectively. The preventive group showed significantly higher activity of ALT when compared with the negative (group 1), positive (group 2), and respective controls (group 4). On the other hand, the therapeutic control (group 6) showed significantly lower activity of ALT when compared with the negative, positive controls and its treated group. Interestingly, the activity of ALT in the positive control was not affected by DEN and 2-AAF when compared with the negative control. The preventive group showed a significant higher activity of ALP when compared with the negative control. However, all other experimental groups showed no significant activity of ALP when compared with either negative or positive controls. The activity of AST was not affected in all the experimental groups, including the positive control and

selenium treated groups when compared with the negative control.

The concentration of globulin was significantly lower in the positive control when compared with the negative control. Also, the therapeutic control showed a significant lower concentration of globulin when compared with the negative control but not with the positive control or its treated group. On the other hand, the preventive group and its respective control showed significantly higher

Table 1. Effect of selenium on ALT, AST, GGT and ALP on chemically induced hepatocarcinogenesis in rats.

Group	ALT (U/L)	AST (g/L)	GGT (U/L)	ALP (U/L)
Group 1	48.73 ± 3.86	107.25 ±41.63	6.33 ± 0.68	3.67 ± 0.52
Group 2	44.67 ± 13.52	94.25 ± 23.04	11.83 ± 2.21 [a]	3.92 ± 1.46
Group 3	59.64 ± 6.53 [a,b,c]	100.29 ± 14.67	12.79 ± 2.02 [a,c]	5.29 ± 1.22 [a]
Group 4	49.23 ± 2.59	98.43 ± 24.57	7.57 ± 2.42 [b]	5.57 ± 2.52
Group 5	47.63 ± 5.66 [d]	92.13 ± 15.14	15.75 ± 1.83 [a,b,d]	4.13 ± 0.74
Group 6	32.07 ± 3.93 [a,b]	75.50 ± 15.32	6.29 ± 2.69 [b]	3.79 ± 0.70

Results are expressed as means ± S.D. Values were analyzed using Student's t and Mann – Whitney's U tests. [a] significantly different from the negative control (1) ($P< 0.05$); [b] significantly different from the positive control (2) ($P< 0.05$); [c] significantly different from the respective group (4) ($P< 0.05$) and [d] significantly different from the respective group (6) ($P< 0.05$).

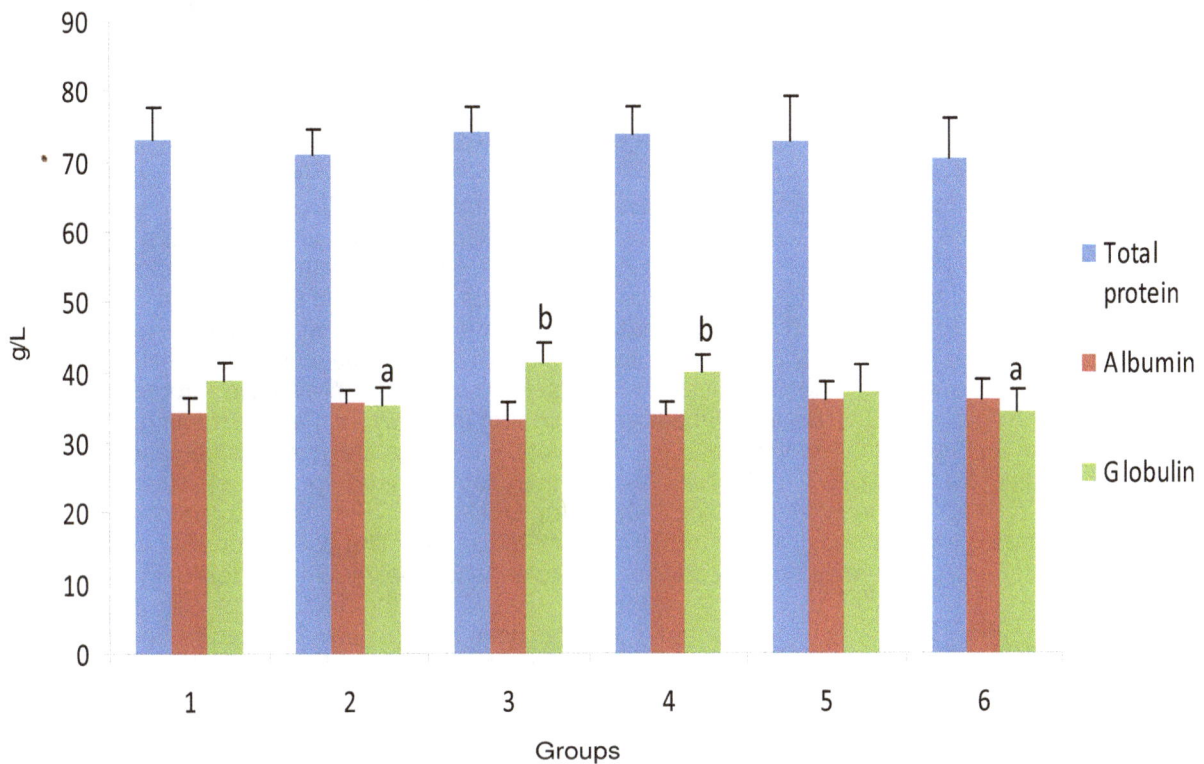

Figure 4. Effect of selenium on total protein, albumin and globulin on chemically induced hepatocarcinogenesis in rats. Results are expressed as means ± S.D. Values were analyzed using Student's t test. [a] Significantly different from the negative control (1) ($P< 0.05$). [b] Significantly different from the positive control (2) ($P< 0.05$).

concentrations of globulin when compared with the positive control. There is no significant concentration of total protein and albumin among all the experimental groups, including the positive control when compared with the negative control (Figure 4).

DISCUSSION

The measurement of concentrations and activities of various biochemical markers and enzymes in the plasma plays a significant role in disease investigation and diagnosis as well as in response to a toxic drug (Malomo, 2000). Previous studies have shown that high levels of AST, ALT and ALP in serum or plasma are usually indicative of liver injury in humans and animals (Spracklin et al., 1996; Yabubu et al., 2003) whereas the lower levels of these enzymes could indicate a degree of liver protection (Manna et al., 1996). This study is in line with other previously reported study (Nakaji et al., 2004), where it was reported that the activities of AST, ALT and ALP were not significantly affected between the control group (rats were given 200 mg/kg of DEN, two weeks later, they were fed via gastric tubes with 10 mg/kg of 2-AAF for two weeks and partial hepatectomy was performed one week later) and Interferon – alpha treatment. Plasma liver enzymes are good indicators of liver injury but they are not always elevated during liver injury. They could be elevated at a particular stage of the disease and return to their normal values at a certain point (Simonsen and Virji, 1984).

Sodium selenite at the dose of 4 mg/L did not maintain the activity of GGT and ALT in the preventive group (selenium for four weeks only, then DEN and 2-AAF) and therapeutic group (four weeks of DEN injection, then selenium until the end of experiments). In fact, selenium significantly increased the activity of GGT in the preventive and therapeutic groups when compared with their respective controls. The elevation in ALT and GGT activities seen in the preventive and therapeutic groups could be due to hepatocellular damage that was initiated by DEN and further damage by 2-AAF. GGT, which is found in all cells of the body except myocytes with particularly high concentrations found in hepatocytes and the kidney, is a liver enzyme involved in the transport of amino acids and peptides into cells (Hanigan and Pitot, 1985). ALT, which is mainly produced in the hepatocytes, is more specific for liver injury (Thomson, 1998). It has been reported that ALT is generally increased in situations where there is damage to the liver cell membrane (Schumann et al., 2002). Thus, when the liver is injured, the levels of ALT in plasma usually rise.

However, the positive control (DEN and 2-AAF only) showed that the ALT activity was not significantly increased when compared with the negative control. This shows that ALT activity maybe worsened by the addition of sodium selenite. On the other hand, the other aminotransferase enzyme (AST) was not affected by either DEN + 2-AAF treatment or selenium supplementation. In addition, other markers such as ALP, total protein and albumin were not affected by the same treatment. This might suggest that the activity of ALT alone is not specific for liver cancer in rat model.

The findings of this study disagree with other previous study (Ozardali et al., 2004). They reported that the activities of plasma AST, ALT and ALP were significantly increased with carbon tetra chloride injection whereas GGT activity was not significantly affected. However, they also reported that selenium treatment showed that the levels of AST, ALT and GGT were decreased to nearly the enzyme values in control group but ALP activity was significantly increased.

Thirunavukkarasu and Sakthisekaran (2003) cited the decreased levels of total protein and albumin and increased levels of globulin in rats treated with DEN and Phenobarbitol when compared to control rats. The findings of this study disagreed with the above mentioned study. It was observed that the concentrations of total protein and albumin were not affected by neither carcinogens nor selenium treatments. In addition, a slight decrease in globulin concentration was observed in the positive control and therapeutic group in comparison with the negative control. Also, it was noted that the globulin concentrations in the preventive group and its respective control were significantly higher when compared with the positive control.

Despite the fact that albumin is a major protein formed by the liver and together with the total protein level in the blood reflect the protein function of the liver (Benoit et al., 2000), the positive control, preventive and therapeutic selenium treated groups showed almost similar concentrations to that of the negative control. This could be due to the long half life of albumin, which has been reported to be around 20 days (Halsted and Halsted, 1991), and a decrease in plasma albumin is usually not seen early in hepatocarcinogenesis (Cheesbrough, 1998). Albumin levels in these groups were normal and this could be also due to the rat's ability to compensate for protein losses.

Recent study on selenium and copper supplementation on blood metabolic markers in male buffalo calves showed that high levels of globulin are beneficial, as it directly correlates with the immune status of the animals (Mudgal et al., 2008). Based on the histopathological examination along with the increased concentration of globulin, it can be concluded that supplementation of selenium in the preventive group and its respective control may have enhanced the immunity status of these rats. On the contrary, supplementation of selenium in the therapeutic control group (selenium for eight weeks but no DEN or 2-AAF) may have reduced the immunity status of those rats. The conclusion that selenium may reduce or enhance the immunity in rats needs to be clarified.

Based on the findings of this study, plasma GGT enzyme marker could be used as an early marker for liver neoplasm in rats. Supplementation of selenium in the preventive and therapeutic experimental groups has increased the activity of GGT and ALT. The effect of selenium on globulin, as an indicator of immunity status, needs to be clarified.

ACKNOWLEDGEMENT

This study was supported by a grant from the Ministry of Science, Technology and Innovation (MOSTI), Malaysia (No 05-01-02-SF0014).

REFERENCES

Benoit R, Breuille D, Rambourdin F, Bayle G, Capitan P, Obled C (2000). Synthesis rate of plasma albumin is a good indicator of liver albumin synthesis in sepsis. Am. J. Physiol. Endocrinol. Metab., 279: 244-255.

Cheesbrough M (1998). District laboratory practice in tropical countries, part 1. pp. 348-361. Cambridge University Press, Cambridge.

Clark L, Combs G, Turnbull B, State E (1996). The nutritional prevention of cancer with selenium 1983 – 1993: a randomized clinical trial. JAMA., 276: 1957-1963.

Halsted JA, Halsted CH (1991). The laboratory in clinical medicine: Interpretation and application. 2nd ed. WB, Saunders Company, Philadelphia, pp. 281-283.

Hanigan M, Pitot H (1985). Gamma-Glutamyl transpeptidase: its role in hepatocarcinogenesis. Carcinogenesis, 6: 165-172.

IOM (Institute of Medicine) (2000). Selenium. In: Dietary Reference Intakes for ascorbic acid, Vitamin E, Selenium, and Carotenoids. Food and Nutrition Board. National academy Press, Washington DC pp. 290-291.

Kensler T, Qian G, Chen J, Groopman J (2003). Translational strategies for cancer prevention in liver. Nat. Rev. Cancer. 3: 321-329.

Kim K, Park J (1994). Antihepatic effect of Artemisia Iwayomogi methanol extract on acute hepatic injury by carbon tetra-chloride in rat. Korean. J. Vet. Res., 34: 619-626.

Mahajan B (1997). Significance of differences in means. In: Methods in Biostatistics for Medical and Research Workers. 6th ed. JAYPEE Brothers Medical Publishers, New Delhi, pp. 130-155.

Malomo S (2000). Toxicological implication of ceftriaxone administration in rats. Nig. J. Biochem. Mol. Biol., 15: 33-38.

Manna Z, Guopei S, Minuk G (1996). Effects of hepatic stimulator substance herbal medicine selenium/vitamin E and ciprofloxacin on cirrhosis in the rat. Gastroenterology. 110: 1150-1155.

Mudgal V, Garg A, Dass R, Varshney V (2008). Effect of Selenium and Copper Supplementation on Blood Metabolic Profile in Male Buffalo (Bubalus bubalis) Calves. Biol. Trace. Elem. Res., 121: 31-38.

Nakaji M, Yano Y, Tinomiya T, Seo Y, Hamano K, Yoon S (2004). INF-alpha prevents the growth of neo-plastic lesions and inhibits the development of hepatocellular carcinoma in the rat. Carcinogenesis. 25: 389-397.

Ozardali I, Bitiren M, Karakilcik A (2004). Effects of selenium on histopathological and enzymatic changes in experimental liver injury of rats. Exp. Toxicol. Pathol., 56: 59-64.

Parkin MD, Bray F, Ferlay J, Pisani P (2005). Global cancer statistics, 2002. Ca. Cancer. J. Clin., 55: 74-108.

Raymond B (2001). Selenium; recent clinical advances. Curr. Opin. Gastroenterol., 17: 162-166.

Schrauzer G, White D, Schneider C (1977). Cancer mortality correlation studies. 111. Statistical association with dietary selenium intakes. Bioinorg. Chem., 7: 35-56.

Schumann G, Bonora R, Ceriotti F (2002). IFCC primary reference procedure for the measurement of catalytic activity concentrations of enzymes at 37°C. Part 4. Reference Procedure for the Measurement of Catalytic Concentration of Alanine Aminotransferase. Clin. Chem. Lab. Med., 40: 718-724.

Simonsen R, Virji M (1984). Interpreting the profile of liver-function tests in pediatric liver transplants. Clin. Chem., 30: 1607-1610.

Solt D, Farber E (1976). New principle for the analysis of chemical analysis. Nature, 263: 701-703.

Spracklin D, Thummel K, Kharasch E (1996). Human reductive halothane metabolism in vitro is catalyzed by cytochrome P450 2A6 and A4. Drug. Meta. Dispos., 24: 976-983.

Tazi A, Galteau M, Siesi G (1980). γ-Glutamyl transferase of rabbit liver: kinetic study of Phenobarbital induction and in vitro solubilization by bile salts. Toxico. Appl. Pharmaco., 55: 18-21.

Thirunavukkarasu C, Sakthisekaran D (2003). D-influence of sodium selenite on glycoprotein contents on normal and N-nitrosodiethylamine initiated and Phenobarbital promoted rat liver tumours. Pharmacological. Res., 48: 167-173.

Thomson CD (1998). Selenium speciation in human body fluids. Analyst, 123: 827-831.

Valkoo M, Rhodes C, Moncol J, Izakovic M, Mazur M (2006). Free radicals, metals and antioxidants in oxidative stress-induced cancer. Chem. Biol. Interact., 160: 1-40.

Yabubu M, Salau I, Muhammad N (2003). Phosphatase activities in selected rat tissues following repeated administration of ranitidine. Nig. J. Biochem. Mol. Biol., 18: 21-24.

6

Lipid profile of a population of diabetic patients attending Nigerian National Petroleum Corporation Clinic, Abuja

Ugwu, C. E[1]*. , Ezeanyika, L. U. S.[2], Daikwo, M. A.[1] and Amana, R.[1]

[1]Department of Biochemistry, Kogi State University, Anyigba, Nigeria.
[2]Department of Biochemistry, University of Nigeria, Nsukka, Nigeria.

This study was conducted to compare the lipid profile of diabetic patients and healthy controls. The lipid profiles and lipoprotein levels of 50 known diabetic patients and 50 healthy subjects were studied. Total cholesterol (TC), Triacylglycerols (TG), Low density lipoprotein-cholesterol (LDL-C), and high density lipoprotein cholesterol (HDL-C) levels were assayed for each group using standard biochemical methods. The mean TC, TG, and low density lipoprotein cholesterol levels were lower in the diabetics than in the control subjects though these were not significant (P > 0.05). The frequency of high TC level was higher in the diabetic group while the frequency of low HDL-C level was higher in the healthy controls. The prevalence of high TG and LDL-C were approximately equal in the two groups. The mean (±SD) HDL- C was significantly lower (P < 0.05) in males compared to the females for both diabetic and control groups. The better lipid profiles in the diabetic patients compared to the controls were apparently due to the regime of management of their condition.

Key words: Diabetes, lipid profile, Nigerians.

INTRODUCTION

Diabetes mellitus (DM) is a common endocrine metabolic disorder and a leading cause of death world wide (Faghilimani et al., 2006). There are more than 154 million diabetics worldwide and its prevalence is on the increase in the developing countries (Bennett, 2000; Sajjadi et al., 2005). Certain racial and ethnic groups have a greater risk of developing diabetes (Manu et al., 2007). Majority of those that suffer from this disease are from Africa and Asia (Manu et al., 2007). This may be due to genetic disposition and life style of people in these areas. Nigeria being the most populous country in Africa may harbor a substantial number of people with this condition.

The disease is accompanied in many cases by secondary alterations of protein and fat metabolism resulting in an array of physical disorders (Welsh et al., 2004). Lipids and lipoproteins abnormalities are well known risk factors for heart disease. Elevated levels of triacylglycerols (TG),

cholesterol, and low density lipoprotein-cholesterol (LDL-C), are documented as risk factors for atherogenesis (LRCP, 1984). The blood level of high density lipoprotein-cholesterol (HDL-C) in contrast bears an inverse relationship to the risk of atherosclerosis and coronary heart disease. The higher the level, the smaller the risk (Khoo et al., 1997; Tao et al., 1992). Lipid abnormalities play an important role in the causation of diabetic atherosclerosis (Easterman and Keen, 1997; Lewis and Steiner, 1996), but the pathophysiology is complex and clearly multifactorial (Gupta and Kapse, 2001), with dysfunction of the fibrinolytic system (Sobel, 1996) pro- oxidative state (Baynes and Thrope, 1996), hyperglycaemia (Malemberg et al., 1995; Lehto et al., 1997) and possibly hyperinsulineamia (Fontbonne et al., 1989) also explaining part of the increased susceptibility of people with diabetes to atherosclerotic complications.

Epidemiological studies have shown that diabetics have 2 - 4 times higher risk of developing cardiovascular diseases (Aboot et al., 1988; Gaillard et al., 1997). Abnormal lipid profiles and lipoprotein oxidation (especially LDL-C)

*Corresponding author. E-mail: ugwuchidiksu@yahoo.com.

Table 1. Biochemical parameters of diabetics and controls.

Parameters	Diabetics(n= 0)	Control(n = 50)	p - value
Total cholesterol (mg/dl)	211.02±59.91	229.77± 33.63	0.057
Triacylglycerols (mg/dl)	142.10± 52.28	148.84 ±105.91	0.687
HDL-C (mg/dl)	44.04± 12.28	44.28± 15.80	0.932
LDL-C (mg/dl)	141.04± 55.15	156.10± 37.63	0.130
Glucose (mg/dl)	197.58± 85.75	85.76± 11.14	0.001

HDL-C = high density lipoprotein – cholesterol, LDL-C = low density lipoprotein cholesterol.

are more common in diabetics and are aggravated with poor glycaemic control. As diabetic patients constitute a unique group with different lifestyles and genetic disposetions, the measurement of their lipid profile is needed to investigate how their lipid metabolism is affected by diabetes. Considering the probable disorders of lipid profile and acceleration of atherosclerotic process in high risk groups, this work assessed the lipid profile of a randomly selected group of adult Nigerian diabetics and compared them with controls.

SUBJECTS AND METHODS

The subjects used in the study were diabetic patients who attend the Nigerian National Petroleum Corporation (NNPC) Group Medical Services Centre, Abuja, Nigeria. A total of 50 diabetic patients (30 males, 20 females) and 50 healthy controls (32 males, 18 females) were randomly selected. Patients with other ailments and metabolic disorders were excluded from the study. Diabetes was ruled out in the control group by asking questions about the clinical signs of diabetes such as polyuria, polydipsia and recent weight loss. Laboratory tests were also used to confirm the absence of diabetes in the control group. Ethical clearance was sought and obtained for the study from the hospital. The aim of the study was explained to the subjects by the physicians and those who gave informed consent were included in the study by the researchers.

In both subjects, venous blood samples were obtained after overnight fast into tubes containing lithium heparin (for lipid profile) and EDTA (for blood glucose) as anticoagulants. The samples were centrifuged at 1500 rpm for 5 min to obtain the plasma. Plasma cholesterol, triacylglycerols, high density lipoprotein – cholesterol (HDL-C), and glucose were assayed using test strips manufactured by Roche Diagnostic Limited (Bell Lane Lewes, East Sussex BN 71LG, UK). All assays were performed on a reflotron machine (Roche, diagnostic GmbH, D-68298 Mannhein, Germany). The machine automatically displays the concentrations of the parameters assayed in mg/dl. The used test strip was then removed from the reflotron machine and disposed off.

LDL-C was calculated indirectly by the method of Friedwald et al. (1972) as shown below.

LDL-C = total cholesterol – HDL-C + TG/ 5

The lipid profile of the subjects was classified based on the ATP III model (NCEP, 2001).

Statistics

The data obtained in this study were presented using descriptive statistics. In order to test whether or not significant differences exist between groups, we analysed the mean values with student t- test. The acceptable level of significance was $P \leq 0.05$.

RESULTS

The mean age of the subjects were 53.14 ± 12 and 49.20 ± 11.10 years for the diabetic and control groups respectively. The sex distribution showed that there were more diabetic males 30(60%), compared to diabetic females 20(40%). There were also more males 32(64%) compared to the females 18(36%) in the control group.

Table 1 shows that the mean total cholesterol, triacylglycerols and LDL-C were lower in the diabetics compared to the controls though these were not significant ($P > 0.05$). The fasting blood sugar level of the diabetics was significantly higher ($P < 0.05$) than those of the controls. Table 2 compared the mean biochemical variables with respect to gender in diabetics and the control respectively. The results shows that the mean HDL-C concentration was significantly lower ($P < 0.05$) in the male diabetics compared to the female diabetics. In the control group (Table 2) the mean HDL-C concentration was also lower ($P < 0.05$) in the males compared to the females.

Table 3 shows the frequency of TC, TG, LDL-C and HDL-C concentrations in both patient and control groups. The results show that the frequency of high TC was higher in the diabetic group (34% Vs 28%). The frequency of high TG and high LDL-C were approximately equal in the two groups. The control group had a higher frequency of low HDL-C than the diabetic group. There was no significant difference in the mean lipid profile of male/female diabetics compared to male/female controls ($P > 0.05$).

DISCUSSION

Patients with diabetes can have many complications including elevated levels of VLDL-C, LDL-C and Triacylglycerols; and low levels of HDL-C (Haffner, 1998). These patients have a preponderance of abnormalities in the composition of smaller, denser particles, which increase atherogenicity even if the absolute concentration

Table 2. Comparison of the biochemical parameters in the males and females in both groups.

Parameters	Diabetics			Controls		
	Male N = 30	Females N = 20	p-Values	Male N = 30	Females N = 20	p-Values
Total cholesterol (mg/dl)	215.03 ± 66.94	205.00 ± 48.49	0.567	230.19 ± 35.96	228.89 ± 29.99	0.897
Triacylglycerols (mg/dl)	143.40 ± 59.74	140.15 ± 39.93	0.832	135.09 ± 70.90	173.28 ±14.93	0.225
HDL-C (mg/dl)	40.53 ± 11.68	49.30 ± 11.14	0.011*	40.31 ± 11.79	51.33 ± 19.58	0.016*
LDL-C (mg/dl)	149.90 ± 60.89	129.05 ± 46.12	0.199	163.47 ± 40.57	143.00 ± 28.21	0.064
Glucose (mg/dl)	206.57 ± 96.96	184.10 ± 65.53	0.369	84.59 ± 11.19	87.83 ± 11.04	0.329

HDL-C: high density lipoprotein–cholesterol; LDL-C: low density lipoprotein-cholesterol.
*significance, $P < 0.05$.

Table 3. Frequency of the biochemical variables in the diabetic and control groups according to the ATP III classification.

	Diabetics (%)	Control (%)
Total cholesterol (mg/dl)		
Desirable(< 200)	21(42)	2(4)
Borderline (200- 239)	12(24)	34(68)
High (≥ 240)	17(34)	14(28)
Triacylglycerols (mg/dl)		
Normal (<150)	32(64)	35(70)
Borderline high (150 – 199)	11(22)	7(14)
High(200 – 499)	7(14)	8 (16)
LDL-C (mg/dl)		
Optimal (<100)	12(24)	3(6)
Near optimal (100 – 129)	7(14)	6(12)
Borderline high (130 – 159)	12(24)	23(46)
High (160 - 189)	10(20)	10(20)
Very high (≥ 190)	9(18)	8(16)
HDL –C (mg/dl)		
Low(< 40)	19(38)	23(46)
High (≥ 60)	7(14)	9(18)
Borderline(40-59)	24(48)	18(36)

of LDL-C is not significantly increased (Haffner, 1998).

In our study, the lipid and lipoprotein profiles of the diabetics were lower than that of the controls. These findings are contrary to previous studies which suggest that lipoprotein abnormalities are higher in diabetics than in non-diabetics (Idogun et al., 2007; Albrki et al., 2007). The study also shows that when the mean (±SD) of the variables are separated for the male and female subjects, HDL-C was significantly lower in both diabetic and non-diabetic males. The higher the level of HDL-C, the lower the risk of developing artherosclerosis (Khoo et al., 1997). The results show no gender difference in the lipid metabolism between the diabetic and non-diabetic males and females.

Vinter-Repalust et al. (2007) reported no significant dif-

ferences in the prevalence of type 2 diabetes mellitus (DM) between males and females. It is known that abnormalities arising from DM are related to gender, duration and drug compliance (Gustafsson et al., 2004).

The patients used in the study were out patients that visit the clinic regularly. The NNPC clinic Abuja is a high profile hospital that caters for mainly workers in the highly paying oil industry in Nigeria. The patients have been on drug for about one year. They were on the average educated and were also properly informed on the management and risks associated with diabetes as against the control group that was drawn from the general population. This could have led to the higher mean lipid abnormalities observed in the control group.

The prevalence rates for high TC, combined high and very high LDL-C and low HDL-C in the diabetic subjects were 34 and 38% respectively. These values are in agreement with the studies of Idogun et al. (2007) and Emile et al. (1993). Lipid abnormality levels in the North American population are approximately 50 and 25% for cholesterol and triacylglycerols respectively (LRCPEC, 1970). The prevalence rates of high TC and triacylglycerol from this study were 34 and 14% respectively.

The study showed combined hyperlipidaemia in the two groups. The diabetic patients had higher prevalence of high serum cholesterol than the controls while the control group had a higher frequency of low level of HDL-C than the patients. The two groups showed approximately equal frequency of high triacylglycerol and LDL-C. This suggests that the two groups could be equally predisposed to cardiovascular diseases. The higher mean levels of some serum lipids in the control group could be due to better nutritional control and drug therapy in the diabetic group. Efforts should therefore be made to continuously educate the populace on diabetes, its management, feeding and life styles.

REFERENCES

Aboot RD, Donhue RP, Kannel NB, Wilson WT (1988).The impact of diabetes on survival following myocardial infection in men Vs. women. J. Am. Med. Ass. 260: 3456 - 3460.

Albrki WM, Elzouki AN Y, EL-Mansoury ZM, Tashani OA (2007).

Lipid profiles in Libian Type 2 Diabtics. J. Sci. Appls. 1(1):18 – 23.

Baynes JW, Thorpe SR (1996). The role of oxidative stress in Diabetic complications of diabetes. Curr Opin Endocr. 3: 277- 284.

Bennett PH (2000). Epidemiology of type 2 diabtes mellitus in: Diabetes Mellitus. Lekoith D, Taylor S. I, Olefsky JM. (eds.) 2nd ed. Wolter, New York. pp. 544 – 557.

Easterman RC, Keen H (1997). The impact of cardiovascular disease on people with diabetes: the potential for prevention. Lancet. 350: S1 – 29 -30.

Emile E, Jorge JG, Jane ER (1993). Lipid level differences and hypertension effect in Blacks and Whites with type 2 Diabetes. Ethnicity Dis.3: 242-249.

Faghilimnai S, Hashemipour M, Kelishadi B (2006). Lipid profile of children with type 1 diabetes compared to controls. ARYA .J. 2(1): 36 – 38.

Friedwal WT, Levy RI, Fredrickson DS (1972). Estimation of the concentration of low density lipoprotein cholesterol in plasma, without use of the preparative ultracentrifuge. Clin. Chem. 18: 499 – 502.

Fontbonne A, Eschwege E, Cambien, F, Richard JL, Ducimetiere P, Thibult N, Warnet JM, Rosselin GE (1989). Hypertriglyceridemia as a risk factor of coronary heart disease mortality in subjects with impaired glucose tolerance or diabetes: Results from the 11 year follow-up of the Paris Prospective study. Diabetogia. 32: 300 – 304.

Gaillard TR, Schutter DP, Bossett BM, Green PA, Osei K (1997). The impact of socioeconomic status on cardiovascular risk factors in Africans at high risk for type II diabetes. Diabetes Care. 20: 745 – 756.

Gupta S , Kapse A (2001). Lipid profile pattern in diabetics from central India. Int. J. Diab. Dev. Ctries. 21: 138 – 145.

Gustafsson I, Brendorp B, Seibaek M, Burchardt H, Hildebrandt P (2004). Influence of diabetes and diabetes – gender interaction on the risk of death in patients hospitalized with congestic heart failure. J. AM. Coll. Cardiol. 43(5): 771 – 777.

Haffner SM (1998). Management of dyslipidemia in adults with diabetes. Diabetes Care. 21: 9(1): 1600 – 1678.

Idogun ES, Unuigbe E.I, Ogunro PS, Akinola OT, Famodu, AA (2007). Assessment of serum lipids in Nigerians with type 2 diabetes mellitus complications. Pak. J. Med. Sci.(Part 1). 23(5): 708 – 712.

Khoo K L, Tan H, Leiw Y M (1997). Serum lipids and their relationship with other coronary risk factors in healthy subjects in a city clinic. Med. J. Malaysia. 52: 38 – 52.

Lehto S, Ronnemaa T, Haffner SM, Pyorala K, Kallio V, Laakso M(1997). Dyslipidemia and hyperglycemia predict coronary heart disease events in middle – aged patients with NIDDM. Diabetes 46: 1354 – 1359.

Lewis GF, Steiner G (1996). Hypertriglyceridemia and its metabolic consequences as a risk factor for atherosclerotic cardiovascular disease in non-insulin-dependent diabetes mellitus. Diabetes Metab Rev. 12: 37 – 56.

Lipid Research Clinical Program (1984). The lipid Research Clinic Coronary Primary prevention trial results II. J. AM. Med. Assoc. 251: 364 – 374.

Malemberg K, Ryden L, Efendic S, Herlitz J, Nicol P, Waldel H, Welin L (1995). Randomized trial of insulin – glucose infusion followed by subcutaneous insulin treatment in diabetic patients with acute myocardial infarction (DIGAMI STUDY) : Effects on mortality at 1 year. J. AM. Coll Cariol. 26(1): 57 – 65.

Manu A, Shyamal K, Sunil G, Sandhu J S (2007). A study on lipid profile and body fat in patients with diabetes mellitus. Anthropologist. (4): 295 – 298.

National Cholesterol Education Program (NCEP) (2001). Expert Panel on Detection, Evaluation, and Treatment of High Blood Cholesterol in Adults (Adult Treatment Panel III).JAMA. 285: 2486 – 2497.

Sajjadi F, Mohammadifard N, Ghaderian N, Alikhasi H, Maghroon M (2005). Clustering of cardiovascular risk factors in diabetic and IGT cases in Isfahan province. 2000 – 2001: Isfan healthy heart program. ARYA .J. 1(2): 94-100.

Sobel BE (1996). Altered fibrinolysis and platelet function in the development of vascular complications of diabetes. Curr. Opin. Endocrinol. 3: 355 – 60.

Tao S, Li Y, Xiao Z, Cen R, Zhang H, Zhuo B, Chen P, Liao Y (1992). Serum lipids and their correlates in Chinese urban and rural population of Beijing and Guangzhou. PRC – USA Cardiovascular and Cardiopulmonary Epidemiology Research Group. J. Epidemiol. 21: 893 – 903.

The lipid Research Clinics Program Epidemiology Committee (1970). Plasma lipid distributions in selected North American populations: the Lipid Research Clinic Program prevalence study. Circulation 60: 42739.

Vinter – Repalust N, Jurkomo L, Katie M, Simunovic R , Petric D(2007). Disease duration, Patient compliance and presence of complications in diabetic patients. Acta. Med. Croatica. 61(1): 57 – 62

Welsh MC, Welsh M, Ekman J, Dixelins R, Hagerkvist R, Anneren C, Akerblom B, Menboobi S, Chandresekharan S, Liu ET (2004). The tyrosine kinase FRK/RAK participates in cytokine-induced islet cytotoxicity. Biochem. J. 382. 15 (pt1): 261 -280.

7

Effect of highly active anti-retroviral therapy (HAART) on lipid profile in a human immunodeficiency virus (HIV) infected Nigerian population

Francis M. Awah[1,2]* and Onyinye Agughasi[2]

[1]Department of Biochemistry, University of Nigeria, Nsukka, Nigeria.
[2]Department of Biochemistry, Madonna University, Elele Campus, Rivers State, Nigeria.

Nigeria remains the country with the third highest number of human immunodeficiency virus (HIV) infected individuals with over 600,000 HIV-infected subjects requiring antiretroviral therapy (ART). Since there is little or no data on the effect of highly active antiretroviral therapy (HAART) on the lipid profile of HIV-infected subjects in Nigeria, this study was aimed at evaluating the dyslipidaemia associated with HAART in HIV/AIDS (acquired immune deficiency syndrome) subjects. The lipid profile of the HIV-infected subjects taking HAART (n=25) was compared to that of the HIV-infected without ART (n=25) and the seronegative control (n=25). Total cholesterol, triglyceride, low density lipoprotein (LDL) and high density lipoprotein (HDL) cholesterol was determined using standard enzymatic methods. The results showed that HAART caused dyslipidaemia in HIV-infected subjects. Total cholesterol, triglycerides and LDL-cholesterol levels significantly increased in those receiving HAART ($p<0.05$) compared to those without treatment and the seronegative subjects. HDL-cholesterol levels were however, not significantly different ($p>0.05$) in the subjects. Triglyceride, LDL and HDL cholesterol levels were not significantly different between the sexes. It could therefore be concluded that HIV-infected subjects on HAART are predisposed to develop hypercholesterolaemia and conditions associated with it like coronary heart diseases. The lipid profile of HIV-infected subjects about to commence ART and those on HAART should be determined prior to and during treatment.

Key words: HIV/AIDS, HAART, lipid profile, dyslipidemia, Nigerian population.

INTRODUCTION

The human immunodeficiency virus (HIV) has thus far infected over 22.4 million people in sub-Saharan Africa and Nigeria remains the country with the third highest number of HIV-infected subjects in the world (WHO/UNAIDS, 2009). Since HAART was introduced in the mid 1990s, as a treatment for HIV/AIDS, the morbidity and mortality associated with HIV/AIDS has reduced considerably. According to the 2005 HIV Sentinel Survey and National HIV Prevalence AIDS Estimates Reports, an estimated 645,810 HIV-infected Nigerians would have been requiring ART by 2010 (FMOH, 2006). HAART regimens typically include a combination of at least three drugs, such as different association of protease inhibitors (PI), non-nucleoside reverse transcriptase inhibitors (NNRTI) and nucleoside reverse transcriptase inhibitors (NRTI). HAART however, has been reported to be associated with a number of side effects in HIV/AIDS subjects among which dyslipidaemia and lipodystrophy are common metabolic disorders with increased risk of cardiovascular diseases and diabetes in the HIV-infected subjects (Carr et al., 1999; Roula et al., 2000; Mulligan et al., 2000; Currier, 2002). Studies have also shown that

---

*Corresponding author. E-mail address: awambuh@yahoo.com.

Abbreviations: HIV, Human immunodeficiency virus; **ART,** antiretroviral therapy; **HAART,** highly active antiretroviral therapy; **AIDS,** acquired immune deficiency syndrome; **HDL,** high-density lipoprotein; **LDL,** low-density lipoprotein; **PI,** protease inhibitors; **NNRTI,** non-nucleoside reverse transcriptase inhibitors; **NRTI,** nucleoside reverse transcriptase inhibitors; **TG,** triglycerides; **SPSS,** statistical package for social sciences; **SEM,** standard error of the mean; **ANOVA,** analysis of variance; **VLDL,** very-low-density lipoprotein.

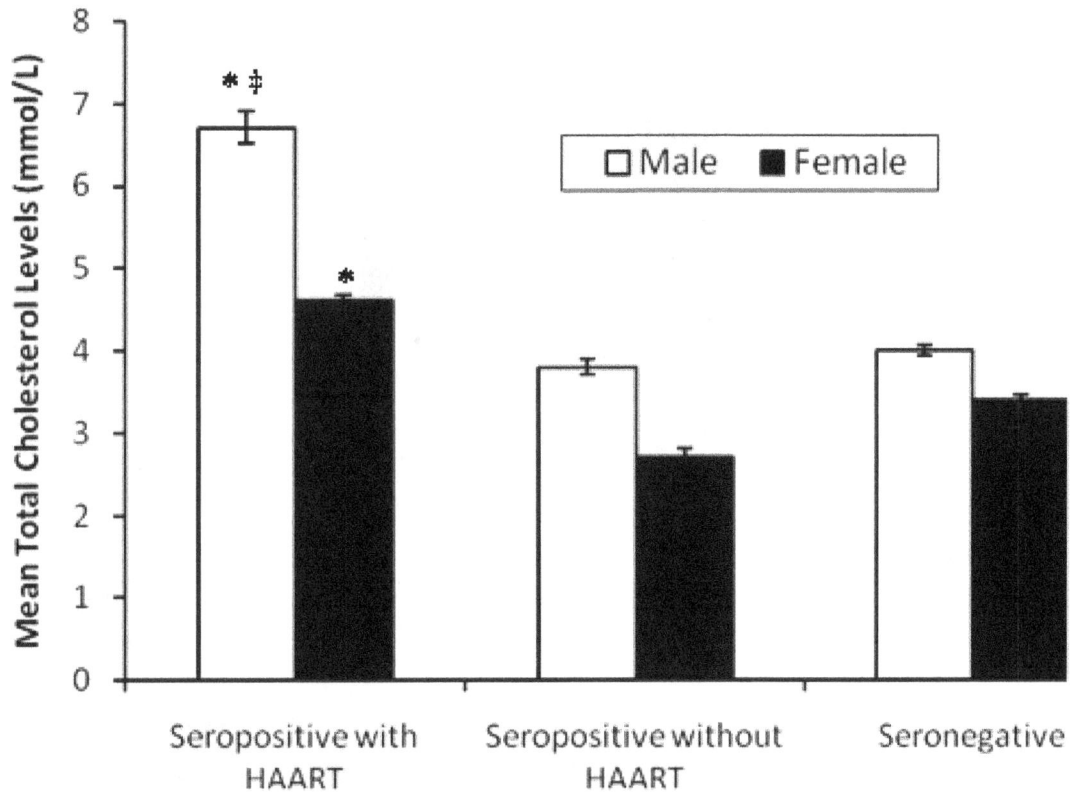

Figure 1. Serum cholesterol level of HIV-infected subjects on HAART compared to those without HAART and the seronegative. *$p < 0.05$, statistical significance of difference from seropositive without HAART and seronegative control. ‡$p < 0.05$, statistical significance of difference from females.

HAART treatment, especially those including protease inhibitors, is associated with hypertriglyceridaemia, hypercholesterolaemia, hypo HDL-cholesterolaemia and hyperinsulinaemia (Miserez et al., 2002; Vigano et al., 2003; Clotet et al., 2003; Palacios et al., 2006; Khiangte et al., 2007). Considering the number of HIV subjects in Nigeria taking HAART, the aim of this study was to evaluate the lipid profile of HIV-infected subject with or without HAART treatment in a bid to assess the potential risks of dyslipidaemia, if any, in male and female subjects who are about to or are enroll on HAART in a Nigerian population.

MATERIALS AND METHODS

Study area and subjects

A total of fifty (50) HIV-infected volunteers aged between 25 to 57 years, attending the General Hospital, Asaba, Delta State, Nigeria between July to October 2008 were surveyed for biochemical investigations and thereafter included in this study. Twenty five (25) of the subjects were on the first line HAART regiment while 25 were not yet on antiretroviral treatment. For both groups there were 13 females and 12 males. Age and sex matched HIV seronegative Madonna University students and staff were used as control (n=25). Selection was based on the subject's availability within the period of study.

Blood collection

Fasting blood samples (5 ml) were collected from all the study subjects by venipuncture into plain sample tubes. The blood samples were allowed to coagulate and spun at 5,000 rpm for 10 min. The serum was collected and stored at 4°C until assayed for biochemical indices within 4 days after collection. Subject's files were checked for their sex, age, stage of infection and antiretroviral therapy usage.

Lipid profile analyses

The serum total cholesterol, triglycerides (TG) and HDLcholesterol were determined in sera using commercial kits supplied by Randox (UK). These analyses were carried out according to the manufacturer's protocol. LDL cholesterol was calculated using the Friedewald's equation.

Statistical analysis

The results were analysed using the Statistical Package for Social Sciences (SPSS) version 10.0 for Windows. All the data are expressed as mean ± standard error of the mean (SEM). Two-way analysis of variance (ANOVA) was used to compare means, and values were considered significant at $p<0.05$.

RESULTS AND DISCUSSION

Metabolic disorders and other adverse drug side effects

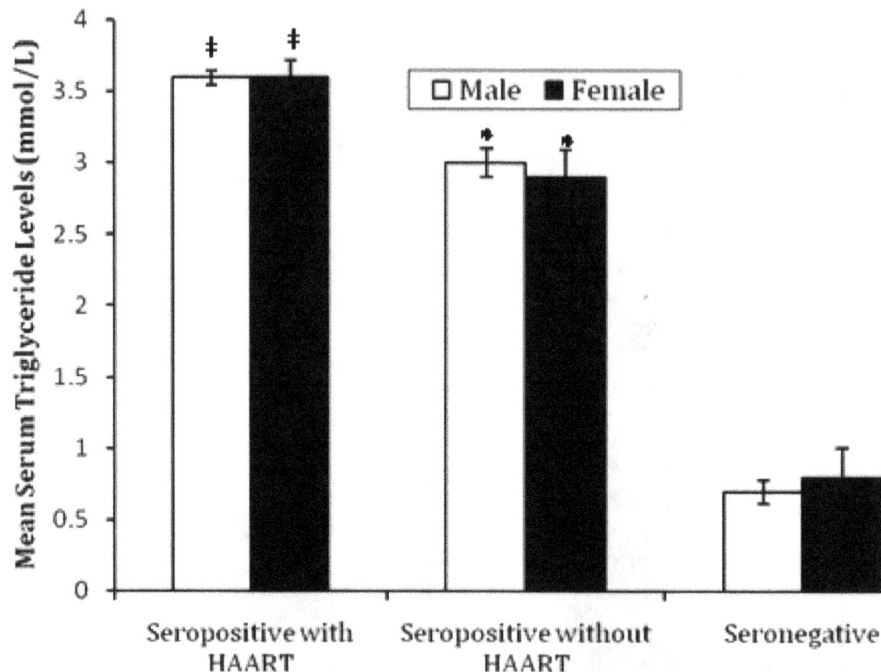

Figure 2. Serum triglyceride level of HIV-infected subjects on HAART compared to those without HAART and the seronegative. *$p < 0.05$, statistical significance of difference from seronegative control. ‡$p < 0.05$, statistical significance of difference from seropositive without HAART and seronegative control.

such as dyslipidaemia are important factors associated with reduced quality of life among HIV infected patients taking HAART (Pujari et al., 2005). The results of this study showed a significant increase in the serum total cholesterol levels ($p<0.05$) of the males and females HIV subjects on HAART compared to the HIV subject without HAART and the seronegative control (Figure 1). Total cholesterol levels were significantly higher in the males HIV-infected subjects compared to the females. These disturbances of cholesterol metabolism upon HAART treatment have been reported by other researchers (Falkenbach et al., 1990). The high levels of oxidative stress and lipid peroxidation associated with HIV/AIDS may, in part, explain the alterations of cholesterol metabolism in HIV-positive patients. Significant higher cholesterol in male subjects on HAART compared to females suggests that males could be more predisposed to hypercholesterolaemia due to HAART treatment compared to females probably due to hormonal differences. Serum triglyceride levels were also significantly higher ($p<0.05$) in the both the male and female seropositive subjects on HAART compared to the seropositive without HAART and the seronegative control. There was no significant difference ($p>0.05$) in the mean triglyceride levels between the male and female HIV-infected subjects. Furthermore, TG levels were significantly higher ($p<0.05$) in the HIV seropositive subject not on HAART compared to the control seronegative subjects (Figure 2). This indicates that HIV-infection may be associated with

a disturbance in TG metabolism and HAART treatment further aggravates the metabolic disturbance. Hypercholesterolemia and hypertriglyceridemia has therefore been seen in HAART users compared to the normal control. Both decreased TG clearance and increased Very-low-density lipoprotein (VLDL) overproduction have been found in HIV-positive patients (Grunfeld et al., 1992) and could be the reason the increase serum TG observed.

There was no significant difference ($p<0.05$) in the means serum HDL-cholesterol ($p>0.05$) levels of the HIV-infected subjects on HAART compared to those without HAART and the seronegative (Figure 3). Compared to the normal HIV seronegative female subjects and HIV-infected female subject not on HAART, HIV-infected female subjects on HAART had significantly higher ($p<0.05$) LDL-cholesterol levels (Figure 4). The HIV-infected male subject on HAART and those not on HAART had no significant difference in their LDL-cholesterol levels, however both groups had significantly higher ($p<0.05$) mean LDL-cholesterol compared to the seronegative control (Figure 4). The mechanisms responsible for metabolic disorders associated with antiretroviral drugs are not fully understood. Lipid disturbances in HIV patients receiving PI treatment are more evident (Leitner et al., 2006). The protease inhibitors could inhibit lipogenesis and impair the activity of adipocyte regulatory proteins (Carr et al., 1998) while NRTI may induce mitochondrial toxicity in subcutaneous fat tissues (Carr

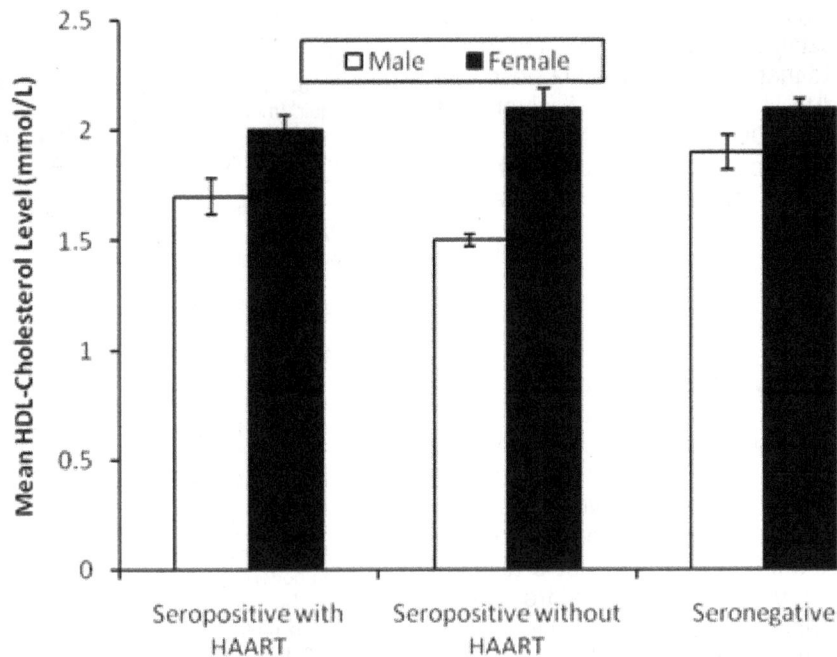

Figure 3. Serum HDL-cholesterol level of HIV-infected subjects on HAART compared to those without HAART and the seronegative.

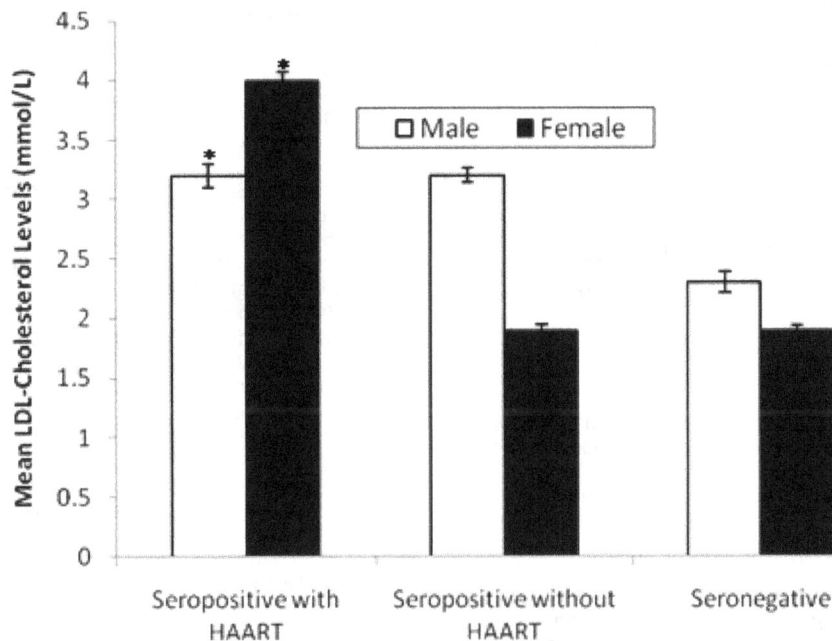

Figure 4. Serum LDL-cholesterol level of HIV-infected subjects on HAART compared to those without HAART and the seronegative. *$p < 0.05$, statistical significance of difference from seronegative control.

and Cooper, 2000). Increase serum concentrations of TG and LDL-cholesterol have been considered independent risk factors for coronary artery diseases and myocardial infarction (Stampfer et al., 1996; Assmann et al., 1998). Abnormalities in lipid metabolism make HIV-positive subjects to be at high risk for the development of coronary heart disease (Asztalos et al., 2006). The observations of this study are in conformity with studies in other countries, which have shown that HIV/AIDS subjects exhibit dyslipidaemia characterized by increase in total cholesterol, triglyceride, and LDL cholesterol (Grover et al., 2005; Asztalos et al., 2006; Obirikorang et al., 2010). This suggests that the dyslipidaemia observed may not be not related to HIV infection itself, given that the total cholesterol and LDL-cholesterol levels of the HIV-infected subjected without HAART treatment was significantly lower than that of HIV-infected subjecting taking HAART.

Conclusion

It is evident from the above findings that long term administration of HAART to HIV-infected subjects could lead to metabolic disorders such as dyslipidaemia which could predispose the patients to high risk of coronary heart diseases. HIV-infected subjects in Nigeria should therefore be assessed for their lipid profile before enrolment for HAART and the lipid profile should also be assessed periodically in the course of the treatment. Elucidating the mechanism via which antiretroviral therapy is associated with metabolic disturbances which could contribute to premature cardiovascular disease is of major importance. It is recommended that further research be carried out with larger sample sizes and additional information regarding other risk factors for dyslipidaemia in Nigerian populations be assessed to better understand the effects of HAART on lipid profile and other metabolic disturbances.

ACKNOWLEDGEMENT

Our profound gratitude goes to the hospital staff in General Hospital, Asaba for their cooperation in this study and those who participated in the study.

REFERENCES

Assmann G, Schulte H, Funke H, von Eckardstein A (1998). The emergence of triglycerides as a significant independent risk factor in coronary artery disease. Eur. Heart J., 19: M8–14.

Asztalos BF, Schaefer EJ, Horvath KV, Cox CE, Skinner S, Gerrior J, Gorbach SL, Wanke C (2006). Protease inhibitor-based HAART, HDL, and CHD-risk in HIV-infected patients. Atherosclerosis, 184(1): 72-77.

Carr A, Samaras K, Thorisdottir A, Kaufmann GR, Chisholm DJ, Cooper DA (1999). Diagnosis, prediction, and natural course ofHIV-1 protease-inhibitor-associated lipodystrophy, hyperlipidaemia, and diabetes mellitus: a cohort study. Lancet, 353: 2093–2099.

Carr A, Cooper D (2000). Adverse effects of antiretroviral therapy. Lancet, 356: 1423-1430

Carr A, Samaras K, Burton S, Law M, Freund J, Chisholm DJ, Cooper DA (1998). A syndrome of peripheral lipodystrophy, hyperlipidaemia and insulin resistance in patients receiving HIV protease inhibitors. AIDS, 12: F51-F58

Clotet B, van der Valk M, Negredo E, Reiss P (2003). Impact of nevirapine on lipid metabolism. J. Acquir. Immun. Defic. Syndr., 34(1): S79-84.

Currier JS (2002). Cardiovascular risk associated with HIV therapy. J. Acquir. Immun. Defic. Syndr., 31: S16–S23.

Falkenbach A, Klauke S, Althoff PH (1990). Abnormalities in cholesterol metabolism cause peripheral neuropathy and dementia in AIDS – a hypothesis. Med. Hypotheses, 33(1): 57-61.

Federal Ministry of Health Nigeria (FMOH) (2006). 2005 National HIV/Syphilis sero-prevalence sentinel survey among pregnant women attending antenatal clinics: technical report, April. Abuja, Federal Ministry of Health.

Grover SA, Coupal L, Gilmore N, Mukherjee J (2005). Impact of dyslipidemia associated with highly active antiretroviral therapy (HAART) on cardiovascular risk and life expectancy. Am. J. Cardiol., 95(5): 586-591.

Grunfeld C, Pang M, Doerrler W, Shigenaga JK, Jensen P, Feingold KR (1992). Lipids, lipoproteins, triglyceride clearance, and cytokines in human immunodeficiency virus infection and the acquired immunodeficiency syndrome. J. Clin. Endocrinol. Metab., 74(5): 1045-1052.

Khiangte L, Vidyabati RK, Singh MK, Bilasini DS, Rajen ST, Gyaneshwar SW (2007). A Study of Serum Lipid Profile in Human Immunodeficiency Virus (HIV) Infected Patients. JIACM. 8: 307-311.

Leitner JM, Pernerstorfer-Schoen H, Weiss A, Schindler K, Rieger A, Jilma B (2006). Age and sex modulate metabolic and cardiovascular risk markers of patients after 1 year of highly active antiretroviral therapy (HAART). Atherosclerosis, 187(1): 177-185

Miserez AR, Muller PY, Spaniol V (2002). Indinavir inhibits sterol-regulatory element-binding protein-1c-dependent lipoprotein lipase and fatty acid synthase gene activations. AIDS, 16(12): 1587-1594.

Mulligan K, Grunfeld C, Tai VW, Algren H, Pang M, Chernoff DN, Lo JC, Schambelan M (2000). Hyperlipidemia and insulin are induced by protease inhibitors independent of changes in body composition in patients with HIV infection. J. Acquir. Immun. Defic. Syndr. 23: 35–43.

Obirikorang C, Yeboah FA, Quaye L (2010). Serum lipid profiling in highly active antiretroviral therapy-naïve HIV positive patients in Ghana: Any potential risk? WebmedCentral Infect. Dis., 1(10): WMC00987.

Palacios R, Santos J, García A, Castells E, González M, Ruiz J, Márquez M (2006). Impact of highly active antiretroviral therapy on blood pressure in HIV-infected patients. A prospective study in a cohort of naive patients. HIV Med., 7(1): 10-15.

Pujari SN, Dravid A, Naik E, Bhagat S, Tash K, Nadler JP, Sinnott JT (2005). Lipodystrophy and dyslipidemia among patients taking fi rst-line, World Health Organization recommended highly active antiretroviral therapy regimens in Western India. J. Acquir. Immun Defic. Syndr., 39(2): 199-202.

Roula BQ, Fisher E, Rublein J, Wohl DA (2000). HIV-Associated Lipodystrophy Syndrome. Pharmacotherapy, 20(1): 13-22.

Stampfer MJ, Krauss RM, Ma J, Blanche PJ, Holl LG, Sacks FM, Hennekens CH (1996). A prospective study of triglyceride level, low-density lipoprotein particle diameter and risk of myocardial infarction. J. Am. Med. Assoc., 276(11): 882-888

Vigano A, Mora S, Testolin C, Beccio S, Schneider L, Bricalli D, Vanzulli A, Manzoni P, Brambilla P (2003). Increased lipodystrophy is associated with increased exposure to highly active antiretroviral therapy in HIV-infected children. J. Acquir. Immun. Defic. Syndr., 32(5): 482-489.

WHO/UNAIDS (2009). Global summary of the HIV/AIDS epidemic. December 2009. www.unaids.org.

8

The use of cholinergic biomarker, cholinesterase activity of blue mussel *Mytilus edulis* to detect the effects of organophosphorous pesticides

K. Yaqin[1]* and P. D. Hansen[2]

[1]Department of Fisheries, Faculty of Marine Science and Fisheries, Hasanuddin University, Makassar, Indonesia.
[2]Technische Universitaet Berlin, FB 7 -Intitute for Ecological Research and Technology, Department of Ecotoxicology, Berlin, Germany.

The aim of the study was to investigate the effects of organophosphorous pesticide on the cholinesterase activity of different organs of *Mytilus edulis*. The mussels were exposed to serial dilutions of the pesticides (0, 50, 100, 200, 500, and 1000 μg/l) for 96 h. A significant inhibition of the cholinesterase activity from gill occurred at the lowest concentration, which indicated that gill was the most sensitive organ. The moderate sensitive organs were foot and mantle, which were inhibited by trichlorfon at 200 μg/l. The least sensitive organs were hemolymph, posterior adductor muscle and digestive gland were inhibited at 1000 μg/l. After incubation of the trichlorfon-exposed mussels for seven days in clean media, the cholinesterase activities from different organs of mussels were not cured completely. The cholinesterase activities from hemolymph, gill, posterior adductor muscle and digestive gland recovered, while the persistent inhibition of cholinesterase activity from foot and mantle were observed.

Key words: Biomarker, cholinesterase, mussel, trichlorfon.

INTRODUCTION

The extensive use of organophosphorous (OP) pesticides in agricultural and other antrophogenic activities causes increase of discharged pesticides wastes in the environment. Although, the occurrence of organophosphorous pesticides in the environment is thought to be unstable compared to organochlorine pesticides, the persistent effects of the pesticides in non-target organisms and ecological system cannot be ruled out (Scholz and Hopkins, 2006). Once organophosphorous pesticides enter the body of organism, most of them are transformed into metabolites which in many cases are more toxic compounds than the parent compounds or induced directly the target enzymes or organs (Belden and Lydy, 2000). Consequently, the effects of most of the pesticides on acetylcholinesterase (AChE) activity are considered as an irreversible action since the time of re-synthesis of the enzyme are naturally longer than the duration of

dissociation of the organophosphorous-complex (Gaglani and Bocquene, 2000). Deteriorate effects of the pesticides become more prominent when de-alkylation or what it called ageing occurs which involves cleavage of an alkyl group of the phosphoryl moiety and the formation of negative charge, which stabilizes the OP-Complex (Ray, 1998). *De novo* synthesis of the enzyme is the only way to recover AChE activity in the synaptic cleft, which goes slower than de-alkylation reaction.

In aquatic ecosystem, the mechanisms of the OP (organophosphorous) actions in aquatic organisms such as *Mytilus* sp in the enzymatic levels, particularly cholinesterase (ChE) activity, provide a comfortable tool as a biomarker for the detection of the pesticide impacts in that ecosystem. Since the works of Grigor'eva et al. (1968), on ChEs from cardiac muscle and hemolymph of *Mytilus edulis* (Moralev and Rozengrat, 2004) and Wachtendonk and Neef (1979) on ChEs from hemolymph, the ChE activity from the mussels has been explored and employed as a biomarker to detect the effects of OP pesticides in laboratory and field studies. Different organs of the mussels have been used to

---

*Corresponding author. Email: khusnul@gmail.com.

evaluate the detrimental effects of the OP pesticides on the ChE (cholinesterase) activity, which showed that gill was more frequently used compared to other organs or whole tissue (Escartin and Porte, 1997; Mc.Henery et al., 1997; Mora et al., 1999a; 1999b; Dizer et al., 2001; Kopecka et al., 2004).

In addition, the correlation between the inhibition of the ChE activity from gill and increasing usage of dichlorvos in marine culture was proved (Mc.Henery et al., 1997). However, compare to the tissue homogenate, hemolymph from *M. edulis* had higher AChE activity (Galloway et al., 2002), and can be used without sacrificing the animals. Hence, the authors claimed that the AChE activity from the mussel hemolymph provides a rapid, relatively cost-effective, reliable, and non-destructive tool to assess the exposure of mussels to OP and carbamate pesticides.

Notwithstanding, transformation of the contaminant induction of ChE activity in hemolymph to ecological levels is more difficult than that in other organs such as gill, mantle, digestive gland, posterior adductor muscle (PAM) and foot. It is because existence of conta-minant-damage in hemolymph is easy to recover due to regeneration of hemolymph in *M. edulis* which occurs quickly (Gosling, 2003). Besides, induction of ChE activity in hemolymph may not be related to nervous system (Brown et al., 2004). Therefore, it is a point of interest to know different responses of different organs of *M. edulis* in term of cholinergic activity which can be useful information to determine which organs of *M. edulis* is sensitive and effective as a target organ in terms of biomarker applications in laboratory and field scales.

The study was carried out to compare the cholinergic responses which are reflected as ChE (cholinesterase) activity in different organs of *M. edulis* after exposure to OP pesticide, trichlorfon. In addition, since induction, adaptation and recovery mechanisms in sentinel organisms are important mechanisms to delineate environmental stress over time (Wu et al., 2005) which are useful factors for applying biomarkers in screening test and field monitoring, the study was also aimed to investigate the recovery of pesticide-induction of ChE activity in the different organs.

MATERIALS AND METHODS

Chemicals

Acethylthiocholine iodide, 5,5'-Dithio-bis-(-2-Nitrobenzoic acid) and γ-globuline were obtained from Sigma, USA. Trichlorfon (PESTANAL®, analytical standard (Riedel-de Haën) was purchased from Sigma-Aldrich, Germany. Bradford-reagent was purchased from BIO-RAD.

In vivo exposure experiment (I)

Mussels, *M. edulis* (6-7 cm) were collected from a clean area of Sylt Island Germany and carried dry to the laboratory. *In vivo* exposure

experiment was conducted for 96 h in duplicate by using trichlorfon as a contaminant model. A stock solution was made by diluting trichlorfon with distilled water. Serial dilutions of trichlorfon were made by adding the trichlorfon stock solution into the glass aquarium that contained 4 l of 3% salinity of artificial sea water (ASW) until it reached the final concentrations (0, 50, 100, 200, 500, and 1000 µg/l). Each experiment media was loaded with 12 mussels. The dissolved oxygen in each aquarium was maintained not less than 80% using an aerator. To assure the desired contaminant concentrations in each aquarium, the media and the contaminant concentrations were replaced everyday 2 h after the mussels were fed by the commercial algae.

After 96 h, the exposure experiment was terminated. Six mussels per aquarium were sacrificed and the intended organs namely, hemolymph, gill, foot, posterior adductor muscle (PAM) and digestive gland were collected and treated as outlined below.

In vivo (II) and *in situ* (III) recovery experiments

The recovery experiment in laboratory scale was conducted by placing the trichlorfon-exposed mussels (experiment I) into 4 l of clean artificial sea water media for seven days. Each aquarium had six mussels. After seven days, the experiment was completed and all the mussels were sacrificed and selected organs were collected and treated as described below.

In situ recovery experiment (III) was performed by transplanting the trichlorfon-exposed mussels into the clean original habitat in Sylt Island. Prior to the transplantation experiment, 12 mussels per aquarium in 4 l of ASW were exposed to 0 and 500 µg/l of trichlorfon for 96 h in duplicate. At the end of the exposure, six mussels per aquarium were sacrificed and the rest were sent dry to Sylt Island. In Sylt Island the mussels were transplanted in sea water using the net. After seven days, the mussels were removed from the sea water and sent back dry to the laboratory and sacrificed for dissecting and collection of the intended organs immediately.

Organs collection and preparation

Hemolymphs of six mussels in each aquarium were sucked from posterior adductor muscle (PAM) using 1 ml syringe and 0.4 mm needle. Immediately, 1 ml of hemolymph was centrifuged at 10000 x g for 10 min at 4°C for the separation of the hemolymph from the hemocytes. The hemocytes-free hemolymph were then harvested, transferred into 1.5 ml eppendorf tube and kept under -80°C prior to ChE activity determination. The intended organs of mussels (gill, foot, mantle, PAM and digestive gland) were dissected out, dam-dried and weighed. A dounce homogenizer was used to homogenize 0.3 g of each tissue in 2 ml of potassium phosphate buffer (0.1 M/pH 8.0). The homogenate obtained was centrifuged at 10000 x g for 10 min at 4°C. The supernatant was removed into 1.5 eppendorf tube and kept under -80°C before ChE activity measurement was conducted.

Cholinesterase activity measurement

The enzyme activity was measured following the modified Ellman method (Ellman et al., 1961), for a 96-well plate and microplate reading (Herbert et al., 1995; Dizer et al., 2001). The enzyme measurement was carried out by placing 50 µl of the diluted sample into each well of the microplate. A blank was made by adding 50 µl of potassium phosphate buffer into a blank section of the microplate wells. The plate was incubated for 5 min at 25°C with 200 µl of 0.75 mM 5,5'-Dithio-bis-(-2-Nitrobenzoic acid) prior to the start of the reaction by the addition of 50 µl of 3 mM a cethylthiocholine iodide.

Accordingly, the plate was read by using a photometer for microtiter plate (Spectra Thermo TECAN) at an interval of 30 s for 5 min at 405 nm. Protein content measurement was carried out using Bradfort reagent solution. Finally, AChE activity was expressed as nmoles of product developed per minute per mg of protein (nmol/min/mg protein).

Statistical analysis

Parametric or non parametric analysis of variant was used to determine the effects of the pesticide for the assays according to data distribution. Distribution and homogeneity of variant of data set were checked firstly. If the data set was not distributed normally and the variant was not homogenous, the data set were log-transformed. Parametric one-way analysis of variant was used on data set, which demonstrated a normal distribution and homogeneity of variant, both before and after transformation. If the means were different significantly, Bonferroni's multiple comparison test was applied to determine the different means between the treatments. Nevertheless, for the data set, which did not show a normal distribution and homogeneity of variant, a non-parametric Kruskall-Wallis test was used to determine the differences between the medians. If the differences were significant, the median values were compared by Dunn's multiple comparison test. $P < 0.05$ was considered as statistically significant. The data set were analyzed by GraphPad prism software program, which expressed mean and standard deviation.

RESULTS

Cholinesterase activity

Hemolymph

Hemolymph of mussels showed the highest ChE activity with mean value 104.70 ± 34.48 nm/min/mg protein. There was a stimulation of the ChE activity on mussel's hemolymph (12% compared to the control) which were exposed to the lowest concentration (50 µg/l), but it was not significant. The effect of the used pesticide on the ChE activity from hemolymph of mussels occurred only significantly when mussels were exposed to 1000 µg/l (Figure 1). At this concentration, the ChE activity of mussel decreased by 39% compared to the control.

After incubation in the clean ASW media for the seven days, hemolymph of mussels in the control showed a slight decrease of the ChE activity (95.12 ± 25.60 nm/min/mg protein) compared to the previous control which were used for the 96 h exposure experiment. The recovery of the ChE activity from hemolymph of mussels, which were exposed to 1000 µg/l of trichlorfon took place after the mussels were incubated in the clean ASW for seven days (Figure 1). The ChE activity of that mussel's hemolymph showed a mean value of 73.17 ± 19.16 nmol/min/mg protein.

Gill

Parametric one-way ANOVA on the transformed data of the ChE activity from gill of mussel showed that significant differences in the ChE activity among the treatments were observed. Further calculation using Bonferroni's multiple comparison revealed that trichlorfon inhibited ChE activity significantly at 50 µg/l (3.44 ± 1.17 nmol/min/mg protein) (Figure 1). At this concentration, the pesticide reduced 29% of the ChE activity from the control level (4.83 ± 1.36 nmol/min/mg protein). The inhibition of the pesticide on the ChE activity persisted significantly at all higher concentrations.

Decreasing activities of the ChE in all treatments were observed when mussels were transferred to the clean condition compared to the control of the 96 h exposure experiment (Figure 1). Furthermore, the differences of the ChE activities of the control and the treatments in the recovery experiment were not confirmed when the data were calculated using Kruskal-Wallis test.

Foot

The ChE activity from foot of mussel was induced significantly by trichlorofon at 200 µg/l (3.10 ± 0.56 nmol/min/mg protein) (Dunn's Multiple Comparation test) (Figure 1). The reduction of the ChE activity at this concentration was 35% compared to the control (4.77 ± 1.19 nmol/min/mg protein). The inhibition of the pesticide persisted significantly when mussels were exposed to other higher concentrations 500 (2.89 ± 0.64 nmol/min/mg protein) and 1000 µg/l (2.10 ± 0.42 nmol/min/mg protein) (Dunn's multiple comparation test).

Transferring the pesticide-exposed mussels to clean condition resulted in relatively slight reduction of the ChE activities for all treatments. However, when the mussels of the control was compared (3.12 ± 0.79 nmol/min/mg protein) to other treatments at clean condition, there were slight increase in the ChE activities of the mussels at 50 (3.40 ± 0.84 nmol/min/mg protein) and 100 µg/l (3.48 ± 1.10 nmol/min/mg protein). Furthermore, insignificant inhibitions of the ChE activities occurred at 200 (2.94 ± 0.63 nmol/min/mg protein) and 500 µg/l (3.11 ± 0.33 nmol/min/mg protein). The inhibition of the ChE activity appeared significantly at 1000 µg/l (2.11 ± 0.188 nmol/min/mg protein) (Dunn's Multiple Comparison test) (Figure 1).

Mantle

Kruskal-Wallis, followed by Dunn's multiple comparison demonstrated that a significant inhibition of ChE activity from mantle by the pesticide occurred at 200 µg/l (Figure 1). Compared to the control (3.82 nmol/min/mg protein), the ChE activity of mussels which were exposed to 200 µg/l of trichlorfon reduced by 26%. Subsequently at 500 µg/l, the inhibition remained (26%) and seemed more distinct at 1000 µg/l (34%).

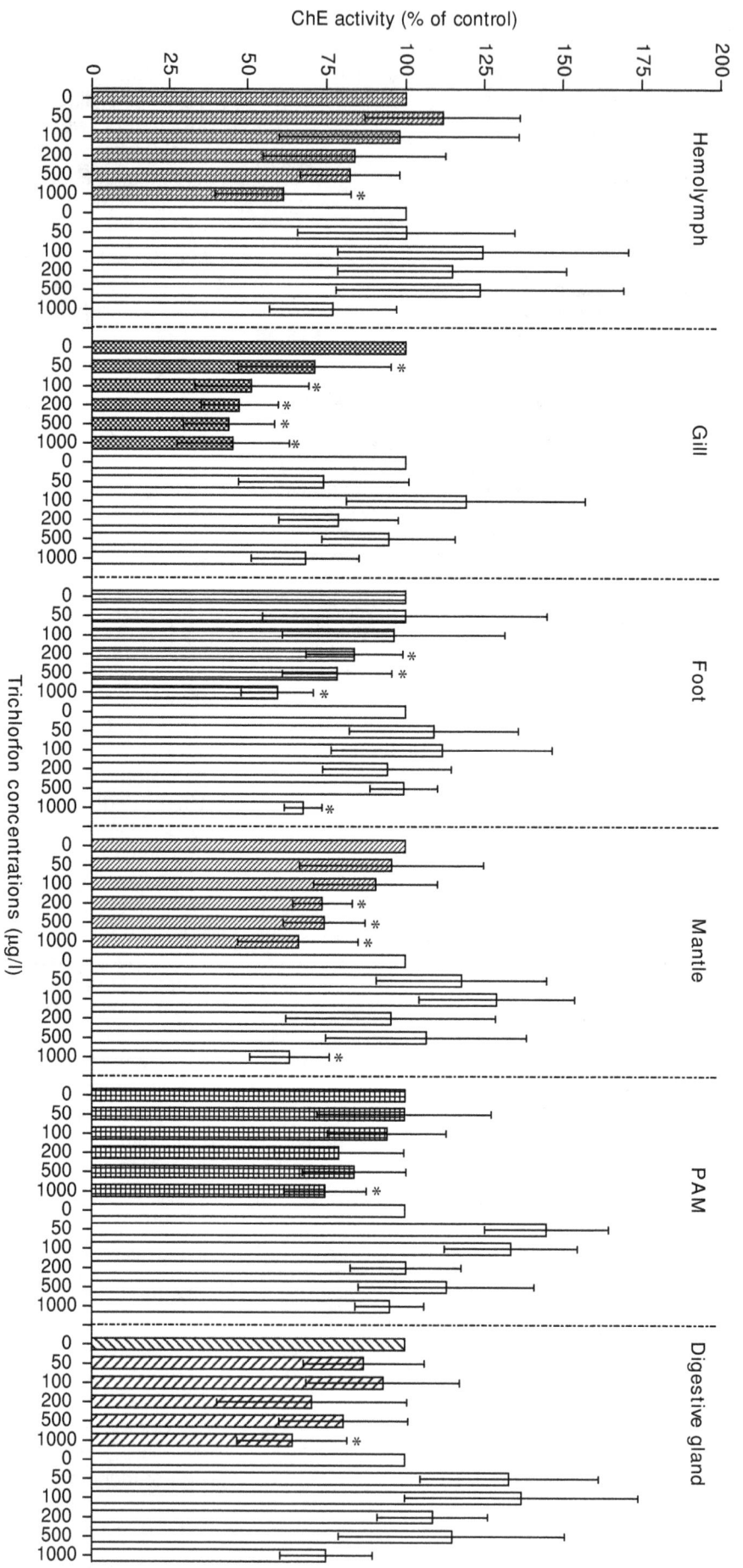

Figure 1. Comparison between pre- and post-incubated mussels *M. edulis* after exposed to trichlorfon concentrations in term of the ChE activity from the six organs. The ChE activity in the six organs of mussels exposed to trichlorfon for 96-h (striated area with ± standard deviation). The ChE activity in the six organs of post-trichlorfon exposed mussels which were incubated in clean ASW for 7 days (empty area with ± standard deviation). * indicate significant difference from control (0 µg/l) (P < 0.05).

After the mussels were replaced to the clean ASW media, the stimulation of the ChE activities occurred seemingly at 50 (3.28 nmol/min/mg protein) and 100 µg/l (3.59 nmol/min/mg protein). Significantly, the level of the ChE activities returned to the level of the control at 200 (2.65 nmol/min/mg protein) and 500 µg/l (2.96 nmol/min/mg protein). In contrast, the effect of the pesticide on the ChE activity of mantle (1.76 nmol/min/mg protein) still appeared significantly (Dunn's Multiple Comparison test) at 1000 µg/l when compared to the control (2.78 nmol/min/mg protein) (Figure 1).

Posterior adductor muscle (PAM)

The data set from the ChE activity of PAM were

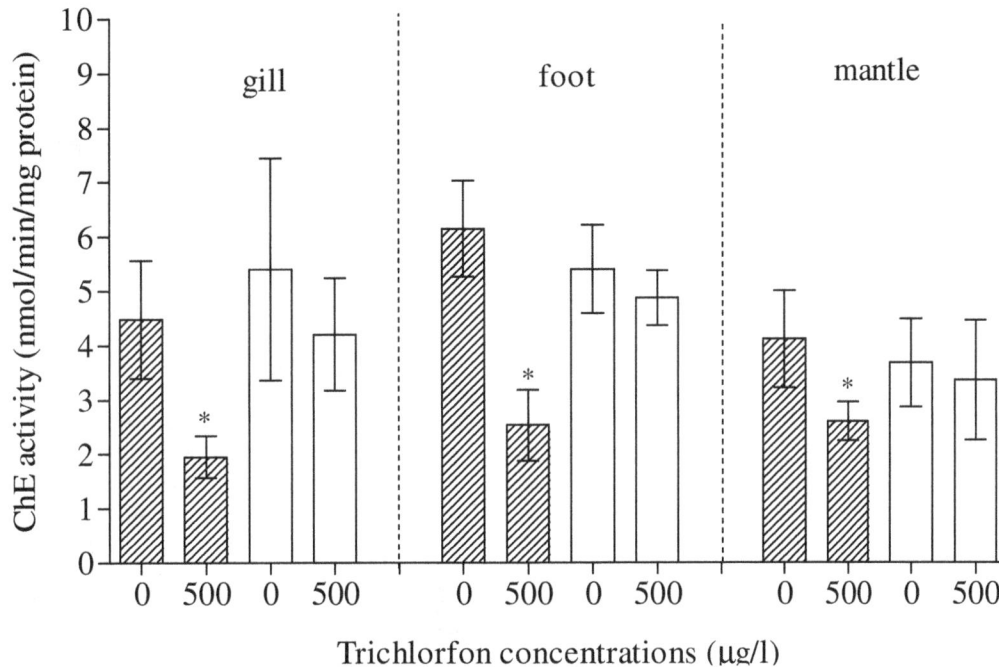

Figure 2. Effects of trichlorfon on the ChE activity from three organs of *M. edulis*. The ChE activity in three organs of mussels exposed to trichlorfon for 96-h (striated area with ± standard deviation). The ChE activity in three organs of post-trichlorfon exposed mussels which were incubated in clean water of original habitat of mussels for 7 days (empty area with ± standard deviation). * indicate significant different from control (0 µg/l) (P < 0.05).

homogenous and normally distributed. Parametric ANOVA showed that the pesticide induced significantly the ChE activity only at 1000 µg/l which produced the lesser ChE activity (1.82 nmol/min/mg protein) from PAM (Figure 2). The percentage of inhibition was 25% compared to the control (2.43 nmol/min/mg protein).

The ChE activity of PAM from the control of the clean incubated-mussels showed a relatively lower level compared to the control of the 96 h pesticide-exposure. Since the data also were distributed normally and the variant was homogenous, parametric ANOVA was used to distinguish the differences among the treatments. The results revealed that there were no differences in ChE activities among the treatments.

Digestive gland

The ChE activity in the digestive gland of mussels in the control showed mean value of 2.57 ± 0.55 nmol/min/mg protein. The ChE activity was only inhibited significantly (Bonferroni's Multiple Comparison test) when mussels were exposed to the highest concentration (1000 µg/l), resulting in 35 % inhibition (Figure 1).

The persistence of the pesticide inhibition on the ChE activity of digestive gland after incubation in the clean ASW media was not observed statistically (Dunn's Multiple Comparison test). However, stimulations of the ChE activities in the control (1.95 ± 0.48 nmol/min/mg

protein) appeared insignificantly at 50 (2.30 ± 0.49 nmol/min/mg protein) and 100 µg/l (2.33 ± 0.37 nmol/min/mg protein). Eventually, the level of the ChE activities returned to the control level at 200 and 500 µg/l and reduced slightly at 1000 µg/l (Figure 1).

In situ recovery experiment

In situ recovery experiment was conducted only by exposing mussels to 500 µg/l of trichlorfon and control. Consequently, data analysis was performed on the organs of mussels which showed inhibition of the ChE activity at concentrations ≤500 µg/l only. The organs were gill, foot and mantle. Repeated *in vivo* experiment for 96 h showed that the ChE activities of the three organs were significantly inhibited by 500 µg/l trichlorfon. After transplanting the mussels to the original habitat in the coastal area of the Island of Sylt, the ChE activities of the organs returned significantly to the activity level of the control (Figure 2).

DISCUSSION

Cholinesterase (ChE) activity of different organs

Although, there are considerable diversity of biochemical properties of ChEs in aquatic organisms (Escartin and

Porte, 1997), ChEs in mussels are generally divided into two main classes; acetylcholinesterase (AChE; EC 3.1.1.7) and butyrylcholinesterase (BuChE; EC 3.1.1.8). The distribution and the properties of ChEs in mussels are organ-dependent (Bocquene et al., 1990; Brown et al., 2004). This study referred to the enzyme as ChEs since they employed acetylthiocholine as non-specific substrate of the enzyme. Therefore, the term of ChE that was used in this study referred to the sum of AChE and BuChE as frequently used as a diagnostic tool for ecotoxicological risk assessment studies in aquatic ecosystem (Torre et al., 2002).

The alteration of ChE activity is a well-known cellular response of marine mussels which is particularly induced by OP and carbamate pesticides. Subsequently, this enzyme activity from the mussels was considered as a respectable tool for detecting the effects of the pesticides (Herbert et al., 1995). Mussel organs such as hemolymph, gill, mantle, PAM, digestive gland and gonad have been employed as target organs for measuring the alteration of ChE activity induced by contaminants. The preference of the used organs usually depends on some factors such as the goal of the study and the availability of the organs that are used for analysis. The ChE activity from hemolymph of mussel is considered to be rapid, inexpensive and a reliable means for measuring the biological impact of pesticide mainly if sacrificing the animals is avoided (Moreira et al., 2001; Galloway et al., 2002). Nevertheless, the link effect of the alteration of the ChE activity from hemolymph induced by contaminants to higher biological organization such as feeding behavior is less relevant compared to the innervated organs like gill, mantle and PAM.

This study evaluated the effects of trichlorfon on ChE activity from various organs from *M. edulis*, viz., hemolymph, gill, foot, mantle, PAM and digestive gland. The highest ChE activity was found in hemolymph and was more than 20 fold when compared to the other organs. These activities from different organs of *M. edulis* were comparable to those which were observed by Herbert et al. (1995).

The results displayed that different organs of *M. edulis* had different sensitivity to trichlorfon. The ChEs from gill was the most sensitive to the effects of trichlorfon followed by foot, mantle, the hemolymph, digestive gland and PAM. The lowest observable effect concentration (LOEC) of gill took place at 50 µg/l. Although, the LOEC of the ChE activity from foot and mantle were similar, the inhibition level of the ChE activity from foot (35%) at 200 µg/l was higher than that from mantle (26%). The significant inhibition of trichlorfon on the ChE activities from the hemolymph, digestive gland and PAM were initiated at 1000 µg/l, even though the ChE activity from the hemolymph experienced greater inhibition (39%) than those from digestive gland (36%) and PAM (25%). These results demonstrated that the threshold of trichlorfon to inhibit the ChE activity from *M. edulis* was organ

dependent. Accordingly, the sensitivity of *M. edulis* organs to trichlorfon in terms of the ChE activity could be divided into three clusters, which were high (gill), moderate (foot and mantle) and low sensitivities (the hemolymph, digestive gland and PAM).

Most studies on ChE activity from mussels were oriented to identify the sensitivity level of ChE activity to pesticides by using single organ. There were some ChEs studies which used multiple organs of mussels in the laboratory scale. Herbert et al. (1995) conducted *in vitro* test to evaluate the sensitivity of ChEs from different organs of *M. edulis* which were hemolymph, gill, PAM, digestive gland and gonad. The experiment revealed that gill and the hemolymph were the most sensitive organs which were followed by digestive gland, PAM and the gonad. Comparison between solid tissue homogenates which were gill, mantle, PAM and whole body showed that gill was the most sensitive organ when the tissue homogenates were exposed to aldicarb. Escartin and Porte (1997) also found that the ChE activity of *M. galloprovincialis* gill tissue homogenate was more susceptible to inhibition by fenitrothion, fenitrooxon and carbofuran compared to that from digestive gland. In addition, a recent study conducted by Canty et al. (2007) recorded that the ChE activity from gill of *M. edulis* was more sensitive than that from hemolymph when exposed to azamethiphos.

Comparison of the sensitivity level of the ChE activities from *M. edulis* organs to the pesticides from different studies was carried out carefully. Different sensitivities of the studies did not merely reflect the sensitivities of the organ *per se*, but mostly they resulted from different methodology and the tested pesticides (Brown et al., 2004).

Recovery of ChE activity from different organs

Recovery mechanisms in sentinel organisms after exposed to environmental stress are point of interest in the use of biomarker in biomonitoring campaigns. To be useful tool in biomonitoring, biomarker must reveal the environmental stress over time so that the knowledge about induction, adaptation and recovery of the stress are required and considered prior to monitoring (Wu et al., 2005).

This study applied two strategies to investigate the recovery of the ChE activity from *M. edulis* after been inhibited by trichlorfon in artificial and natural conditions. The results demonstrated that before transferring to the clean media, the percentages of inhibitions of the ChE activities in gill, foot, and mantle from the trichlorfon exposed mussels, which were used in laboratory scale recovery were 55.94, 39.31 and 25.65%, respectively. Furthermore, for the mussels that were used in the recovery experiment and transplanted in natural sea water, the percentages of inhibitions induced by trichlorfon on the ChE activity in gill, foot, and mantle were 56.51, 58.84 and 36.89% compared to the control respectively. After incubation for seven days, in both laboratory and natural conditions, the ChE activities from the three organs returned to the level of control statistically. This indicated that both artificial and natural conditions served as a suitable media for mussels to recover from ChE activity inhibition. Besides, the result also indicated that the artificial media used in the laboratory scale was sufficient to mimic natural sea water for the mussels to recover from the induction caused by 500 µg/l of trichlorfon for 96 h.

In the laboratory scale, mussels were exposed to serial concentrations of trichlorfon. After been replaced and incubated in clean media for seven days, the ChE activities of the tested organs increased to the level of control. This recovery occurred in mussels which were exposed to the pesticide in concentrations from 50 to 500 µg/l. Nonetheless, the exposed mussels to 1000 µg/l of trichlorfon did not show increase in the ChE activities from all studied organs.

After transferring the animals to the clean media only the ChE activities from gill, hemolymph, digestive gland and PAM that returned to the level of the control, but the evidence were not observed in foot and mantle. These evidences suggested that recovery mechanisms on trichlorfon inhibited ChE activity from mussels were organs specific. In other words, the complete recovery of the ChE activities in the whole organs of mussels after been exposed to 1000 µg/l of trichlorfon was not confirmed. Gill illustrated a sensitive and quick recovery organ regarding the pesticide effects. Mc.Henery et al. (1997) also observed that fast induction and recovery of the ChE activity from gill of M. edulis occurred when the animals were exposed to serial concentrations of dichlorvos (10, 100 and 1000 µg/l) and replaced in clean media for seven days. In contrast, this study demonstrated foot and mantle as moderate sensitive organs, which could retain pesticides effects for such periods. The lowest sensitive organs, that is, the hemolymph, digestive gland and PAM elucidated fast recovery responses. On that account, it is suggested that the use of foot and mantle from M. edulis in studies on induction of neurotoxic contaminants on the ChE activity and the recovery mechanism is recommended.

The application of these organs viz. foot and mantle as target organs for measuring the ChE activity from mussels could be considered as counterparts of common employment organ such as gill to elucidate more comprehensive understanding on neurotoxic xenobiotic effects in mussels when the ChE activity would be applied as biomarker. As demonstrated in the study the sensitivity and the recovery response of the ChE activity in mussels differed from organ to organ. Combining all types of inductions and recoveries of target organs is needed to reduce potential false positive or negative on assessing the impacts of the pesticides on mussels.

Conclusion

The study depicted that the sensitivity of ChE activity from M. edulis to trichlorfon is organ dependent. The sensitivity of ChE activity in M. edulis organs to trichlorfon can be discriminated into three clusters, which are high (gill), moderate (foot and mantle) and low sensitivities (hemolymph, digestive gland and PAM). Moreover, the recovery of the ChE activity induction from different organs is also organ specific which can be divided to three clusters as well. The most sensitive (gill) and insensitive organs, (hemolymph, digestive gland and PAM) delineated the recovery from pesticide induction, while the moderate sensitive organs (foot and mantle) still retained the effect of the pesticide during the experiment. The results of the study demonstrated that the use of battery of target organs in screening test using ChE activity as a biomarker in laboratory scale is recommended.

ACKNOWLEDGEMENT

This research was made possible by DAAD (Deutche Akademische Austauschdient/German Academic Exchange Service).

REFERENCES

Belden JB, Lydy MJ (2000). Impact of atrazine on organophosphate insecticide toxicity. Environ. Toxicol. Chem., 19: 2266–2274.

Bocquene G, Gaglani F, Truquet P (1990). Characterization and assay conditions for use of AChE activity from several marine species in pollution monitoring. Mar. Environ. Res., 30:75–89.

Brown M, Davies IM, Moffat CF, Redshaw J, Craft JA (2004). Characteristic of choline esterases and their tissue and subcellular distribution in mussel (Mytilus edulis). Mar. Environ. Res., 57:155-169.

Canty MN, Hagger JA, More RTB, Cooper L, Galloway TS (2007). Sublethal impact of short term exposure to the organophosphate pesticide azamethiphos in the marine Mollusc Mytilus edulis. Mar. Poll. Bull., 54: 396-402.

Dizer H, Fischer B, Harabawy ASA, Hennion M-C,Hansen P-D (2001). Toxicity of domoic acid in the marine mussel Mytilus edulis. Aquat. Toxicol., 55: 149-156.

Ellman GL, Courtney KD, Andres VJr, Featherstone RM (1961). A new and rapid colorimetric determination of acetylcholinesterase activity. Biochem. Pharmocol., 7: 88-95.

Escartin E, Porte C (1997). The use of cholinesterase and carboxylesterase activities from Mytilus Galloprovincialis in pollution monitoring. Environ. Toxicol. Chem., 16: 2090-2095.

Gaglani F, Bocqueme G (2000). Molecular biomarkers of exposure of marine organisms to organophosphorus pesticide and carbamates. In: Use of biomarkers for environmental quality assessment, Lagadic L, Caquet Th, Amiard J-C, Ramade F. (eds). Science Publishers, Inc, USA. pp. 113 – 137.

Galloway TS, Millward N, Browne MA, Depledge MH (2002). Rapid assessment of organophosphorous/carbamate exposure in the bivalve mollusc Mytilus edulis using combined esterase activities as biomarkers. Aquat. Toxicol., 61:169-180.

Gosling E (2003). Bivalve Molluscs; Bilogy, Ecology and Culture. Oxford. Fishing News Books.

Herbert A, Guilhermino L, de Asis HCS, Hansen P-D (1995). Acetylcholinesterase activity in aquatic organisms as pollution biomarker. Z. Angewandte Zoo, 3: 1-15.

Kopecka J, Rybakowas A, Barsiene J, Pempkowiak J (2004). AChE levels in mussels and fish collected off Lithuania and Poland (Southern Baltic). Oceanologia, 46: 405-418.

Mc.Henery JG, Linley-Adams GE, Moore DC, Rodger GK (1997). Experimental and field study of effects of dichlorvos exposure on acetylcholinesterase activity in the gills of the mussel, *Mytilus edulis* L. Aquat Toxicol., 38: 125-143.

Mora P, Michel X, Narbonne JF (1999a). Cholinesterase activity as potential biomarker in two bivalves. Environ. Toxicol. Pharmacol., 7: 253-260.

Mora P, Fournier D, Narbonne JF (1999b). Cholinesterases from the marine mussels *Mytilus galloprovincialis* Lmk. and *M. edulis* L. and from the freshwater bivalve *Corbicula fluminea* Müller. Comp. Biochem. Physiol. C., 122: 353–361.

Moralev SN, Rozengart EV (2004). Comparative sensitivity of cholinesterases of different origin to some irreversible inhibitors. J. Evol. Biochem. Physiol. 40,1-17.

Moreira SM, Coimbra J, Guilhermino L (2001). Acetylcholinesterase of *Mytilus galloprovincialis* LmK. Hemolymph: A suitable environmental biomarker. Bull. Environ. Contam. Toxicol., 67: 470 – 475.

Ray D (1998). Organophosphorus esters : an evaluation of chronic neurotoxic effects. Report. Institute for Environment and Health University of Leicester. Leicester UK.

Scholz NL, Hopkins WA (2006). Ecotoxicology of anticholinesterase pesticides: data gaps and research challenges. Environ. Toxicol. Chem., 25: 1185–1186.

Torre DFR, Ferrari L, Salibian A (2002). Freshwater pollution biomarker: response of brain acetylcholinesterase activity in two fish species. Comp. Biochem. Physiol. C., 131: 271-280.

Wachtendonk von D, Neef J (1979). Isolation, perufication and molecular properties of an acetylcholinesterase (E.C. 3.1.1.7) from the haemolymph of the sea mussel *Mytilus edulis*. Comp. Biochem. Physiol., 63C: 279 – 286.

Wu RSS, Siu WHL, Shin PKS (2005). Induction, adaptation and recovery of biological responses: Implications for environmental monitoring. Mar. Poll. Bull., 51: 623–634.

9

Reduction of platelet and lymphocyte counts and elevation of neutrophil counts in rats treated with aqueous leaf extract of *Ocimum gratissimum*

Okon U. A.*, Ita S. O., Ekpenyong C. E., Davies, K. G. and Inyang, O. I.

Department of Physiology, College of Health Sciences, University of Uyo, Nigeria.

Ocimum gratissimum (Linn.) has been widely used for food and medicinal purposes. The effect of aqueous leaf extract of *O. gratissimum* on haematological parameters of albino Wistar rats was studied. Acute toxicity study showed a median lethal dose (LD_{50}) value of 2121.32 mg/kg body weight for intraperitoneal (i. p) route in mice. Twenty rats' weights ranging between 85 to 115 g were used. The rats were divided into four groups; with group 1 as the control group. Increasing doses (212, 424 and 636mg/kg body weight) of extract were administered orally to the three groups for a period of 21 days. Significant ($p<0.001$ and $p<0.01$) increases of WBC in groups II and III and significant ($P<0.05$) decreases in platelet count and lymphocyte levels were observed. Similarly, a significant ($P<0.05$) increase in the neutrophil levels was observed. There were no significant changes in red blood cell (RBC), packed cell volume (PCV), mean corpuscular volume (MCV), mean corpuscular haemoglobin (MCH) and mean corpuscular haemoglobin concentration (MCHC). *O. gratissimum* reduces platelet and lymphocyte counts, while it increases total WBC and neutrophil levels. It is evident from this study that there is need for caution with the consumption or administration of excessive dosage of *O. gratissimum*.

Key words: *Ocimum gratissimum*, toxicity, platelet, lymphocyte, neutrophil.

INTRODUCTION

Man throughout the ages has depended on his immediate environment for food and medication. Most especially, man has consistently resorted to plants for solution to the myriad of health problems challenging him. Traditional herbal practitioners have made several claims on numerous herbal preparations with specific claim on the efficacy of *Ocimum grastissimum* in the treatment of several disease conditions including infections, oncogenic and neurological disorders.

There is widespread usage and consumption of *O. gratissimum* globally. It is used as spice and is known to have enormous nutritional values. The leaves of the plant which is highly appreciated for its pleasant aroma is used for seasoning food and as vegetable. It contains high moisture, low protein, ash, vitamins A, B_2, and D, calcium, phosphorus, selenium, iron, zinc and magnesium (Oboh et al., 2009).

Phytochemical screening reveals that the volatile aromatic oil from the leaves of *Ocimum gratissimum* consists of thymol, Eugenol, terpenes, xanthones and lactones (Oboh, 2008).

Other reports had showed that this plant contains; α–pinene, β –pinene. 1, 8 – cineole, β – caryophyllene, a murolene and sehirene (Sainsbury and Sofowora, 1971). Other constituents include germacrene, x-copaene, humutene, β -elemene, β – bourbonenem and serinerel (Pande and Pathak, 2009). Regional variation in the phytochemical constituent of *Ocimum gratissimum* has been reported (Oboh et al., 2009).

The widespread usage of this plant is informed by its peculiar pharmacological properties such as its ability to scavenge free radical as antioxidant (Odukoya et al., 2005; Akinmoladun et al., 2007; Aprioku and Obianime, 2008), antidiarrhoeal (Orafidiya et al., 2000; Adebolu and Salau, 2005), antihelmintic (Fakae et al., 2000; Pessoa et al., 2002). Its content of thymol makes it a more

---

*Corresponding author. E-mail: chairmo2010@rocketmail.com.

acceptable antiseptic agent (Agnaniet et al., 2005) particularly by the local folks.

Blood is a specialized connective tissue that contains cells suspended in a fluid extracellular matrix called plasma, making it the main target for any detrimental effect(s) of *O. gratissimum* extract. The RBC plays a primary role of transporting substances including nutrients, respiratory gases and other waste materials throughout the body. The WBC defends the body against pathogens and other foreign bodies. The platelets play the role of preventing blood loss. Therefore, severe alteration in the concentration of any of these haemopoietic components may be detrimental.

Day after day, the consumption of *O. gratissimum* increases globally in view of its many uses. However, the usage of any plant, either as food or medication should be weighed against its possible detrimental effect(s) on the physiochemical integrity of the body. The fact that traditional medicine practitioners prescribe and administer decoctions of the leaves to clients regardless of the possible adverse effect of administration of unverified dosages on the body system and also considering the strategic place of the haemopoietic system in animal physiology. This study was conducted to investigate the possible effects of *O. grastissimum* on some of the haematological parameters as well as its safety for usage as food or medication by man.

MATERIALS AND METHODS

Collection and identification of plant materials

Fresh leaves of *O. gratissimum* were collected from Ukanafun Local Government Area in Akwa Ibom State, Nigeria. The leaves were later identified and authenticated by a senior herbarium officer in the Department of Pharmacognosy, Faculty of Pharmacy, University of Uyo, Uyo.

Preparation of extract

Fresh leaves of *O. gratissimum* were air-dried and ground into coarse powder. 450 g of the powdered leaves was macerated in 2000 ml of distilled water. The mixture was allowed for 24 h before filtration through a fine sieve. The filtrate was evaporated to dryness at 45°C using rotations evaporator. The extract was kept in a refrigerator at -4°C until use.

Acute toxicity study

The median lethal dose (LD_{50}) of the plant extract was determined by method of Lorke (1983). The rats were divided into 4 groups and treated with dosage of 250, 500, 750, 1000 mg/kg body weight in the 1st phase and dosage of 1500, 1750, 2000 and 2250 mg/kg body weight intraperitoneally in the 2nd phase. The median lethal dose (LD_{50}) was calculated using the second phase, according to the formula:

Lethal dose (LD_{50}); $LD_{50} = \sqrt{(D_0 \times D_{100})}$

Where D_0 = Dosage of 0% mortality. D_{100} = Dosage of 100% mortality.

Experimental design and treatment of animals

The rats were weighed and kept in wooden cages of 50 to 80 cm dimension. They acclimatized for 1 week before the commencement of the experiment. The animals were fed with rat chow and allowed free access to water. Both sexes of rats were randomly assigned into 4 groups with group 1 as the control while group 2 to 4 were administered with the extract dosage of 212, 424, 636 mg/kg, respectively. The experimental procedures involving the animals and their care were conducted in conformity with the approved guidelines by the local Research and ethical Committee.

Sample collection

After daily extract administration for 21 days, the rats were suffocated in chloroform on the 22nd day. Incisions were made into the ribs with a sterile pair of scissors to expose the heart. A sterile syringe with needle was used for collection of blood directly from the heart of each of the rats, by cardiac puncture. The blood sample was transferred into properly-labeled sample bottles with anticoagulant.

Blood analysis

The blood samples were analyzed using an automated machine. Each of the samples collected were ran sequentially using the standard procedure for blood analysis. KX-21 haematological analyzer made by Symex Kobe Japan was used. Data obtained were analyzed using ANOVA and student t-test. Level of significance was pre-determined as $P < 0.05$.

RESULTS AND DISCUSSION

Acute Toxicity Study

While animals that received very high dosage died, other signs of toxicity were noticed 2 to 4 h after extract administration. There was decrease locomotion, wrighting, constipation and decreased in sensitivity to touch. After 15 h of extract administration, the median lethal dose was calculated to be 2121.3 mmg/kg body weight.

Analysis of blood parameters

The results of the different dosages of aqueous extract of *O. gratissimum* on the WBC, platelets, lymphocyte and neutrophils are shown in Table 1. The extract-treated groups showed significant ($p < 0.001$ and $p < 0.01$) increases for test groups II and III, respectively. The results also showed significant decrease in total platelet count ($P < 0.001$) for groups II and IV, $p < 0.05$ for group III when compared with the control group as shown in Table 1. Also, in the experimental groups, the mean value for lymphocytes decreased significantly ($P < 0.001$) when compared with animals in the control group. The animals

Table 1. Effect of aqueous leaf extract of *Ocimum gratissimum* on white blood cells, platelets, lymphocytes and neutrophils.

Group/treatment	WBC (x10³/µL)	Platelets (x10³/µL)	Lymphocytes (%)	Neutrophils (%)
I (Control)	16.90 ± 0.24	945.20 ± 1.96	81.12 ± 0.20	18.68 ± 0.22
II (212 mg/Kg)	19.34 ± 0.18***	744.20 ± 1.00***	77.04 ± 0.56***	22.96 ± 0.28***
III (424 mg/Kg)	18.64 ± 0.69**	963.40 ± 1.93*	69.66 ± 0.28***	20.34 ± 0.22*
IV (636 mg/Kg)	16.90 ± 0.24	783.40 ± 1.03***	71.58 ± 0.49***	28.42 ± 0.25***

*** = Significantly different from group I (Control) at $P < 0.001$. ** = Significantly different from group I (Control) at $P < 0.01$. * = Significantly different from group I (Control) at $P < 0.05$.

Table 2. Effect of aqueous leaf extract of *Ocimum gratissimum* on RBC, PCV, Hb and haematological indices.

Group/Treatment	RBC (x10³/µL)	PCV (%)	Hb (g/dL)	MCV (fL)	MCH (pg)	MCHC (g/dL)	PCV (%)	Hb (g/dL)
I (Control)	7.21 ± 0.28	44.16 ± 1.17	12.50 ± 0.36	61.18 ± 0.96	17.40 ± 0.41	28.02 ± 0.35	44.16 ± 1.17	12.50 ± 0.36
II (212 mg/Kg)	7.11 ± 0.22	46.68 ± 0.92	12.40 ± 0.32	61.84 ± 0.62	17.44 ± 0.15	28.06 ± 0.43	46.68 ± 0.92	12.40 ± 0.32
III (424 mg/Kg)	7.22 ± 0.18	45.44 ± 0.92	12.52 ± 0.35	61.74 ± 0.70	17.38 ± 0.39	27.48 ± 0.30	45.44 ± 0.92	12.52 ± 0.35
IV (636 mg/Kg)	7.55 ± 0.12	47.16 ± 1.61	13.06 ± 0.25	62.06 ± 1.28	17.06 ± 0.22	28.36 ± 0.34	47.16 ± 1.61	13.06 ± 0.25

All values are the mean ± SEM.

in the extract-treated groups showed a significant ($p<0.001$) increase in the mean value of neutrophils of groups II and IV when compared to the control group but was significantly ($p<0.05$) higher in group III.

The administration of *O. gratissimum* to rats did not show any significant change in RBC, PCV, Hb, MCV, MCH and MCHC, as shown in Table 2.

Administration of *O. gratissimum* (OG) to rats did not have any significant effect on RBC, PCV, Hb, MCV, MCH and MCHC as reported in this study. However, the WBC counts showed significant higher values in the tested groups II and III when compared with the control group; these data are in agreement with the data reported by Ephraim et al. (2000). These increases might be attributed to the homeostatic response by the endogenous defense system to the adverse effects of *O. gratissimum*.

The significant decrease in platelets count recorded in this study may be due to the toxic effect of the chemical properties of *O. gratissimum*. *O. gratissimum* is reported to reduce platelets count most probably because of its phytochemical saponins and cardiac glycosides content (Tohti et al., 2006).

The lymphocyte level at the end of the experiment was 81.12% on the average for the control group and 77.04, 69.66 and 71.58% on the average for the test groups. These significant reductions in the lymphocyte level of the tested groups as compared to the control group are in agreement with the findings of Ephraim et al. (2000) and Jimoh et al. (2008).

Contrary to the reports of Ephraim et al. (2000) and Jimoh et al. (2008), the neutrophil level significantly increased in the test groups when compared to the control group. On the basis of the present study, the increase neutrophil level re-emphasize its anti-bacterial and anti-fungal properties and justifies the use of the plant by traditional medicine practitioners; considering that neutrophils constitute the first line of defense. However, it is feared that prolonged consumption of *O. gratissimum* could result in thrombocytopenia, leading to bleeding disorders, considering its ability to decrease platelet aggregation and platelet count. Meanwhile, increased neutrophils could offer some physiological advantage in neutrophil-mediated defense mechanisms. Further studies may be necessary to give more information about the effects on lymphocytes and neutrophils.

Obaji et al. (2009) had reported generalized anti-haematinic effects of *O. gratissimum* and cautioned on the adverse consequences of its prolonged usage. It is possible that *O. gratissimum* could induce haemolysis and can even suppress haemopoiesis because of its phytochemical constituent of saponin, but the detail specific mechanism of action is not clear. Nevertheless, it appears that the anti-haematinic effect of saponins to some extent could be cell linage selective. There is also a possibility that other constituent of *O. gratissimum* could stimulate the activity of some haemopoietic growth factors while inhibiting others. It is known that some growth factors caused maturation of a single linage progenitor cells e.g. erythropoietin (EP) for erythrocytes, thrombopoietin (TPO) for thrombocytes and interleukin 9 (IL-9) for lymphoid cells (Reddy, 2008). It is therefore possible that some of the active agent in *O. gratissimum* could promote the action of growth factors related to neutrophil production, while at the same time inhibiting growth factors associated with platelets and lymphocytes production.

Contrasting results reported on similar studies in various regions tends to support the notion that, there may be regional differences in the phytochemical constituent of *O. gratissimum*. Again, there may be need to consider the dosage and duration of administration in each study, as this is likely to contribute to the conflicting results obtained from the different experimentation.

Conclusion

It is evident that oral administration of aqueous leaf extract of OG reduces platelets and lymphocyte counts, but increases total WBC counts and neutrophil levels in wistar albino rats. If this result were to be extrapolated to human beings, then caution would be needed not to consume quantities of *O. gratissimum*. However, pharmacological studies are required to confirm the effect of *O. gratissimum* extract, without serve side effects.

REFERENCES

Adebolu TT, Salau AO (2005). Antimicrobial activity of leaf extracts of *Ocimum gratissimum* on selected diarrhea causing bacteria in South western Nigeria. Afr. J. Biotechnol., 4: 682-684.

Agnaniet H, Arguillet J, Bessieve JM, Menut C (2005). Aromatic plant of tropical central Africa. Part XL-VIL. Chemical and Biological investigation essential of oil *ocimum* species from Gabon. J. Ess. Oil. Res., Abstract.

Akinmoladun AC, Ibukun EO, Afor E, Obuotor EM, Farombi EO (2007). Phytochemical constituent and antioxidant activity of extract from the leaves of *Ocimum gratissimum*. Scientific Res. Essay, 2: 163-166.

Aprioku JS, Obianime AW (2008). Antioxidant activity of the aqueous crude extract of *Ocimum gratissimum* Linn. Leaf on basal and cadmium-induced serum levels of phosphatases in male guinea-pigs. JASEM, 12: 33-39.

Fakae BB, Campbell AM, Barrett J, Scott IM, Teesdale-Spittle PH, Liebau E, Brophy PM (2000). Inhibition of glutathione-S-trasferases (GSTs) from parasitic nematodes by extracts from traditional Nigerian medicinal plants. Phytother. Res., 1148: 630-634.

Ephraim KD, Salami HA, Osewa TS (2000). The Effect of Aqueous Leaf Extract of *Ocimum Gratissimum* on Haematological and Biochemical Parameters in Rabits. Afr. J. Biomed. Res., 3: 175-199.

Jimoh OR, Olaore J, Olayaki L, Olawepo A, Bilimaminu S (2008). Effect of aqueous Extract of *Ocimum gratissimum* on Haematological Parameters of Wistar rats. Biokemistri, 20(1): 33-37.

Lorke D (1983). A New Approach to practical Acute Toxicity Test. Arch. Toxicol., pp. 275-287.

Obaji NN, Egwurugwu BI, Uche A, Nwafor CS, Ufearo RC, Uchefuna DC, Nwaorah OM, Adienbo OM, Olorunfemi OJ (2009). Effects of *Ocimum gratissimum* on the Hematological Parameters of Albino Rats. Int. J. Trop. Agric. Food Sys., 3(4): 283-286.

Oboh G (2008). Antioxidative potentials of *Ocimum gratissum* and ocimum canum leaf polyphenols and protective Effects on some pro-oxidants. Induced Lipid peroxidation in Rat Brain: An *in vitro* study. Am. J. Food Technol., 3(5): 325-334.

Oboh F, Masodje H, Enabulele S (2009). Nutritional and antimicrobial properties *of Ocimum gratissimum* Leaves. J. Biol. Sci., 9(4): 377-380.

Odukoya OA, Ilori OO, Sofidiya MO, Aniunoh OA, Lawal BM, Tade IO (2005). Antioxidant activity of Nigerian dietary spices. Elect. J. Environ. Agric. Food Chem., 4: 1086-1093.

Orafidiya OO, Elujoba AA, Iwalewa FO, Okeke IN (2000). Evaluation of antidiarrheal properties of *Ocimum gratissimum* volatile oil and its activity against enteroaggregative *Escherichia coli*. Pharm. Pharmacol. Lett., 10: 9-12.

Pande M, Pathank A (2009). Effect of Ethanolic Extract of *Ocimum Gratissimum* on sexual Behavious in Male Mice. Int. J. Pharm. Tech. Res., 1(3): 468-473.

Pessoa LM, Morais SM, Bevilaqua CML, Luciano JHS (2002). Antihelmintic activity of essential oil of *Ocimum gratissimum* Linn. and eugenol against. *Haemonchus contortus*. Vet. Parasitol., 9: 59-63.

Reddy LP (2008). Haemopoiesis-Erythropoiesis-Haemoglobin-Anaemia in Fundamentals of Medical Physiology, (4th Ed). Hyderabad, India: Para Medical Publisher, pp. 207-210.

Sainsbury M, Sofowora EA (1971). Essential Oils from the leaves and inflorescence of *Ocimum gratissimum*. Phytochemistry, 10: 3309-3310.

Tohti I, Tursun M, Umar A, Imin H, Moore N (2006). Aqueous extracts of *Ocimum basilicum* L. (Sweet basil) decrease platelet aggregation induced by ADP and thrombin *in vito* and rats arterio-venous shunt thrombosis *in vivo*. Thromb. Res., 118(6): 733-739.

10

Asymmetrical dimethylarginine (ADMA) and nitric oxide as potential cardiovascular risk factors in type 2 diabetes mellitus

Mohamed H. Mahfouz[1]*, Ibrahim A. Emara[1], Mohamed S. Shouman[2] and Magda K. Ezz [3]

[1] Biochemistry and [2] Internal Medicine Departments, National Institute of Diabetes and Endocrinology (NIDE), Cairo, Egypt.
[3] Biochemistry Department, Faculty of Science. Ain Shams University.

Hyperglycemia affects biochemical parameters and influences the progression of coronary heart disease and mortality rates in diabetic patients. L-arginine is the substrate used by NO synthase to produce the vasodilator NO. However, in patients with type 2 diabetes mellitus (T2DM), there is an increase in serum levels of methylated L-arginines, such as ADMA, which is a recently identified potent cardiovascular risk factor. The aim of this study was designed to determine both risk factors (ADMA and NO) in type 2 diabetic patients with and without cardiovascular disease and to evaluate whether there is an association between ADMA and glycosylated hemoglobin (HbA1c) on the one hand and nitric oxide on the other hand. The study included 3 groups of subjects; Group I (Control group); comprising 20 healthy subjects; the mean age 48 ± 1.6 years; Group II: 20 diabetic patients without cardiovascular complications; the mean age 51.0 ± 1.96 years and Group III; 20 diabetic patients with evidence of cardiovascular complications; the mean age 54.0 ± 2.1 years. Fasting and postprandial serum glucose, HbA1c, lipid profile (total cholesterol, triacylglycerol, HDL-c and LDL-c), ADMA and serum NO metabolite level, were determined. Serum glucose (fasting and postprandial), HbA1c and ADMA levels showed significant increase in diabetic patients type 2 with and without cardiovascular complications compared to healthy normal control. Total cholesterol, triacylglycerol and LDL-c manifested significant elevations, while HDL-c level showed insignificant change in both groups in compared to non diabetic healthy subjects. Serum NO metabolite level was significantly reduced in the both diabetic patient groups compared with controls. No correlation between ADMA level and studied parameters in diabetic patients without evidence of cardiovascular complications, whereas in cardiovascular complications group, the ADMA level was positively correlated with both postprandial serum glucose and HbA1c, but there was a negative correlation between ADMA levels and NO. Also, NO was negatively correlated with postprandial serum glucose and HbA1c. In conclusion, ADMA and NO may serve as predictors for future cardiovascular events in type 2 diabetic patients. So, early diagnosis and good glycemic control are more effective in reducing the cardiovascular complications.

Key words: Type 2 diabetes, cardiovascular disease, ADMA, nitric oxide.

INTRODUCTION

Type 2 diabetes mellitus (T2DM) is a progressive and complex metabolic disorder characterized by chronic hyperglycemia resulting from impaired insulin secretion and/or insulin action, the lack of effective insulin leads to disturbances in carbohydrate, lipid and protein metabolism. It is a proinflammatory, hypercoagulable state that predisposes patients to develop cardiovascular disease. It is also associated with risk factors for atherosclerosis, including dyslipidemia, hypertension, inflammation and altered hemostasis, (Granberry and Fonseca, 2005). Patients with T2DM tend to have a characteristic dyslipidemia (increased concentrations of

---

*Corresponding author. E-mail: mhesham5@yahoo.com.

LDL-c and decreased concentrations of HDL-c), likely responsible for their being 2 to 4 times more inclined to develop cardiovascular disease than those without T2DM (Haffner et al., 1998). In fact, patients with T2DM are twice as likely as those without T2DM to have elevated triacylglycerol levels and decreased HDL-c concentrations (Garg and Grundy, 1990).

Cardiovascular disease remains the most important cause of morbidity and mortality in diabetes mellitus (Haffner et al., 1998), accounting for approximately 65% of total mortality in patients with type 2 diabetes mellitus (Stamler et al., 1993), in whom hyperglycaemia is one of the main metabolic abnormalities (Yasuda et al., 2006). Aggressive risk factor management is an important mean for reducing the cardiovascular morbidity in this patient group, which implies accurate risk stratification. Poorly controlled blood pressure and hyperglycemia seem to be significantly involved in the development process of cardiovascular disease in patients with type 2 diabetes (Selvin et al., 2004).

Beside traditional risk markers, the nitric oxide (NO), which is synthesized by NO synthases, is an important antiatherogenic molecule (Moncada, 1999). As initially described by Vallance et al. (1992) an endogenous inhibitor of NO synthase, asymmetric dimethylarginine (ADMA), occurs in significant amounts in peripheral blood. Since this discovery, numerous clinical studies have been performed that found elevated circulating ADMA concentrations in humans suffering from diseases associated with increased cardiovascular risk. Different prospective cohort studies in selected patient groups suggested ADMA as an independent predictor of cardio-vascular morbidity (Mittermayer et al., 2006). In a cross-sectional study, ADMA was associated with macrova-scular disease in patients with type 2 diabetes (Krzyzanowska et al., 2006). However, the predictive role of ADMA for the occurrence of cardiovascular events in type 2 diabetes has not been examined prospectively until now.

The aim of this study was designed to determine both risk factors (ADMA and NO) in type 2 diabetic patients with and without cardiovascular disease and to evaluate whether there is an association between ADMA and glycosylated hemoglobin (HbA1c) on the one hand and nitric oxide on the other hand.

SUBJECTS AND METHODS

Subjects

The study was performed on forty outpatients with type 2 diabetes and 20 healthy subjects. All patients were selected from outpatients Clinic of National Institute of Diabetes and Endocrinology (NIDE), Cairo, Egypt. Type 2 DM was diagnosed according to the Report of the Expert Committee on the Diagnosis and Classification of Diabetes Mellitus (2006). Cardiovascular disease was defined as a positive medical history for myocardial

infarction, angina, coronary artery bypass graft and stroke, and ischemic changes in electrocardiogram. Type 2 diabetic patients were taking the same oral therapy. Patients with any history of smoking and alcohol habits, respiratory disorder and showed any clinical or laboratory signs of liver disease, or thyroid function impairments, renal dysfunction, chronic inflammatory and clinically significant infectious diseases were excluded. The healthy control subjects, which matched with age and sex as patients with type 2 diabetes, had no recognizable diseases and clinically free from any abnormality. They were not receiving any medications and represented the control group. Demographic data was recorded for each subject using self-made questionnaire. Approval had been taken from the research ethics committee of General Organization of Teaching Hospitals and Institutes. An informed consent was obtained from all patients and healthy subjects that described the aim of the study and the procedures that would be required from them. The study included 3 groups of subjects, Group I: control group (n = 20), the mean age 48 ± 1.6 years; Group II: type 2 diabetic patients without cardiovascular complications (n = 20), the mean age 51.0 ± 1.96 years, the mean duration of the disease was 5.1 ± 0.37 years and Group III: type 2 diabetic patients with evidence of cardiovascular complications (n = 20), the mean age 54.0 ± 2.1 years, the mean duration of the disease was 8.1 ± 0.75 years.

Methods

Blood samples were collected into vacutainer tube without additive after 12 h overnight fasting from healthy subjects and diabetic patients, this was followed by the ingestion of meals and a further blood samples were collected 2 h later (postprandial). Blood was then centrifuged at 3000 rpm for 10 minutes at 4°C. Serum was rapidly separated; subdivided into aliquots and were stored at -80°C until the measurements of lipid profile, ADMA and nitrite /nitrate metabolites (NO). Another part of collected blood was taken on EDTA for determination of HbA1c level. Hemolysed samples were excluded. Serum glucose concentration (fasting and postprandial) was assayed at once by glucose oxidase method according to Barham and Trinder (1972). Serum total cholesterol was determined by the enzymatic method (Allain et al., 1974). Triacyl-glycerol was assayed by peroxidase-coupled method (McGowan et al., 1983). HDL-c was measured by enzymatic method after precipitation of other lipoproteins with $MgCl_2$ and dextran sulphate (Finley et al., 1978), LDL-c was calculated according to Friedewald et al. (1972) and HbA1c level was done according to the method of Grey et al. (1996) using an immunoturbidimetric assay on Dimension RxL Max (Dade Behring). Serum ADMA concentration was measured using commercially available enzyme-linked immunosorbent assay (ELISA) kits based on the method of Schulze et al. (2004). Nitric oxide decomposes rapidly in aerated solutions to form stable nitrate/nitrite products, so NO production was evaluated by measuring the serum concentration of nitrate/nitrite products via a Griess reaction in accordance with Moshage et al. (1995).

Statistical analysis

Data was expressed as the mean ± S.D. Statistical analysis was performed with Statistical Package for the Social Science for Windows (SPSS, version 10.0, 1999, Chicago, IL, USA). Differences between groups were analyzed by one-way analysis of variance (ANOVA). Post-hoc testing was performed by the Bonferroni test to compare the difference among the groups studied as reported by Altman (1991). Pearson's correlation analysis was performed to determine the relationships between

Table 1. Demographic and biochemical characteristics of control and diabetic patients (Mean ± SD).

Parameters	Group I	Group II	Group III
Age (Years)	48.0 ± 1.6	51.0 ± 1.96	54.0 ± 2.1
Gender (F/M)	12/8	14/6	11/9
Duration of diabetes (Years)	-------	5.1 ± 0.37	8.1 ± 0.75 [c]
BMI (Kg/m^2)	26.19 ± 0.3	25.87 ± 0.37	25.14 ± 0.44
SBP (mmHg)	116.6 ± 9.7	114.5 ± 11.3	124.4 ± 4.1
DBP (mmHg)	69.1 ± 4.8	74.0 ± 6.7	74.2 ± 7.4
Fasting serum glucose (mg/dl)	80.4 ± 9.1	191.9 ± 32.0 [a]	219.4 ± 45.5 [a,c]
Postprandial serum glucose (mg/dl)	109 ± 9.9	269.9 ± 41.6 [a]	350.7 ± 66.9 [a,d]
HbA$_{1c}$ (%)	5.9 ± 0.47	9.17 ± 1.07 [a]	11.37 ± 0.94 [a,d]

SBP, Systolic blood pressure; DBP, diastolic blood pressure; BMI, body mass index
[a]$p < 0.0001$ Vs. Group I, [c]$p < 0.029$ Vs. group II, [d]$p < 0.0001$ Vs. group II.

Table 2. Serum lipid profile in control and diabetic patients (Mean ± SD).

Parameters	Group I	Group II	Group III
Cholesterol (mg/dl)	177.3 ± 40.0	202.7 ± 30.3 [a]	279.4 ± 25.8 [b, c]
Triacylglycerol (mg/dl)	121.1 ± 29.6	209.9 ± 39.1 [e]	283.9 ± 87.9 [b, d]
HDL-c (mg/dl)	44.7± 16.6	40.2 ± 12.16	42.6 ± 11.6
LDL-c (mg/dl)	113.0 ± 31.2	134.1 ± 24.14 [a]	190.0 ± 24.5 [b, c]

[a]$p < 0.05$ group 1, [b]$p < 0.0001$ Vs. group I, [c]$p < 0.0001$ Vs. group II, [d]$p < 0.001$ Vs. group II,
[e]$p < 0.0001$ Vs. group I.
Group I: control group, Group II: type 2 diabetic patients without cardiovascular complications.
Group III: type 2 diabetic patients with evidence of cardiovascular complications.

ADMA, NO and other cardiovascular risk variables. P-value < 0.05 was accepted to indicate statistical significance.

RESULTS

Demographic and some biochemical characteristics of the patients and controls are presented in Table 1. No differences were found between diabetic groups and controls with respect to age, gender and BMI (P > 0.05). In the diabetic group, the mean duration of diabetes with and without vascular complications was 5.1 ± 0.37 and 8.1 ± 0.75 respectively. Serum glucose (fasting and postprandial) and glycosylated Hemoglobin (HbA1c) showed significant increase in diabetic patients type 2 with and without cardiovascular complications compared to healthy normal control. Vascular complications in type 2 diabetic patients produced pronounced increase in serum fasting glucose, postprandial serum glucose and HbA1c concentrations when compared to diabetic patients without vascular complications (p < 0.029, p < 0.0001, p < 0.0001 respectively).

Table 2 demonstrates the changes of lipid profile of diabetic patients type 2. Total cholesterol (TC), triacylglycerol (TG) and LDL-c manifested significant elevations (p < 0.05, p < 0.0001, p < 0.05 respectively) without cardiovascular complications and with cardiovascular complications (p < 0.0001 for each) when compared to

normal control subjects, while HDL-c level showed insignificant change in both groups in compared to non diabetic healthy subjects. TC, TG and LDL-c represented pronounced increases in diabetic patients with vascular complications (p < 0.0001, p < 0.001, p < 0.0001 respectively) compared to diabetic patients without vascular complications.

Table 3 illustrates the effects of diabetes mellitus type 2 on asymmetrical dimethylarginine (ADMA) and nitric oxide (NO). The level of ADMA represented significant elevation (P < 0.0001 for each) in both groups in comparison to normal subjects but the vascular complications in type 2 diabetic patients produced pronounced increase (P < 0.0001 for each) when compared to diabetic patients without vascular complications.

Serum NO metabolite level (nitrate/nitrite) was significantly reduced (P < 0.0001) in the both diabetic patient groups compared with controls. However the diabetic patients with vascular complications showed pronounced increase of NO (P < 0.0001) when compared to diabetic patients without vascular complications (Table 3).

Pearson's correlation analyses revealed that, no correlation was observed between ADMA level and studied parameters in diabetic patients without evidence of cardiovascular complications, whereas in cardiovascular complications group, the ADMA level was positively correlated with both postprandial serum glucose and HbA1c

Table 3. Serum levels of asymmetrical dimethylarginine (ADMA) and nitric oxide (NO) in control and diabetic patients (Mean ± SD).

Parameters	Group I	Group II	Group III
ADMA (µmol/L)	0.41± 0.092	1.31 ± 0.45[a]	2.34 ± 0.59 [a, b]
NO (µmol/L)	274.4 ± 34.3	222.2 ± 8.8 [a]	125.9 ± 48.3 [a, b]

[a]$P < 0.0001$ Vs. group I, [b]$P < 0.0001$ Vs. group II.
Group I: control group, Group II: type 2 diabetic patients without cardiovascular complications.
Group III: type 2 diabetic patients with evidence of cardiovascular complications.

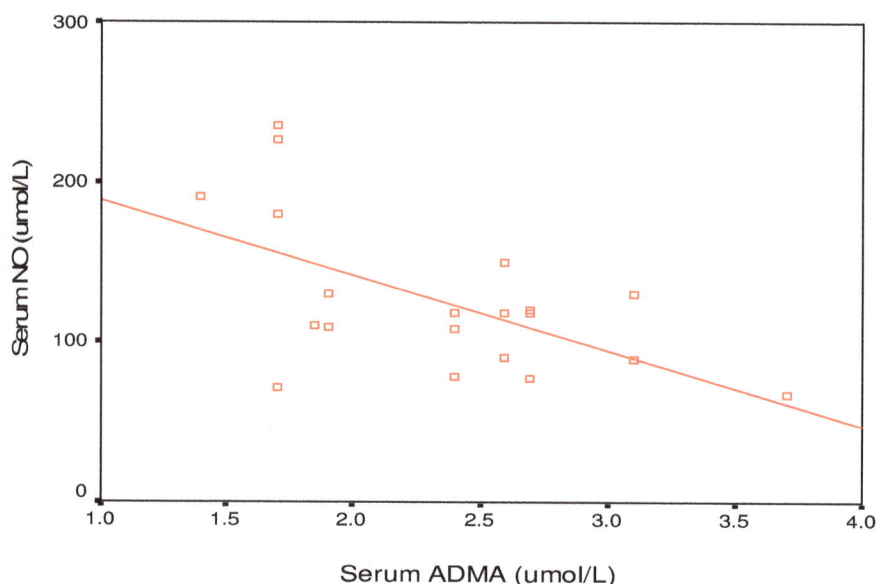

Figure 1. Correlation between serum ADMA and NO in type 2 diabetic patients with evidence of cardiovascular complications ($r = - 0.580$, $p = 0.007$).

($r = 0.593$, $p = 0.006$; $r = 0.496$, $p = 0.026$, respectively) as represented in Figures 2 and 3. On the other hand, there was a negative correlation between ADMA levels and NO ($r = - 0.580$, $p = 0.007$) (Figure 1). Also NO was negatively correlated with postprandial serum glucose and HbA1c ($r = - 0.497$, $p = 0.026$; $r = - 0.460$, $p = 0.041$, respectively) (Figures 4 and 5).

DISCUSSION

Cardiovascular disease is the major cause of morbidity and mortality in patients with type 2 diabetes mellitus T2DM (Haffner et al., 1998), in whom hyperglycaemia is one of the main metabolic abnormalities (Yasuda et al., 2006). Blood glucose control occupies the centre stage in T2DM management (Grundy et al., 1999). Patients with diabetes can have many complications including elevated levels of VLDL-c, LDL-c triacylglycerol and low levels of HDL-c (Haffner, 1998). These patients have a preponderance of abnormalities in the composition of smaller, denser particles, which increase atherogenicity even if the absolute concentration of LDL-c is not significantly increased (Haffner, 1998). In our study, Table 2 elucidates that total cholesterol, triacylglycerol and LDL-c levels showed pronounced increases in diabetic patients with evidence of cardiovascular complications compared with control group, while no change in HDL-c level in diabetic groups compared with control group. These findings are in agreement with the previous studies which suggest that lipoprotein abnormalities are higher in diabetics than in non-diabetics (Idogun et al., 2007; Albrki et al., 2007).

Our finding that increased ADMA and decreased NO in type 2 diabetic subjects is significant because endothelial dysfunction associated with increased ADMA concentrations seems to begin before the detectable vascular damage in type 2 diabetic patients. Additionally, type 2 diabetic patients had increased levels of ADMA leading to endothelial damage, even if hypertension or hyperlipidemia does not exist. Measurement of ADMA and NO as markers of endothelial dysfunction may

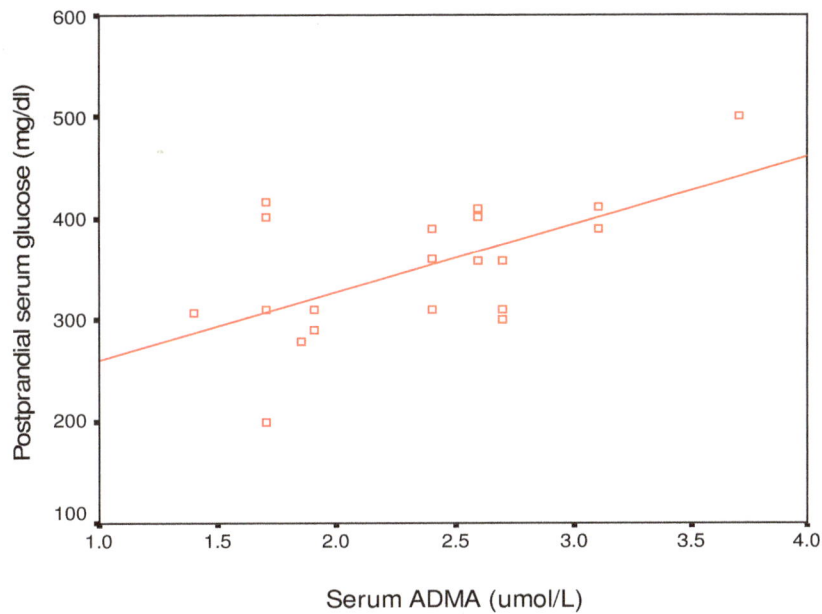

Figure 2. Correlation between serum ADMA and postprandial serum glucose in type 2 diabetic patients with evidence of cardiovascular complications (r = 0.593, p = 0.006).

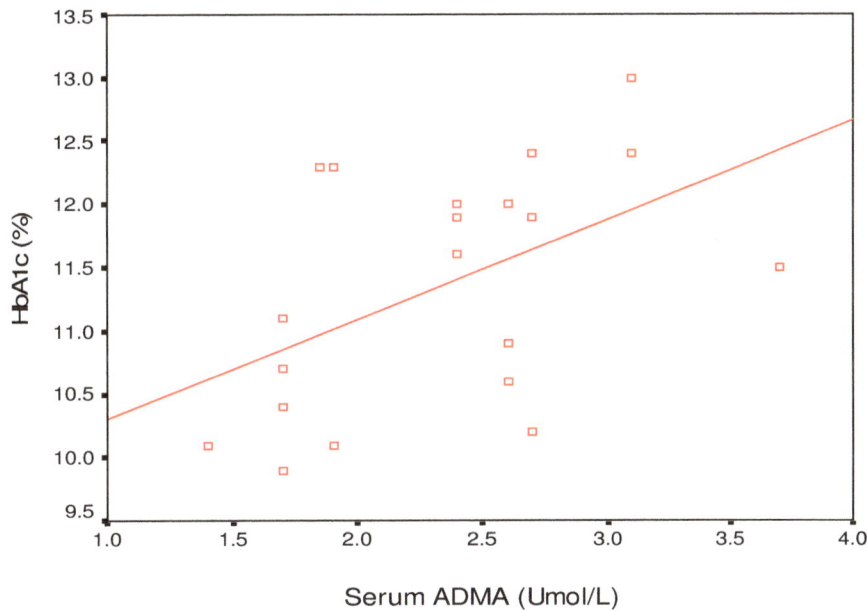

Figure 3. Correlation between serum ADMA and HbA1c (%) in type 2 diabetic patients with evidence of cardiovascular complications (r = 0.496, p = 0.026).

provide an opportunity for the prevention of irreversible endothelial damage in these patients (Altinova et al., 2007).

There are some explanations about the interaction between hyperglycemia and the L-arginine-NO system.

Hyperglycemia-induced activation of protein kinase C, increased superoxide anion production from glucose autoxidation and accumulation of advanced glycation end product due to nonenzymatic cross-linking of proteins via oxidative stress can reduce the bioavailability of

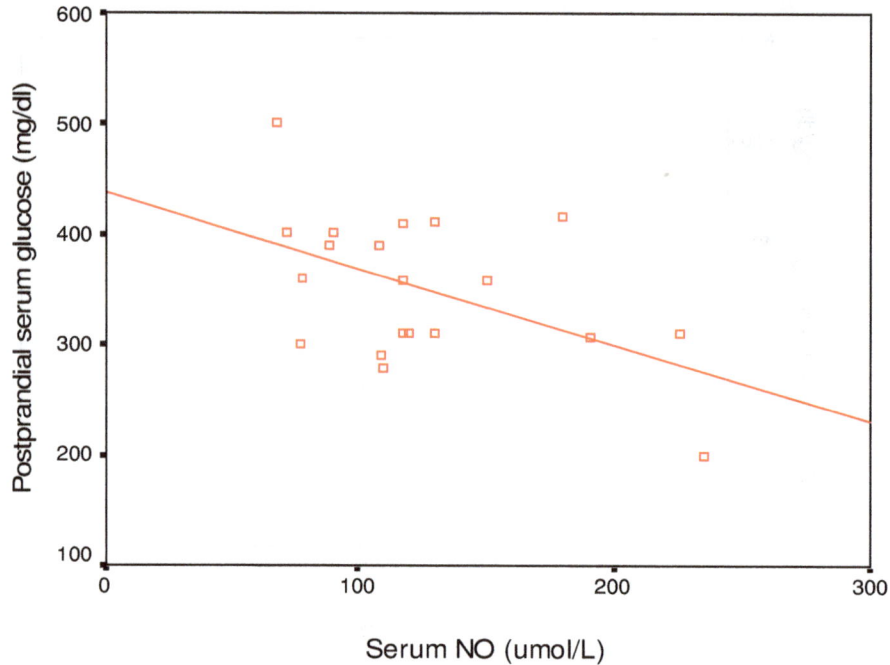

Figure 4. Correlation between serum NO and postprandial serum glucose in type 2 diabetic patients with evidence of cardiovascular complications ($r = -0.497$, $p = 0.026$).

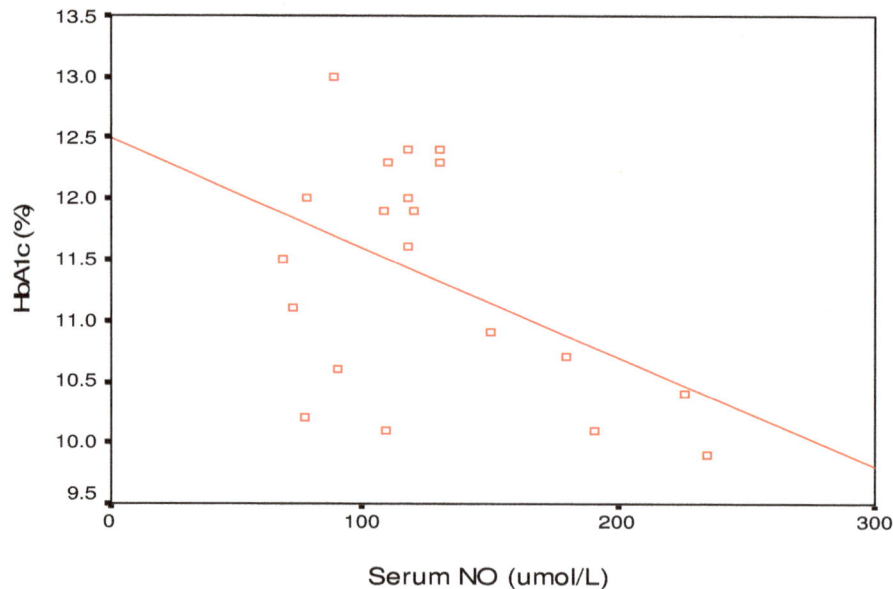

Figure 5. Correlation between serum NO and HbA1c (%) in type 2 diabetic patients with evidence of cardiovascular complications ($r = -0.460$, $p = 0.041$).

NO and activation of the polyol pathway, which increases the use of nicotinamide adenine dinucleotide phosphate can reduce the biosynthesis of NO (Chan and Chan, 2002). However, the exact mechanism of how hyperglycemia influences circulating ADMA concentrations in T2DM is not fully known. One possible mecha- nism has been suggested in an animal study that hyperglycemia-induced oxidative stress increases ADMA by impairing the dimethylarginine dimethylaminohydro- lase (DDAH), which is involved in the metabolic degra- dation of ADMA (Lin et al., 2002). Furthermore, Sorrenti et al. (2006) reported that exposure to high glucose in endothelial cells increases oxidative stress, reduces DDAH-2 and leads to a NOS imbalance. Although renal clearance is the first mechanism for the elimination of ADMA (Zoccali et al., 2001), enzymatic degradation of ADMA by DDAH has recently gained substantial importance. DDAH degrades ADMA to dimethylamine and L-citrulline and DDAH activity is found in almost all tissues, especially in kidney and liver (Kimoto et al., 1995). One of the allelic isoforms of this enzyme, DDAH-2, is mainly present in vascular tissues that coexpress endothelial NOS (Leiper, et al., 1999). Another mechanism for the increase in ADMA concentrations in hyperglycemic media may be associated with the enzyme arginine methyltransferase, which synthesizes ADMA, because hyperglycemia-induced oxidative stress up-regulates the expression of arginine methyltransferases (Maas, 2005).

One of the most critical vasoactive mediators synthesized by the vascular endothelium is nitric oxide (NO), previously known as endothelium-derived relaxing factor (Palmer et al., 1987). Nitric oxide is synthesized from L-arginine by NO synthase. Endothelium-derived NO is a powerful endogenous vasodilator and also has important roles in the maintenance of vascular homeostasis. For instance, NO inhibits platelet aggregation, leucocyte migration and cellular adhesion to the endothelium and attenuates vascular smooth muscle cell proliferation and migration. In addition, it inhibits activation and expression of adhesion molecules and the production of superoxide anions (Vallance and Chan, 2001). In type 1, type 2 diabetes and insulin resistance syndrome, there is evidence that the release and/or bioavailability of NO are diminished (Chan et al., 2000). Despite the heterogeneous nature of these conditions, they all share the same feature of increased plasma glucose concentrations which could affect the L-arginine: NO pathway. Nitric oxide bioavailability can be reduced due to increased oxidative stress which can result from increased superoxide anions production from glucose autoxidation. Hyperglycaemia-induced activation of protein kinase C (PKC) followed by that of phospholipase A2, results in increased production of arachidonic acid metabolites which also have potent oxidizing effects. In contrast, reduced NO synthesis can result from activation of the polyol pathway which increases the utilization of nicotinamide adenine dinucleotide phosphate (NADPH), an important cofactor in the biosynthesis of NO. Furthermore, accumulation of advanced glycation end product AGE due to non-enzymatic crosslinking of proteins, could quench NO, further reducing its bioavailability. In addition to these mechanisms, the endogenous NO synthase inhibitor, asymmetric dimethylarginine (ADMA), has emerged as a key factor in determining NO biosynthesis (Chan et al., 2000).

ADMA is derived from the catabolism of proteins containing methylated arginine residues and is released as the proteins are hydrolysed. These proteins are predominately found in the nucleus and are involved in RNA processing and transcriptional control (Najbauer et al.,

1993). The synthesis of ADMA and NG-monomethyl-L-arginine (L-NMMA) requires the enzyme protein arginine methyltransferase type I (PRMT I) which methylates arginine residues (Ghosh et al., 1988). Protein arginine methyltransferase type II (PRMT II) forms symmetric dimethylarginine (SDMA). SDMA is a stereoisomer of ADMA and has no direct inhibitory effect on NO synthase. All three methylarginines (ADMA, SDMA and L-NMMA) enter endothelial cells through the cationic amino acid transporters known collectively as the y^+ transporter. The activity of this transporter was found to co-locate with caveolin-bound NO synthase which suggests that the y^+ transporter activity could be an important determinant of the local concentrations of methylarginines (McDonald et al., 1997). The three methylarginines compete with each other and with arginine for transport into the cell (Leiper and Vallance, 1999). Hence high concentrations of ADMA could potentially interfere with intracellular transport of L-arginine resulting in a decrease in NO synthesis. The transport system concentrates methylarginines within the endothelial cells such that intracellular concentrations are greater than circulating concentrations. In this regard, a defective y^+ transporter system could result in a higher concentration of circulating ADMA leading to decreased NO biosynthesis. Hence the y^+ transporter system could be a potential site of defect in disease states (Toutouzas et al., 2008).

We observed that our diabetic patients did not show a good glycemic control. Therefore, the present study showed a positive correlation between serum ADMA and HbA1c levels. It is likely to be a major effect of tight blood glucose control on increased serum ADMA concentrations in diabetic patients.

Our study showed no correlation between ADMA level and studied parameters in diabetic patients without evidence of cardiovascular complications, whereas in patients with cardiovascular complications, the ADMA level was positively correlated with both postprandial serum glucose and HbA1c, but there was a negative correlation between ADMA levels and NO. Also, NO was negatively correlated with postprandial serum glucose and HbA1c. This agrees with the previously reported data by Schernthaner and Krzyzanowska (2008) that indicate ADMA is directly related to the serum glucose level. Besides, Yasuda et al. (2006) reported that intensive control of serum glucose levels led to a decrease in ADMA level in hospitalized patients with type 2 DM. No correlation was observed between ADMA and lipid parameters including HDL-c and LDL-c in patients with type 2 diabetes. This is in accordance with previous researches (Krzyzanowska et al., 2006; Paiva et al., 2003).

In conclusion, ADMA and NO may serve as predictors for future cardiovascular events in type 2 diabetic patients.

So, early diagnosis and good glycemic control are more effective in reducing the cardiovascular complications. Further studies would be required to clearly establish the utility of decreasing ADMA levels or increasing NO in the treatment of type 2 diabetic patients.

ACKNOWLEDGEMENT

The authors would like to acknowledge the members of Internal Medicine Department in National Institute of Diabetes and Endocrinology (NIDE), Cairo, Egypt for their help in selection and clinical diagnosis of patients.

REFERENCES

Albrki WM, Elzouki ANY, EL-Mansoury ZM, Tashani OA (2007). Lipid profiles in Libian Type 2 Diabtics. J. Sci. Appls. 1(1): 18- 23.

Allain CC, Poon LS, Chan CS, Richmond W, Pu FC (1974). Enzymatic determination of total serum cholesterol. Clin. Chem. 20: 470- 475.

Altinova AE, Arslan M, Sepici-Dincel A, Akturk M, Altan N, Toruner FB (2007). Uncomplicated Type 1 Diabetes Is Associated with Increased Asymmetric Dimethylarginine Concentrations. J. Clin. Endocrinol. Metab. 92: 1881-1885.

Altman DG (1991). Practical Statistics for Med. Res. Chapman & Hall, London pp. 179- 223.

Barham D, Trinder P (1972). An important colour reagent for the determination of blood glucose by the oxidase system. Analyst. 97: 142 -145.

Chan NN, Chan JC (2002). Asymmetric dimethylarginine (ADMA): a potential link between endothelial dysfunction and cardiovascular diseases in insulin resistance syndrome? Diabetologia 45: 1609-1616.

Chan NN, Vallance P, Colhoun HM (2000). Nitric oxide and vascular responses in Type I diabetes. Diabetologia 43: 137-147.

Expert Committee on the Diagnosis and Classification of Diabetes Mellitus (2006). Report of the Expert Committee on the Diagnosis and Classification of Diabetes Mellitus. Diab. Care 29(1): 43-48.

Finley PR, Schifman RB, Williams RJ, Lichhti DA (1978). Cholesterol in high- density lipoprotein: use of Mg $^{2+}$/dextran sulphate in its enzymatic measurement. Clin. Chem. 24: 931- 933.

Friedewald WT, Levy RI, Fredrickson DS (1972). Estimation of the concentration of low-density lipoprotein cholesterol in plasma, without use of the preparative ultracentrifuge. Clin. Chem. 18: 499-502.

Garg A, Grundy SM (1990). Management of dyslipidemia in NIDDM. Diabetes Care. 13(2): 153-169.

Ghosh SK, Paik WK, Kim S (1988). Purification and molecular identification of two protein methylases I from calf brain: myelin basic protein- and histone-specific enzyme. J. Biol. Chem. 263: 19024-19033.

Granberry MC, Fonseca VA (2005). Cardiovascular risk factors associated with insulin resistance: effects of oral antidiabetic agents. Am. J. Cardiovasc. Drugs 5(3): 201-209.

Grey V, Perlas M, Aebi C (1996). Immunoturbidimetric method for determination hemoglobin A1c. Clin. Chem. 42(12): 2046-2047.

Grundy SM, Benjamin IJ, Burke GL, Chait A, Eckel RH, Howard BV, Mitch W, Smith SC Jr, Sowers JR (1999). Diabetes and cardiovascular disease: a statement for healthcare professionals from the American Heart Association. Circulation 100: 1134-1146.

Haffner SM (1998). Management of dyslipidemia in adults with diabetes. Diab. Care 21: 9(1): 1600- 1678.

Haffner SM, Lehto S, Ronnemaa T, Pyorala K, Laakso M (1998). Mortality from coronary heart disease in subjects with type 2 diabetes and in nondiabetic subjects with and without prior myocardial infarction. N. Engl. J. Med. 339:229-234.

Idogun ES, Unuigbe EI, Ogunro PS, Akinola OT, Famodu AA (2007). Assessment of serum lipids in Nigerians with type 2 diabetes

mellitus complications. Pak. J. Med. Sci. (Part 1). 23(5): 708-712.

Kimoto M, Whitley GS, Tsuji H, Ogawa T (1995). Detection of NG, NG-dimethylarginine dimethylaminohydrolase in human tissues using a monoclonal antibody. J. Biochem. 117: 237-238.

Krzyzanowska K, Mittermayer F, Krugluger W, Schnack C, Hofer M, Wolzt M, Schernthaner G (2006). Asymmetric dimethylarginine is associated with macrovascular disease and total homocysteine in patients with type 2 diabetes. Atherosclerosis. 189: 236-240.

Leiper JM, Santa Maria J, Chubb A, MacAllister RJ, Charles IG, Whitley GS, Vallance P (1999). Identification of two human dimethylarginine dimethylaminohydrolases with distinct tissue distributions and homology with microbial arginine deiminases. Biochem. J. 343: 209-214.

Leiper J, Vallance P (1999). Biological significance of endogenous methylarginines that inhibit nitric oxide synthases. Cardiovasc. Res. 43: 542-548.

Lin KY, Ito A, Asagami T, Tsao PS, Adimoolam S, Kimoto M, Tsuji H, Reaven GM, Cooke JP (2002). Impaired nitric oxide synthase pathway in diabetes mellitus: role of asymmetric dimethylarginine and dimethylarginine dimethylaminohydrolase. Circulation 106: 987-992.

Maas R (2005). Pharmacotherapies and their influence on asymmetric dimethylarginine (ADMA). Vasc. Med. 10(1): S49-S57.

McDonald KK, Zharikov S, Block ER, Kilberg MS (1997). A caveolar complex between the cationic amino acid transport I and endothelial nitric oxide synthase may explain the 'arginine paradox.' J. Biol. Chem. 272: 31213-31216.

McGowan MW, Artiss JD, Strandberrgh DR, Zak BA (1983). Peroxidase-coupled method for the colorimetric determination of serum triglycerides. Clin. Chem. 29: 538- 542.

Mittermayer F, Krzyzanowska K, Exner M, Mlekusch W, Amighi J, Sabeti S, Minar E, Muller M, Wolzt M, Schillinger M (2006). Asymmetric dimethylarginine predicts major adverse cardiovascular events in patients with advanced peripheral artery disease. Arterioscler. Thromb. Vasc. Biol. 26: 2536-2540.

Moncada S (1999). Nitric oxide: discovery and impact on clinical medicine. J. R. Soc. Med. 92: 164-169.

Moshage H, Kok B, Huzenenga R, Jansen P (1995). Nitrite and nitrate determination in plasma: A critical evaluation. Clin. Chem. 41: 892-896.

Najbauer J, Johnson BA, Young AL, Aswad DW (1993). Peptides with sequences similar to glycine, arginine-rich motifs in proteins interacting with RNA are efficiently recognized by methyltransferase(s) modifying arginine in numerous proteins. J. Biol. Chem. 268(14): 10501-10509.

Palmer RMJ, Ferrige AG, Moncada S (1987). Nitric oxide release accounts for the biological activity of endothelium derived relaxing factor. Nature. 327: 524-526.

Paiva H, Lehtimaki T, Laakso J, Ruokonen I, Rantalaiho V, Wirta O, Pasternack A, Laaksonen R (2003). Plasma concentrations of asymmetric-dimethylarginine in type 2 diabetes associate with glycemic control and glomerular filtration rate but not with risk factors of vasculopathy. Metabolism. 52: 303-307.

Schernthaner G, Krzyzanowska K (2008). Role of asymmetric dimethylarginine in cardiovascular disease and diabetes. Biomarkers. Med. 2(4): 317-320.

Schulze F, Wesemann R, Schwedhelm E, Sydow K, Albsmeier J, Cooke JP, Böger RH (2004). Determination of asymmetric dimethylarginine (ADMA) using a novel ELISA assay. Clin. Chem. Lab. Med. 42(12): 1377-1383.

Selvin E, Marinopoulos S, Berkenblit G, Rami T, Brancati FL, Powe NR, Golden SH (2004). Meta-analysis: glycosylated hemoglobin and cardiovascular disease in diabetes mellitus. Ann. Intern. Med. 141: 421-431.

Sorrenti V, Mazza F, Campisi A, Vanella L, Li Volti G, Di Giacomo C (2006). High glucose-mediated imbalance of nitric oxide synthase and dimethylarginine dimethylaminohydrolase expression in endothelial cells. Curr. Neurovasc. Res. 3: 49-54.

Stamler J, Vaccaro O, Neaton JD, Wentworth D (1993). Diabetes, other risk factors, and 12-yr cardiovascular mortality for men screened in the Multiple Risk Factor Intervention Trial. Diab. Care 16(2): 434-444.

Toutouzas K, Riga M, Stefanadi E, Stefanadis C (2008). Asymmetric dimethylarginine (ADMA) and other endogenous nitric oxide synthase (NOS) inhibitors as an important cause of vascular insulin resistance. Horm. Metab. Res. 40(9): 655-659.

Vallance P, Chan NN (2001). Endothelial function and nitric oxide: clinical relevance. Heart. 85:342–350.

Vallance P, Leone A, Calver A, Collier J, Moncada S (1992). Accumulation of an endogenous inhibitor of nitric oxide synthesis in chronic renal failure. Lancet. 339: 572-575.

Yasuda S, Miyazaki S, Kanda M, Goto Y, Suzuki M, Harano Y, Nonogi H (2006). Intensive treatment of risk factors in patients with type-2 diabetes mellitus is associated with improvement of endothelial function coupled with a reduction in the levels of plasma asymmetric dimethylarginine and endogenous inhibitor of nitric oxide synthase. Eur. Heart. J. 27: 1159-1165.

Zoccali C, Bode-Boger SM, Mallamaci F, Benedetto FA, Tripepi G, Malatino L, Cataliotti A, Bellanuova I, Fermo I, Frolich JC, Boger R (2001). Asymmetric dimethylarginine (ADMA): an endogenous inhibitor of nitric oxide synthase predicts mortality in end stage renal disease (ESRD). Lancet. 358:2113–2117.

11

Degradation of pyrimidine ribonucleosides by extracts of *Aspergillus terreus*

Osama M. Abdel-Fatah*, Maysa A. Elsayed and Ali M. Elshafei

Department of Microbial Chemistry, National Research Centre, Dokki, Cairo, Egypt.

Cell-free extracts of nitrate-grown mycelia of *Aspergillus terreus* could catalyze the hydrolytic deamination of cytidine to uridine and ammonia followed by the hydrolytic cleavage of the N-glycosidic bond of the produced uridine to the corresponding base (uracil) and ribose. The same extracts could not catalyze the hydrolytic deamination of cytosine. Addition of inorganic arsenate to the reaction mixture containing cytidine or uridine did not affect the amount of ribose liberated indicating the absence of pyrimidine ribonucleosides phosphorylase in the extracts. Cytidine deaminase showed an optimum activity at pH 7.0 and 60°C and stability to high degrees of temperature. Uridine hydrolase activity was optimized at pH 8.0 and 55°C. Incubation of the extracts at 55°C for 60 min showed no effect on uridine hydrolase activity whereas incubation of the extracts at 60 and 70°C for different interval times caused a gradual decrease in activity and the enzyme lost its activity completely by incubation at 80°C for 15 min. Dialyzing the extracts showed no effect on cytidine deaminase activity and a decrease in uridine hydrolase activity. Addition of EDTA at a concentration of 5×10^{-3} M and 10^{-2} M caused an inhibition to the two enzymes activities. The presence of $MgSO_4$ in the reaction mixture seems to activate greatly both enzymatic cytidine deamination (225 and 128% increases) and uridine hydrolysis (22 and 77% increases) at final concentrations of 5×10^{-3} M and 10^{-2} M respectively. However $HgCl_2$ and $CuSO_4$ were found to be potent inhibitors for both activities at the two concentrations.

Key words: Pyrimidine ribonucleosides, cytidine, uridine, cytidine deaminase, uridine hydrolase, *Aspergillus terreus*.

INTRODUCTION

Most of the studies concerning the cleavage of N-glycosidic bond of ribonucleosides in filamentous fungi have been reported mainly on purine ribonucleosides (Elzainy et al., 1978; Hassan et al., 1979; Elzainy et al., 1990; Abu-Shady et al., 1994; Elshafei et al., 1995; Abdel-Fatah et al., 2003). However the cleavage of N-glycosidic bond of pyrimidine ribonucleosides in filamentous fungi was very rare (Hassan et al., 1983; Allam et al., 1987). Cleavage of the amino group of cytidine was studied in extracts of *Penicillium citrinum* (Elzainy et al., 1990), *Aspergillus niger* (Ali, 1998) and *Aspergillus phoenicis* (Abdel-Fatah, 2005). A highly specific uridine hydrolase was firstly demonstrated from autolysates of baker's yeast (Carter, 1951). The presence of specific uridine hydrolase was also reported by Grishchknov et al., (1978) in extracts of *Corynebacterium glutamicus*. Uridine and cyti-

dine hydrolase were also plant pathogenic bacterium namely, *Xanthomonas phaseoli* (Hochster and Nozzolillo, 1961). Uridine ribohydrolase from *Saccharomyces cereviciae* was identified cloned and characterized the corresponding URH1 gene and its physiological function was determined by the measurement of metabolic fluxes in several mutants im-paired in the pyrimidine salvage pathway (Kurtz et al., 2002). Pyrimidine nucleoside hydrolase from *Escherichia coli* showed high catalytic efficiency toward flourouridine which could be exploited for suicide gene therapy in cancer treatment (Giabbai and Degano, 2004). The availability of information about the degradation of pyrimidine ribonucleosides and their derivatives from different sources is the basis for studies on the use of these unique enzymes in biotechnology and in medical therapy (Johansson and Karlsson, 1997; Hocek et al., 2005). The aim of the present work was to investigate the presence of enzymatic cleavage of pyrimidine ribonucleosides in *A. terreus* and study the properties of the enzymes encountered in this degradation.

*Corresponding author: E - mail: abdelfatahom@yahoo.com.

Table 1. Degradation of pyrimidine ribonucleosides by extracts of *A. terreus*.

| pH | Products (μmoles) formed from | | |
| | Cytidine | | Uridine |
	Ammonia	Ribose	Ribose
5.0	2.00	0.0	0.0
7.0	2.27	1.4	1.8
9.0	2.12	0.9	1.6

Reaction mixture contained: cytidine or uridine, 5 μmoles Tris-acetate buffer pH as indicated, 50 μmoles; extracts protein, 3.96 mg; total volume, 1.0 ml; reaction temperature, 40°C; reaction time, 60 min.

MATERIALS AND METHODS

Microorganism

Aspergillus terreus NRRL 265 was obtained from the culture collection of the Northern Utilization Research and Development Division, United States, Department of Agriculture, Peoria, Illinois, U.S.A.

Media: Aspergillus terreus was cultivated and kept on slants of modified solid Czapek-Dox medium (Difco Manual, 1972). The composition of this medium is as follows (g/l): D-glucose, 20; $NaNO_3$, 2.0; KH_2PO_4, 1.0; $MgSO_4$. $7H_2O$, 0.5; KCl, 0.5 and agar 20. The liquid medium was sterilized by autoclaving under 1.5 atmosphere for 20 min. In case of solid media, the sterilization time was prolonged for 30 min under the same pressure.

Cultivation of organism

Conidia were scraped from mycelia, which were grown on slants for 7 days at 30°C and suspended by hand shaking, in sterile distilled water. Two ml aliquots of this suspension were used to inoculate, under aseptic conditions, 250 ml Erlenmeyer flasks each containing 50 ml of sterile medium. The inoculated flasks were incubated statically at 30°C for 4 days.

Preparation of cell-free extracts

After four days incubation period the mycelia were collected and harvested by filtration, then washed thoroughly with cold distilled water and finally blotted dry with absorbent paper. The blotted-dry mycelia were ground with approximately twice its weight of washed cold sand in a cold mortar and extracted with 0.1 M Tris-HCl pH 8.0. The slurry so obtained was centrifuged at 5500 rpm for 5 min. The supernatant was used as the crude enzyme preparation.

Dialysis of the extracts

Dialysis of the extracts was made against 200 fold its volume of cold 0.01 M Tris-HCl pH 8.0 for a period of 24 h at 4°C using dialysis bags (cellulose tubing 21mm dia., Sigma Diagnostic. St Louis. MO 631 78 USA).

Colorimetric determination

Determination of ribose was made by the method described by Ashwell, (1957). Ammonia was determined by Nessler's reagent described by Schramm and Lazorik, (1975). Protein was determined by the method of Lowry et al., (1951).

Enzyme assay

Cytidine deaminase activity was assayed by measuring the appearance of ammonia when the enzyme preparation was incubited with cytidine. This was accompanied by chromatographic identification of the formed products. Uridine hydrolase activity was determined by measuring the reducing sugar as ribose formed from the pyrimidine ribonucleosides, as previously described by Elzainy et al., (1990). This was accompanied by chromatographic identification of the base. One unit of enzyme is defined as the amount that formed one micromole of ribose or ammonia under the standard assay conditions. Specific activities are expressed as units/ml/mg protein. The estimation was repeated for three times and the data given represents the mean value of the three repetitions.

Identification of ribose: Ribose was identified using the ascending paper chromatographic (Whatman $N^o.1$) technique according to the method of Smith and Seakins, (1976). Two solvent systems were used; solvent 1 consists of n-propanol -ethyl acetate - water (70:10:20) and solvent 2 consists of isopropanol - water (160: 40). The developed brown spots of the identified and authentic ribose were located by using aniline oxalate reagent.

Identification of the pyrimidine ribonucleosides and their bases

Chromatographic identification of the pyrimidine ribonucleosides and their bases was made using chromatographic Whatman N^o3 MM filter paper and two solvent systems. Solvent 1 consists of n-butanol - glacial acetic acid - water (120:30:50) and solvent 2 consists of n-butanol - formic acid - water (154:20:26) (Smith and Seakins, 1976). The spots were located with an ultraviolet lamp.

RESULTS

Enzymatic degradation of pyrimidine ribonucleosides by extracts of *A. terreus*

An experiment was made to test the ability of cell-free extracts of nitrate-grown mycelia of *A. terreus* to catalyze the degradation of the pyrimidine ribonucleosides namely, cytidine and uridine. Each substrate was added to the reaction mixture containing Tris-acetate buffer at pH 5.0, 7.0 and 9.0. The results obtained indicate that ammonia was detected in the reaction mixtures containing cytidine adjusted at pH 5.0, 7.0 and 9.0 whereas ribose was detected in the reaction mixtures containing cytidine or uridine adjusted at pH 7.0 and 9.0 with different proportions. The results cited in Table 1 indicated that extracts of *A. terreus* could catalyze the deaminating and hydrolytic degradation of cytidine and uridine respectively.

Chromatographic identification of the products

Ribose was chromatographically identified in the reaction mixture containing cytidine or uridine adjusted at pH 8.0. The developed brown spots of the identified and authentic ribose had the same R_f values of 0.51 and 0.72 in solvent 1 and 2 respectively (materials and methods). Uri-

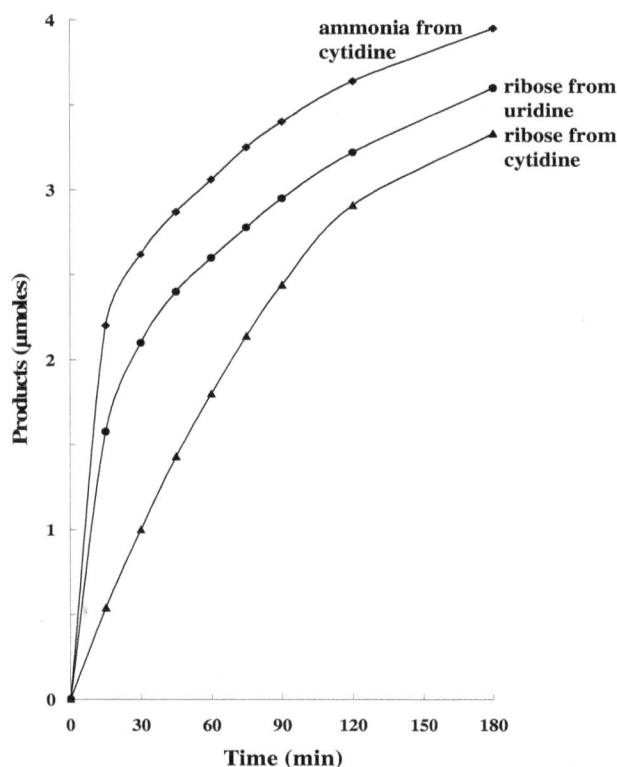

Figure 1. Degradation of cytidine and uridine as a function of time. Reaction mixture contained: Cytidine or uridine, 5 μmoles; Tris-HCl buffer pH 8.0, 50 μmoles; extract protein, 3.96 mg; total volume, 1.0 ml; temperature, 55°C; and reaction time as indicated.

dine and uracil were detected in the reaction mixture containing cytidine and uracil in the reaction mixture containing uridine. Cytosine could not be detected in the reaction mixture containing cytidine indicating that cytidine was degraded to uridine due to the action of cytidine deaminase. N-glycosidic bond of the formed uridine was then cleaved to produce uracil and ribose. The R_f values of the identified and authentic uridine were identical and were found to be 0.38 in solvent 1 and 0.25 in solvent 2. The expected base, uracil had R_f coincide with the authentic sample (R_f 0.49 in solvent 1 and 0.4 in solvent 2) (not shown).

Degradation of cytidine and uridine as a function of time

Figure 1 shows that the amount of ammonia formed in the reaction mixture containing cytidine was higher than the amount of the formed ribose. This result suggests that the degradation process starts by the deamination of cytidine producing ammonia and uridine which in turn could be cleaved by a hydrolase to produce uracil and ribose. The previously mentioned assumption was confirmed by the formation of considerable amounts of ri-

bose from a reaction mixture containing uridine as a substrate (Figure1).

Inability to detect pyrimidine ribonucleoside phosphorylase

Investigating this point was achieved through comparing the activities of the extracts in the presence and absence of inorganic arsenate with the two substrates cytidine and uridine. Results obtained show that the amount of ribose in presence of arsenate equals to that formed in the absence of arsenate indicating that phosphorolytic activity could not be detected in the extracts under these experimental conditions (not shown).

Substrate specificity of cytidine deaminase of *A. terreus*

This experiment was carried out to investigate substrate specificity of the enzyme that catalyzes the hydrolytic deamination of cytidine. Equimolar amount of a specific substrate namely cytidine, cytosine, adenosine, adenine, guanosine and guanine was incubated with Tris-HCl buffer at pH 8.0 at 60°C for 120 min after which ammonia was determined in all reaction mixtures. Results obtained indicate that from the six tested substrates, only cytidine was degraded by hydrolytic deamination, as 3.64 μ moles of ammonia was formed from 5 μmoles of cytidine indicating its specificity for this substrate (not shown).

pH value dependence of pyrimidine ribonucleosides degradation

Figure 2 shows the effect of pH values of the reaction mixtures on ammonia formation due to the action of cytidine deaminase. Optimum pH value for ammonia formation from cytidine occurred at pH 7.0, while as pH 8.0 was the optimum value for ribose formation from uridine and cytidine by cell-free extracts of *A. terreus* (Figure 3).

Effect of temperature

Maximal deaminating and hydrolytic activities occurred at 60 and 55°C respectively (Figures 4 and 5). This can be explained by stability of cytidine deaminase and uridine hydrolase towards high degrees of temperature.

Thermal stability behaviour

Cell-free extracts of *A. terreus* was subjected to different degrees of temperature to test the stability of cytidine deaminase and uridine hydrolase against high degrees of temperature. Results cited in Figures 6 and 7 show that cytidine deaminase has a thermal stability higher than uridine hydrolase where cytidine deaminase was still active at high temperatures (33% of activity remained when the extracts was

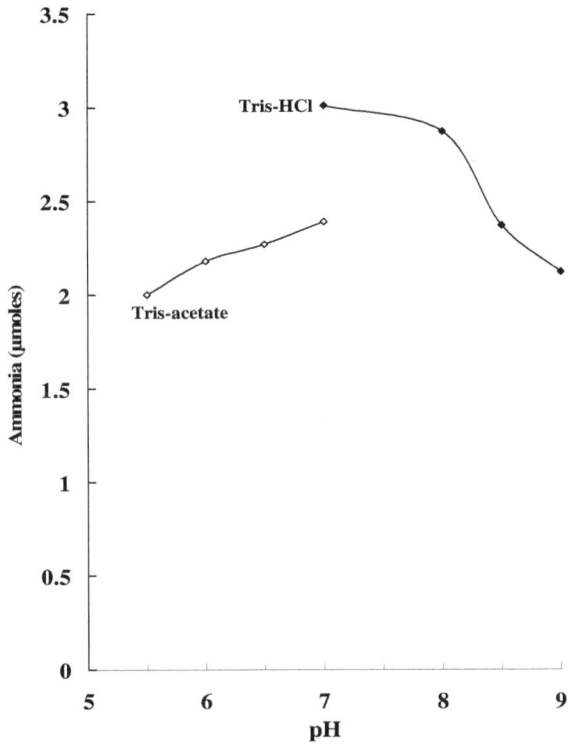

Figure 2. Effect of pH on cytidine deaminase activity. Reaction mixture contained: Cytidine, 5 μmoles; (•) Tris-acetate or (○) Tris-HCl buffer pH as indicated, 50 μmoles; extract protein, 3.52 mg; total volume, 1.0 ml; temperature, 60°C and reaction time, 60 min.

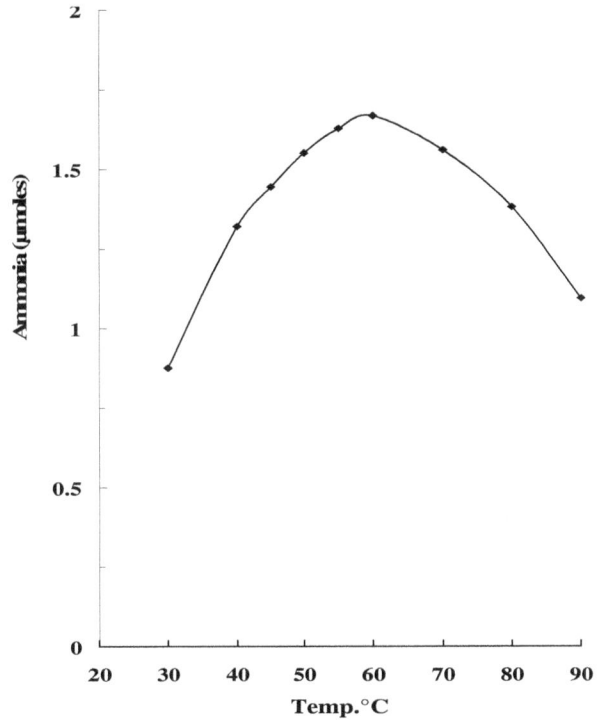

Figure 4. Effect of temperature on cytidine deaminase activity. Reaction mixture contained: Cytidine, 5 μmoles; Tris-HCl buffer pH 7.0, 50 μmoles; extract protein, 3.52 mg; total volume, 1.0; temperature, as indicated and reaction time, 60 min.

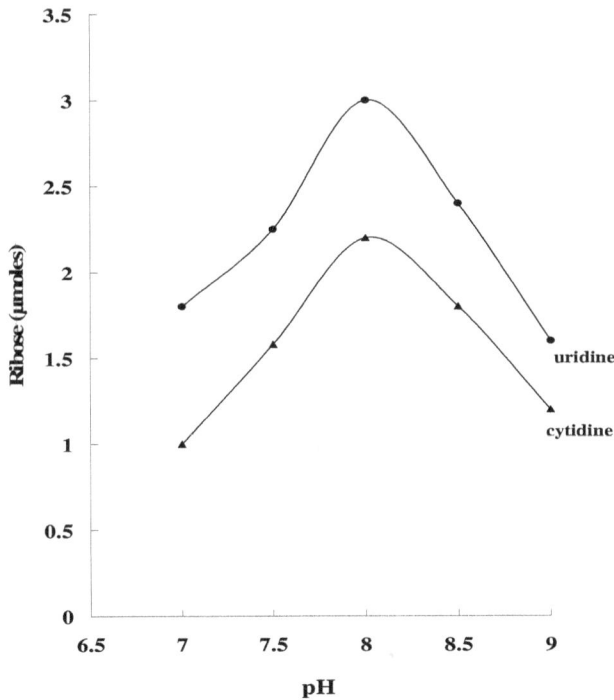

Figure 3: Ribose formation from cytidine and uridine as a function of pH value. Reaction mixture contained: Cytidine or uridine, 5 μmoles; Tris-HCl buffer pH as indicated, 50 μmoles; extract protein, 3.96 mg; total volume, 1.0 ml; temperature, 55°C; and reaction time, 60 min.

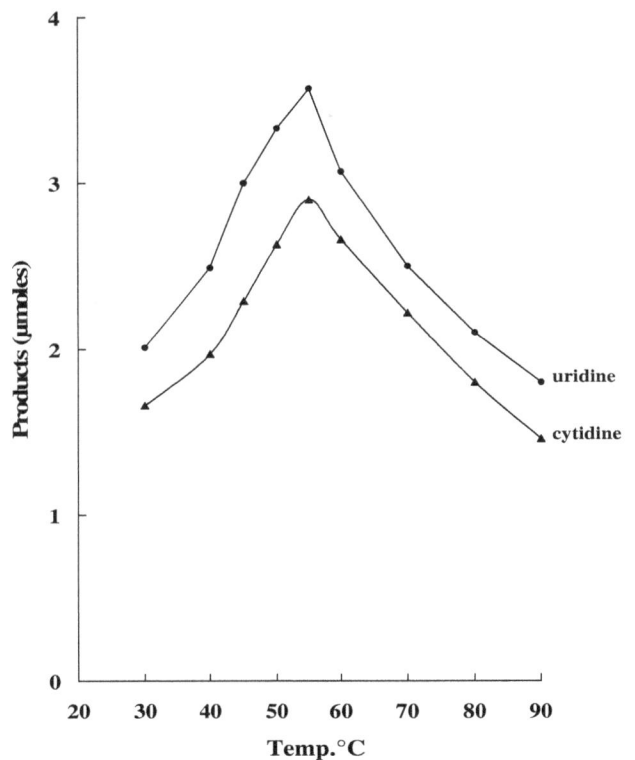

Figure 5. Effect of temperature on the formation of ribose from cytidine and uridine. Reaction mixture contained: Cytidine or uridine, 5 μmoles; Tris-HCl buffer pH as indicated, 50 μmoles; extract protein, 3.96 mg; total volume, 1.0 ml; temperature, 55°C; and reaction time, 60 min.

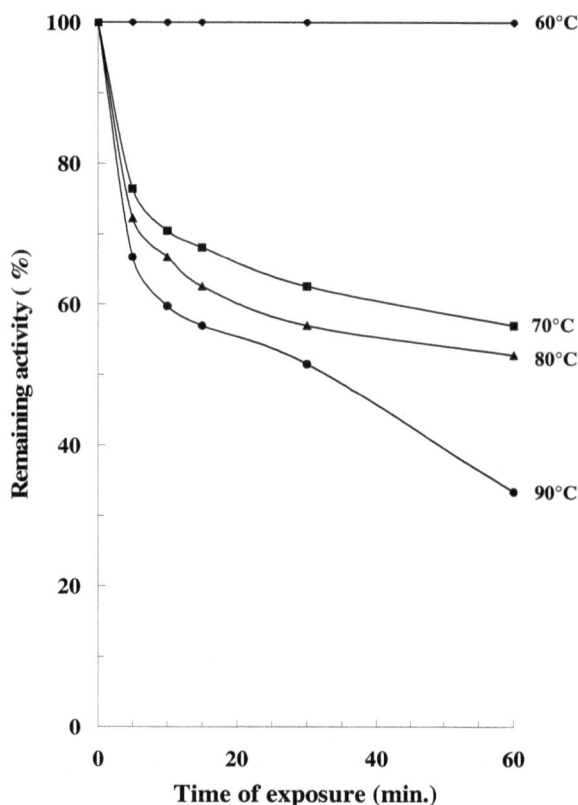

Figure 6. Heat inactivation kinetics of cytidine deaminase. Reaction mixture contained: Cytidine, 5 µmoles; Tris-HCl buffer pH 8.0, 50 µmoles; extract protein, 3.03 mg; total volume, 1.0 ml; reaction temperature, 60°C; time of exposure, as indicated and reaction time, 60 min.

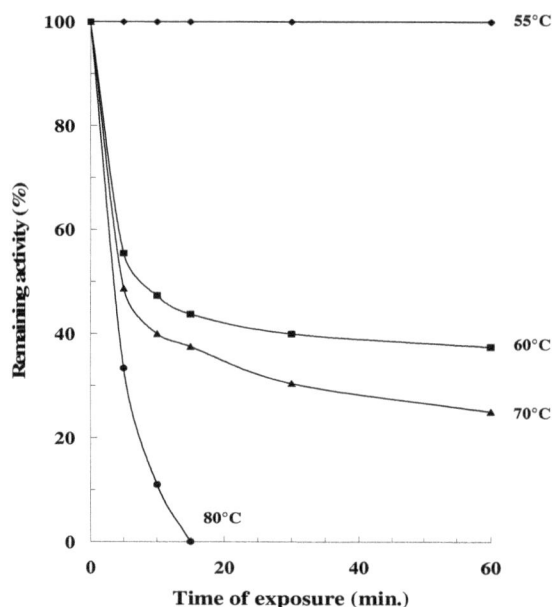

Figure 7. Heat sensitivity of uridine hydrolase. Reaction mixture contained: uridine, 5 µmoles; Tris-HCl buffer pH 8.0, 50 µmoles; extracts protein, 3.03 mg; total volume, 1.0 ml; reaction temperature, 55°C; time of exposure, as indicated and reaction time, 60 min.

Table 2. Effect of different buffer systems on pyrimidine ribonucleosides degradation

Type of buffer (pH 8.0)	Products (µmoles) formed from		
	Cytidine		Uridine
	Ammonia	Ribose	Ribose
Tris-HCl	4.9	3.8	4.8
Tris-acetate	4.0	1.6	1.8
Boric-Borax	2.9	1.4	2.4
Phosphate	4.5	3.6	3.8

Reaction mixture contained: cytidine or uridine, 5 µmoles; different buffers as indicated at pH 8.0, 50 µmoles; extract protein, 4.08 mg; total volume, 1.0 ml; temperature, 55°C and reaction time, 60 min.

exposed to 90°C for 60 min), while uridine hydrolase lost its activity completely when the crude extracts were exposed to 80°C for 15 min.

Best buffer system

Four reaction mixtures were made containing the same amount of substrate (cytidine or uridine) and received equimolar amount of Tris-HCl, Tris-acetate, Boric-Borax and phosphate buffer (0.05 M) at pH 8.0. Results obtained in Table 2 indicate that the activity of each of the two enzymes in Tris-HCl buffer is the highest when compared with the analogous activities obtained from the other three tested buffers.

Effect of additives

Addition of ethylenediamine-tetraacetate (EDTA), a metal chelating agent, at a concentration of 5×10^{-2} and 10^{-2} M caused about 50 and 43% inhibition on cytidine deaminase activity. The same concentrations caused about 54 and 74% inhibition on uridine hydrolase activity (Table 3). The previously mentioned results can be interpreted by the fact that cytidine deaminase and uridine hydrolase may require some metal cation(s) or activator(s) in the enzyme catalysis process. Addition of iodoacetate or 2-mercaptoethanol at a concentration of 5×10^{-3} and 10^{-2} M had no effect on cytidine deaminase and uridine hydrolase activities. The previous findings indicate that sulfhydryl group has no role in deaminating and hydrolytic catalysis process. These results were supported by the addition of reduced glutathione at the same concentrations which had no effect on the activity of the two enzymes under study (Table 3).

Effect of dialyzing the extracts

To study the effect of dialyzing cell-free extracts of *A. terreus* on the formation of ammonia and ribose from cytidine and uridine respectively, dialyzed extracts were tested and the obtained activity was compared with the analogous activity of nondialyzed extracts. The data presented in Table 5 show that the dialyzing process has no

Table 3. Effect of the addition of some compounds on the activity of cytidine and uridine-degrading enzymes.

Additive compounds	Concentration (M)	Enzyme activity (%)	
		Cytidine deaminase	Uridine hydrolase
---	---	100	100
Iodoacetate	5×10^{-3}	100	96
	10^{-2}	100	94
Mercaptoethanol	5×10^{-3}	100	97
	10^{-2}	100	95
Reduced glutathione	5×10^{-3}	100	100
	10^{-2}	100	100
EDTA	5×10^{-3}	50	46
	10^{-2}	57	26

Reaction mixture contained: cytidine or uridine, 5 µmoles; Tris-HCl buffer pH 8.0, 50 µmoles; extract protein, 2.62 mg; total volume, 1.0 ml; reaction temperature, 55°C and reaction time, 60 min

Table 4. Effect of dialysis process on the activity of cytidine and uridine- degrading enzymes.

Type of extract	Cytidine deaminase		Uridine hydrolase	
	Ammonia (µmoles)	Sp. activity	Ribose (µmoles)	Sp. activity
Crude	3.40	1.14	3.0	1.01
Dialyzed	2.74	1.2	0.6	0.26

Reaction mixture contained: cytidine or uridine, 5 µmoles; Tris-HCl buffer pH 8.0, 50 µmoles; extract protein, 2.97 mg; dialyzed extract protein, 2.25; total volume, 1.0 ml; reaction temperature, 55°C; reaction time, 60 min.

no effect on cytidine deaminase activity while as, uridine hydrolase lost about 75% of its activity due to the dialyzing process. The decrease in the enzymatic activity in case of dialyzed extracts can be explained by the release of certain dialyzable cofactors or activators that are present in the non-dialyzed extracts and required for the catalysis process (Table 4).

Effect of addition of some metal salts

From the previous experiment it was shown that dialyzing the extracts caused a decrease in the activity of uridine hydrolase. This finding may indicate the importance of certain dialyzable cofactors or activators that are present in the non-dialyzed extracts and required for the catalysis process. The presence of $MgSO_4$ in the reaction mixture seems to activate greatly both enzymatic cytidine deamination (225 and 128% increases) and uridine hydrolysis (22 and 77% increases) at a final concentration of 5×10^{-3} M and 10^{-2} M respectively. However $HgCl_2$ and $CuSO_4$ were found to be potent inhibitors for both activities at the two concentrations. $CuSO_4$ was also found to inhibit completely uridine hydrolase and partially cytidine deaminase (32 and 42% decreases in activity at both concentrations respectively) (Table 5).

Effect of freezing and thawing

The enzyme preparation was assayed for activity in the usual manner, stored in the refrigerator at about -5°C, left

for 24 h, after which it was thawed and an aliquot was withdrawn for assay of activity under the same experimental conditions. This process was repeated every 24 h for 3 days. Results of the analysis of products (ammonia and ribose) cited in Table 6 show that the amounts of products remain more or less the same after 72 h (3 cycles) indicating the stability of cytidine deaminase and uridine hydrolase towards this treatment.

DISCUSSION

Studies cited in this work revealed that extracts of A. terreus catalyzed the hydrolytic deamination of cytidine to uridine and ammonia and could not catalyze the deamination of cytosine .The formed uridine or uridine itself as a substrate could be cleaved to produced uracil and ribose. Cytidine deaminase of A. terreus reached its maximum activity in Tris-HCl buffer at pH 7.0 which resembles cytidine deaminase of A. niger (Ali, 1998) and differ from cytidine deaminase of A. phoenicis in having its maximum activity in citrate buffer at pH 6.0 (Abdel-Fatah, 2005). Cytidine deaminase of A. terreus showed a relatively high optimum temperature (60°C) and thermal stability behaviour compared to that obtained with cytidine deaminase of P. citrinum and A. niger at 50°C (Allam et al., 1991 and Ali, 1998 respectively) and A. phoenicis at 45°C (Abdel-Fatah, 2005). Uridine hydrolase of A. terreus showed an activity at a rather extreme pH value (8.0) compared to that reported with purine ribonu-

Table 5. Effect of addition of some metal salts on the activity of cytidine and uridine-degrading enzymes.

Metal salts	Concentration (M)	Relative activity (%)	
		Cytidine deaminase	Uridine hydrolase
---	---	100	100
CuSO$_4$	5×10^{-3}	0.0	0.0
	10^{-2}	0.0	0.0
CaCl$_2$	5×10^{-3}	100	178
	10^{-2}	100	122
KCl	5×10^{-3}	118	0.0
	10^{-2}	106	0.0
HgCl$_2$	5×10^{-3}	0.0	0.0
	10^{-2}	0.0	0.0
CoSO$_4$	5×10^{-3}	68	0.0
	10^{-2}	58	0.0
MgSO$_4$	5×10^{-3}	325	122
	10^{-2}	228	177
ZnSO$_4$	5×10^{-3}	100	0.0
	10^{-2}	100	0.0
FeSO$_4$	5×10^{-3}	100	0.0
	10^{-2}	100	0.0
MnCl$_2$	5×10^{-3}	100	55
	10^{-2}	100	0.0
NaCl	5×10^{-3}	113	200
	10^{-2}	124	150

Reaction mixture contained: cytidine or uridine, 5 µmoles; Tris-HCl buffer pH 8.0, 50 µmoles; dialyzed extract protein, 2.25 mg; total volume, 1.0 ml; reaction temperature, 55°C and reaction time, 60 min.

Table 6. Effect of freezing and thawing on the activity of cytidine and uridine- degrading enzymes.

Time of storage at -5°C	Relative activity (%)	
(hr)	Cytidine deaminase	Uridine hydrolase
--	100	100
24	100	100
48	98	97
72	97	97

Reaction mixture contained: cytidine or uridine, 5 µmoles; Tris-HCl buffer pH 8.0, 50 µmoles; extracts protein, 3.96 mg; total volume, 1.0 ml; reaction temperature, 55°C and reaction time, 60 min.

cleoside hydrolase of *P. citrinum* and *A. terricola* at pH 4.0 (Elzainy et al., 1990; Abu-Shady et al., 1994 respectively) and *A. phoenicis* at pH 3.5 (Abdel-Fatah et al., 2003). On the other hand the cleavage of N-glycosidic bond of guanosine, inosine, adenosine, cytidine anduridine by extracts of *Fusarium moniliforme* occurred at pH 6.0 (Allam et al., 1987). Results obtained in this work indicate the absence of pyrimidine ribonucleosides phosphorylase while as phosphorolytic cleavage of the pyrimidine ribonucleosides by extracts of *P. oxalicum* occurred at pH 6.2 (Hassan et al., 1983). In conclusion the cleavage of cytidine and uridine by extracts of *A. terreus* revealed the following facts:

1. Presence of a constitutive cytidine deaminase which is responsible for the deamination of cytidine to uridine and ammonia.

2. Presence of a constitutive uridine hydrolase which catalyzes the cleavage of uridine to uracil and ribose.

The way through which cytidine and uridine were degraded can be summarized as follows:

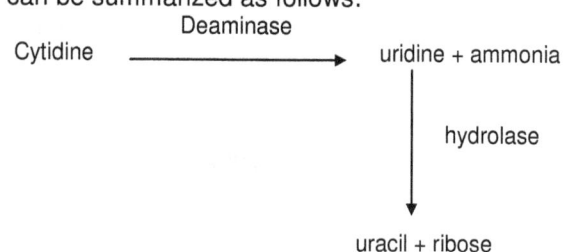

Cytidine —— Deaminase ——→ uridine + ammonia
                                    | hydrolase
                                    ↓
                            uracil + ribose

The cessation of further breakdown of uracil is not unexpected, as the denovosynthesis of this molecule requires a lot of energy. So it is much logical for the cell not to degrade this preformed molecule. Furthermore what has been actually found in some biological systems is that the cell invented very short pathway for its reutilization as:

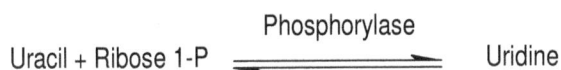

$$\text{Uracil + Ribose 1-P} \xrightleftharpoons{\text{Phosphorylase}} \text{Uridine}$$

REFERENCES

Abdel-Fatah OM (2005). Some properties of cytidine deaminase of *Aspergillus phoenicis*. N. Egypt. J. Microbiol. 10: 242-254.

Abdel-Fatah OM, Elsayed MA, Elshafei AM (2003). Hydrolytic cleavage of purine ribonucleosides in *Aspergillus phoenicis*. J. Basic. Microbiol. 43: 439-448.

Abu-Shady MR, Elshafei AM, El-Beih FM, Mohamed LA (1994). Hydrolytic cleavage of purine ribonucleosides by extracts of *Aspergillus terricola*. Acta Microbiologica Polonica, 43: 297-304.

Allam AM, Hassan MM, Ghanem BS, Elzainy TA (1987). Nature of enzymes that catalyze the cleavage of the N-glycosidic bond of pyrimidine ribonucleosides in some filamentous fungi. Biochem. Syst. Ecol. 15: 515-517.

Allam AM, Elzainy TA, Elsayed MA, Elshafei AM (1991). Some kinetic studies on ribonucleosides degradation by extracts of *Penicillium citrinum*. Acta. Biotechnol. 11: 227-233.

Ali TH (1998). Some kinetic studies on cytidine aminohydrolase activity from *Aspergillus niger* NRRL3. Acta Microbiologica Plonica. 47: 365-372.

Ashwell G (1957). Colorimetric analysis of sugars. In: Methods in Enzymology (Colowick SP, Kaplan NO, eds). Academic Press, New York. 3: 73,

Carter CE (1950). Paper chromatography of purine and pyrimidine derivatives of ribonucleic acid. J. Am. Chem. Soc. 72: 1466-1471.

Difco Manual (1972). Difco Manual of Dehydrated Culture, Media and Reagents 9th ed: 245. Difco Laboratories, Detroit, Michigan, USA.

Elshafei AM, Abu-Shady MR, El-Beih FM, Mohamed LA (1995). Mode and extent of degradation of adenosine and guanosine by extracts of *Aspergillus terricola*. Microbiol. Res. 150: 291-295.

Elzainy TA, Elsayed MA, Elshafei AM (1990). Mode and extent of degradation of ribonucleosides by extracts of *Penicillium citrinum*. Ann. Microbiol. 40: 159-169.

Elzainy TA, Hassan MM, Allam AM (1978). Hydrolytic cleavage of purine nucleoside in *Aspergillus niger*. Egypt. J. Bot. 21: 53-60.

Giabbai B, Degano M (2004). Crystal structure to 1.7 a of the *Esherichia coli* pyrimidine nucleoside hydrolase Yeik, a novel candidate for cancer therapy. Structure, 12: 739-749.

Grishchenkov VG, Sukhodolets W, Smirnov YN (1978). Characteristics of the metabolism of nucleosides and pyrimidine bases in *Corynebacterium glutamicus*. Microbiologiya. 47: 693.

Hassan MM, Elzainy TA, Allam AM (1979). Studies of some of the properties of purine nucleoside hydrolase from *Aspergillus niger*. Egypt. J. Chem. 22: 189-196.

Hassan MM, Ghanem BS, Elzainy TA, Allam AM (1983). Hydrolytic and phosphorylytic activities for the cleavage of purine and pyrimidine nucleosides in *Penicillium oxalicum*. Ann. Microbiol. 33: 83-91.

Heppel LA, Hilmoe RJ (1952). Phosphorolysis of purine riboside by enzyme from yeast. J. Biol. Chem. 198: 683-694.

Hocek M, Nauš P, Pohl R, Votruba I, Furman PA, Tharnish PM, Otto MJ (2005). Cytostatic 6-arylpurine nucleosides. 6. SAR in anti-HCV and cytostatic activity of extended series of 6-hetarylpurine ribonucleosides. J. Med. Chem. 48: 5869-5873.

Hochster RM, Nozzolillo CG (1961). The breakdown of ribonucleosides in extracts of phytopathogenic organism *Xanthomonas phaseoli*. Can. J. Microbiol., 7: 389.

Johansson M, Karlsson A (1997). Cloning of the cDNA and chromosome localization of the gene for human thymidine kinase 2. J. Biol. Chem. 272: 8454-8458.

Kurtz JE, Exinger F, Erbs P, Jund R. (2002). The URH1 uridine ribohydrolase of *Saccharomyces cerevisiae*. Curr Genet. 41: 132-141.

Lowry OH, Rosebrough NJ, Farrand AL, Randall RJ (1951). Protein measurement with the folin phenol reagent. J. Biol.Chem. 193: 265-275.

Schramm VL, Lazorik FC (1975). The pathway of adenylate catabolism in *Azotobacter vinelandii*. J. Biol. Chem. 250: 1801-1808.

Smith I, Seakins JWT (1976). Chromatographic and electrophoretic techniques. William Heinemann Medical Book Ltd. 1: 153-159.

12

Chelating ability of sulbactomax drug in arsenic intoxication

V. K. Dwivedi[1]*, A. Arya[1], H. Gupta[1], A. Bhatnagar[2], P. Kumar[1] and M. Chaudhary[1]

[1]Pre Clinical Division, Venus Medicine Research Centre, Baddi, H.P. India 173205, India.
[2]Analytical Division, Venus Medicine Research Centre, Baddi, H.P. India 173205, India.

The aim of this study was to determine the chelating ability of Sulbactomax drug in arsenic pre-exposed rats. 24 rats were divided into I to III groups. Group I was control while Groups II to III were arsenic exposed and arsenic plus Sulbactomax treated group. Arsenic toxicity was induced via intraperitoneally administration of arsenic trioxide (As_2O_3) 10 mg/kg body weight/day for three weeks. Toxicity was confirmed by decreased body weight, increased body temperature, loss of appetite and decreased hemoglobin levels in all groups except Group I. After confirmation of these symptoms, drug was administered via intravenous route for three week treatment. At the end of the study, blood samples were collected and the arsenic, zinc concentration, hemoglobin level and δ -aminolevulinic acid dehydrates activity were measured in the blood while other parameters were measured in the plasma sample. Our results revealed that zinc concentration, hemoglobin level, δ -aminolevulinic acid dehydratase, catalase and superoxide dismutase (SOD) enzyme activities significantly increased, while arsenic concentration, lipid peroxidation level, myloperoxidase enzyme activity, tumor necrosis factor TNF-α and IL-6 levels significantly reduced in Group III as compared to Group II. So, the findings concluded that administration of Sulbactomax drug act as an antioxidant and chelating agent that reduce arsenic metal and free radical mediated tissue injury.

Key words: Arsenic, Sulbactomax, antioxidant enzymes, cytokines parameters, arsenic, zinc.

INTRODUCTION

Arsenic (As) is a member of Group V in the periodic table of elements along with nitrogen, phosphorus, antimony and bismuth. The chemistry of arsenic is rather complex, and the compounds it forms are numerous. This is largely because arsenic possesses several different valence or oxidation states, which markedly result in the different biological behavior of its compound. It is a ubiquitous element which is present in low amount, soil and water. Arsenic compounds may represent a concern to environmental and occupational health become concentrated in the environmental as result of natural anthropogenic source. High doses of inorganic arsenic lead to neurologic, muscular, renal and gastrointestinal manifestation which may be responsible for a fatal outcome (Von, 1958). Long term administration of arsenic can cause hepatocellular carcinoma (Weisberger and Williams, 1975) in experimental animals, angiosarcoma of the liver (Ishak, 1976) and carcinoma of skin and lung (Jackson and Grainge, 1975). Arsenic is methylated by alternating reduction of pentavalent arsenic to trivalent and addition of a methyl group from S-adenosylmethionine (Singh and Rana, 2007). Glutathione and possibly other thiols, serve as reducing agents. Arsenic also inhibits several other cellular enzymes, especially those involved in cellular glucose uptake, gluconeogenesis and fatty acid oxidation etc. Exposure of arsenic causes melanosis, depigmentation and various organ failure such as kidney and hepatic. The heme metabolism pathway is known to be highly susceptible to alterations induced by heavy metals and environmental

*Corresponding author. E-mail: vivekdwivedi@venusremedies.com.

Abbreviation: As_2O_3, Arsenic trioxide; SOD, superoxide dismutase; CAT, catalase; MDA, malonaldialdehyde; GSH, reduced glutathione; MPO, myloperoxidase, TNF, tumor necrosis factor, IL, interlukin; FDC, fixed dose combination.

chemicals, offering the chance to use these changes as indicator of damage caused by arsenic (Flora et al., 2005).

A fixed-dose combination (FDC) is a formulation of two or more active ingredients combined in a single dosage form available in certain fixed doses. Sulbactomax is a novel fixed dose combination of ceftriaxone and sulbactam antibiotics along with third vector VRP1034. Ceftriaxone is third generation class of antibiotic and sulbactam is β-lacatamase inhibitor. VRP1034 (Trade secrete) is having potent chelating ability which reduce oxidative stress along with removal of heavy metal ions from the body. Dwivedi et al. (2009) reported that a novel fixed dose combination of Sulbactomax drug also play significant role in various bacterial infection. Beside their antibacterial property, authors have tried to determine whether Sulbactomax drug act as an antioxidant and has metal chealtion ability in arsenic pre-exposed rat's model.

MATERIALS AND METHODS

Chemicals

All the chemicals and biochemicals used in the present study were procured from Sigma, St. Louis, MO, USA. Δ-aminolevulinic acid (δ-ALAD) was purchased from Sigma Chemical, St Louis, MO, USA. Arsenic trioxide was purchased from CDH laboratories Ltd, New Delhi. Other chemicals, purchased locally, were of analytical grade. Ketamine hydrochloride was purchased from Samarth Life Science Pvt. Ltd. Mumbai. Other commercial kits were procured from Erba diagnostics Mannheim Gmb, Germanny. ELISA kits were procured from *invitrogen* for the measurment of cytokines parameters.

Drugs

Sulbactomax (ceftriaxone plus sulbactam + VRP1034), was obtained from Venus Medicine Research Centre, Baddi, H.P. The ratio of ceftriaxone and sulbactam in Sulbactomax is 2:1 respectively, along with VRP1034.

Animal groups and treatments

The animals were obtained from the animal house facility of Venus Medicine Research Centre, Baddi, and H.P. The experiment was carried out after approval from Institutional ethics committee. The study was performed on male Wistar rats weighing 145 ± 10 g housed in polypropelene cages in an air-conditioned room with temperature maintained at 25 ± 2°C and 12 h alternating day and night cycles. Animals were allowed standard rat chow diet and sterile distilled water. Twenty four rats were selected and divided into three groups of eight rats each which is given as follow:

Group I: Control normal saline treated group;
Group II: Arsenic trioxide induced group (10.0 mg/kg body weight/day);
Group III: Arsenic trioxide + Sulbactomax treated group (1.5 mg/kg body weight/day).

Arsenic trioxide was administered by *i.p* route to Groups II and III for 21 days except Group I. Body weight, body temperature, food intake and water intake were measured per day for 21 days. After clear symptoms, such as loss of appetite, increased body temperature and decreased body weight, appeared in all arsenic induced groups, treatment was started for 21 days. Sulbactomax drug were given to Group III according to their body weight via intravenous route for 21 days. At the end of experiment, blood samples were collected from each group in 3.8% sodium citrate containing vials. Immediately 1.5 ml blood samples were transferred into other tubes for measurement of hemoglobin, δ-ALAD enzyme activity and arsenic and zinc metal estimation.

Plasma preparation

Blood was centrifuged at 6000 rpm for 15 min and supernatant was carefully taken into other polypropelene tube and stored at 2°C to 8°C for the measurement of antioxidant enzymatic and biochemical parameters.

Determination of hemoglobin level

Hemoglobin level was measured according to Sahli haemoglobinometer.

Blood δ -aminolevulinic acid dehydratase (ALAD) activity

The ALAD activity in blood was assayed according to method of Berlin and Schaller (1974). For assay δ-ALAD activity, take 0.2 ml of blood sample and mixed with 1.3 ml double distilled water and incubate the test tubes for 10 min at 37°C for complete hemolysis. After incubation of all test tubes, added 1.0 ml of δ-ALA standard solution and further incubate all test tubes for one hr at 37°C. The reaction was stopped after one hr by adding 1.0 ml of 10% TCA solution. All test tubes were centrifuged at 600 g for 5 to 10 min. After centrifuge, 1.5 ml of the supernatant was taken in clean test tubes and equal amount of Ehrlich reagent was added and absorbance was recorded at 555 nm wavelength after 5 min. The molar extension coefficient 6.1×10^4 was used for calculation.

Measurement of intracellular and extracellular antioxidant enzymes activities

Superoxide dismutase (SOD) assay

SOD activity was determined by the method of Misra and Fradovich (1974). The reaction mixture consisted of 1.0 ml carbonate buffer (0.2 M, pH 10.2), 0.8 ml KCl (0.015 M), 0.1 ml of plasma sample and water to make the final volume to 3.0 ml. The reaction was started by adding 0.2 ml of epinephrine (0.025 M). The change in absorbance was recorded at 480 nm at 15 second interval for one min at 25°C. Suitable control lacking enzyme preparation was run simultaneously.

One unit of enzyme activity is defined as the amount of enzyme causing 50% inhibition of auto oxidation of epinephrine.

Catalase (CAT) activity

Catalase activity was measured according to procedure of Aebi (1984) at room temperature with slight modification. 100 µl plasma samples were placed on ice bath for 30 min at room temperature. 10 µl Triton-X was added in each plasma containing test tube. In a cuvette, 200 µl phosphate buffer (0.2 M; pH 6.8), 20 µl of sample and 2.53 ml distilled water was added. The reaction was started by adding 250 µl of H_2O_2 (0.066 M in phosphate buffer) and decrease in optical density was recorded at 240 nm wave length at every 15 second for one min. The molar extinction coefficient of 43.6 M Cm^{-1} was used for determination of catalase activity.

One unit of enzyme activity was defined as the amount of

enzyme that liberates half of the peroxide oxygen from H_2O_2 in one minute at 25°C.

Reduced glutathione (GSH) estimation

Reduced glutathione was estimated by the method of Eillman (1959). 0.5 ml plasma sample was mixed with equal amount of 5% (w/v) TCA reagent and kept for 10 min at room temperature, proteins were precipitated and filterate was removed carefully after centrifuge at 3500 rpm for 15 min. 0.25 ml filtrate was taken and added to 2.0 ml of Na_2HPO_4 (4.25%) and 0.04 ml of DTNB (0.04%). A blank sample was prepared in similar manner using double distilled water in place of the filtrate. The pale yellow color was developed and optical density was measured at 412 nm by spectrophotometer.

Estimation of total thiol content

Total thiol content was analyzed by the method of Hu (1994). 0.2 ml plasma sample was taken in test tubes and added 0.6 ml of Tris EDTA buffer (Tris 0.25 M , EDTA 20 mM; pH 8.2) followed by addition of 40 µl of 10 mM of 5,5' dithionitrobis 2-nitrobenzoic acid (DTNB in methanol) and make the total reaction volume up to 4.0 ml by adding 3.16 ml of methanol. All test tubes were sealed and the color was developed for 15 to 20 min, followed by centrifugation at 3000 g for 10 to 15 min at room temperature. The absorbance of the supernatant was measured at 412 nm wavelength.

Measurement of myleoperoxidase enzyme

Myeloperoxidase enzyme was determined by O-dianisidine method with slight modification (Kurutas et al., 2005). The assay mixture consisted of 0.3 ml of sodium phosphate buffer (0.1 M; pH 6.0), 0.3 ml of H_2O_2 (0.01 M), 0.2 ml of O-dianisidine (0.02 M) (freshly prepared in distilled water) and make final volume up to 3.0 ml with distilled water. The reaction was started by the addition of 0.025 ml plasma. The change in absorbance was recorded at 460 nm wavelength. All measurements were carried out in duplicate.

One unit of enzyme activity is defined as that giving an increase in absorbance of 0.001 min^{-1}.

Measurement of lipid per-oxidation level

Free radical mediated damage was assessed by the measurement of the extent of lipid peroxidation in the term of malonaldialdehyde (MDA) formed, essentially according to method of Ohkawa et al. (1979). It was determined by thio barbituric reaction. The reaction mixture consisted of 0.25 ml of plasma preparation, 0.20 ml of 8.1% sodium dodecyl sulphate (SDS), 1.5 ml of (20%, pH 3.5) acetic acid, 1.5 ml of 0.8% thio barbituric acid (TBA) and 0.6 ml distilled water and make the volume upto 4.0 ml. The tubes were kept in boiled water at 95°C for one hour and cooled immediately under running tap water. Added 1.0 ml of water and 5.0 ml of mixture of n-butanol and pyridine (15:1 v/v) and was vortexed. The tubes were centrifuged at 3500 rpm for 15 to 20 min. The upper layer was aspirated out and optical density was measured at 532 nm. The molar extension coefficient 1.56×10^5 was used for calculation.

Determination of biochemicals and cytokines parameters

The hepatic and renal enzymes serum glutamic oxaloacetic transaminase (SGOT), serum glutamic pyruvic transaminase (SGPT), Creatinine and alkaline phosphatase (ALP) levels) were measured in the plasma sample by standard kit method. Cytokine parameters (TNF α and Interlukin-6) were assayed by ELISA Reader, (Merck, Serial No- 21041098, MIOS -junior) according to manufacturer's instruction.

Metal estimation

For estimation of arsenic and zinc concentration in the blood, 0.5 ml of blood samples were directly mixed with 4.5 ml of acidic glycerol (1% HNO_3 and 5% glycerol mixture). Arsenic and zinc metal estimation were measured by using flame atomic absorption spectrophotometer (Analytikjena, model No contra A300, Germany) with hallow cathode lamp at wave length 193.7 and 213.9 nm respectively. The direct absorption of solution was determined by atomic absorption spectrophotometer and suitable standard curves of each metal were prepared using 20 to 100 µg/ml. All chemical used for metal estimation were Merck grade.

Statistical analysis

The resulting data were analyzed statistically. All values were expressed as mean ± SD. One-way analysis of variance (ANOVA) followed by Newman-Keuls comparison test was used to determine statistical difference between control vs. arsenic pre-exposed group and arsenic pre-exposed group vs. Sulbactomax treated groups. P values < 0.05 were considered statistically significant.

RESULTS

There was no mortality found during the experimental period. The clear symptoms such as anemia, loss of hair, loss of body weight and diarrhea were observed in arsenic pre-exposed group.

There was statistically significant decreased body weight (P<0.01), food intake (P<0.001) and water intake (P<0.01), along with significant (P<0.001) increase in body temperature in arsenic pre-exposed group after 21 days of intraperitoneally administering arsenic trioxide as compared to the control group. After treatment with Sulbactomax drug for 21 days, these physiological parameters were significantly improved in treated group as compared to the arsenic pre-exposed group and these physiological parameters almost reached near to the control group (Table 1). The arsenic concentration and δ-ALAD enzyme activity are major parameters in the arsenic toxicity. These parameters were affected by arsenic which is involved in the heme synthesis path way, due to inhibition of this enzyme heme synthesis is inhibited. So in the present study, the concentartion of arsenic (P<0.001) was significantly increased along with significantly (P<0.001) decreased zinc concentration in the blood of pre-exposed group as compared with control normal saline group. These metal concentration were significantly (P<0.001) improved in the blood of Sulbactomax treated group as compared with arsenic pre-exposed group after 21 days treatment.

There was statistically (P<0.001) significant decreased hemoglobin level and δ-ALAD enzyme activity in arsenic pre-exposed group as compared with control group. After

Table 1. Status of body weight, body temperature, food and water intake in arsenic pre-exposed group and Sulbactomax treated group.

Groups	Body weight (g)	Body temperature (°C)	Food intake (g)	Water intake (ml)
Control normal saline group	145.5 ± 3.5	35.53 ± 0.28	74.6 ± 2.68	75.2± 6.88
As pre-exposed group	120 ± 17.8**	38.01± 0.96***	53.1± 8.49***	60.0± 5.37**
Sulbactomax treated group	134.4 ± 12.7*	36.72± 0.59**	68.1± 6.64***	72.5± 9.45**

All data are expressed as mean± SD of each group. Newman keuls test was performed for statistical significance between control group vs. arsenic pre exposed group and arsenic exposed group vs. Sulbactomax treated group. Where ***p<0.001 (highly significant), **p<0.01(significant), *p<0.05 (significant), ^{Ns}p>0.05 (non significant).

Table 2. Status of biochemical parameters, arsenic and zinc levels in pre-exposed group and Sulbactomax treated group.

Groups	Hemoglobin level (mg/dl)	δ-ALAD activity (nmole/min/ml blood)	Blood arsenic (μgm/ml)	Blood zinc (μgm/ml)
Control group	11.8 ± 0.67	7.46 ± 0.62	0.065 ± 0.01	5.53 ± 0.98
As pre-exposed group	8.7 ± 0.71***	5.07± 0.11***	2.56± 0.42***	3.45± 0.20***
Sulbactomax treated group	9.9 ± 0.52**	6.61± 0.81***	1.01± 0.10***	4.57± 0.44***

All data are expressed as mean± SD of each group was performed for statistical significance between control group vs. arsenic pre exposed group and arsenic exposed group vs. Sulbactomax treated group. Where ***p<0.001 (highly significant), **p<0.01(significant), *p<0.05 (significant), ^{Ns}p>0.05 (non significant).

intravenous treatment of Sulbactomax drug (ceftriaxone plus sulbactam with VRP1034) for 3 weeks, the level of hemoglobin (P<0.01) and δ-ALAD enzyme activity (P<0.001) were significantly increased in drug treated group as compared to arsenic pre-exposed group and these parameters were significantly increased and reached near to control level when compared with Sulbactomax treated group (Table 2). There was significant increase in the endogenous antioxidant enzymes activities (SOD, P<0.001; CAT, P<0.001; total thiol, P<0.001; and GSH, P<0.001) along with significant decrease MDA level (P<0.001) and myloperoxidase enzyme activity (P<0.01) in the plasma of arsenic plus Sulbactomax treated group when compared with the arsenic pre-exposed group, in that these parameters almost reached those of the control group after treatment with

Sulbactomax drug for 21 days (Table 3).
The hepatic enzymes (serum glutamic oxaloacetic transaminase, serum glutamic pyruvic transaminase and alkaline phosphatase) and renal enzymes (Creatinine) were (P<0.001) significantly reduced in the plasma of the Sulbactomax treated group after 21 days treatment as compared to the arsenic pre-exposed group. These enzymatic parameters almost reacted like those of the control group after treatment with the drug (Table 4).
The levels of cytokines (TNF– α and IL-6) were (P<0.001) significantly increased in the plasma of arsenic pre-exposed group as compared with control group. These levels were found to be significantly (P<0.001) lowered in the plasma of Sulbactomax treated group after 21 days treatment as compared with arsenic pre-exposed

group (Figures 1 and 2).

DISCUSSION

Arsenic (As) is a well known heavy metal that causes tissues damage, including the immune system. Arsenic is an uncoupler of mitochondrial oxidative phosphorylation that induces generation of reactive oxygen species. Various studies have reported that arsenic is immunotoxic (Fuente et al., 2002; Hornhardt et al., 2006). Arsenic metal interferes with the antigen-presenting function of splenic macrophages that is able to alter the response of IgM and IgG antibody-forming cells to sheep erythrocytes, and the proliferative response of lymphocytes to phytohaemagglutinin (Sikorski et al., 1989; Sikorski et al., 1991; Savabieasfahani

Table 3. Status of antioxidant enzymatic and free radical mediated damage parameters in pre-exposed group and Sulbactomax treated group.

Groups	SOD (nmole/min/ml plasma)	CAT (nmole/min/ml Plasma)	Total thiol (mg/dL)	GSH (mg/dL)	MDA (µmole/ml plasma)	MPO (µmole/min/ml plasma)
Control group	3.47 ± 0.22	61.93 ± 8.49	0.487 ± 0.07	7.95± 0.64	7.30 ± 0.42	8.45± 1.23
As pre-exposed group	0.985 ± 0.011***	36.31± 3.71***	0.167± 0.04***	3.28± 0.98***	11.37± 1.61***	21.56± 3.11***
Sulbactomax treated group	2.23 ± 0.10***	53.0± 6.22***	0.342± 0.01***	4.59± 1.02***	7.75±0.35***	12.99± 2.85***

All data are expressed as mean± SD of each group. Newman keuls test was performed for statistical significance between control group vs arsenic pre exposed group and arsenic exposed group vs. Sulbactomax treated group. SOD; superoxide dismutase, CAT; catalase, GSH; reduced gluathione, MDA; malonaldialdehyde, MPO; myloperoxidase. Where ***p<0.001 (highly significant), **p<0.01(significant), *p<0.05 (significant), ^{Ns}p>0.05 (non significant).

Table 4. Status of hepatic and renal enzymetic parameters in pre-exposed group and Sulbactomax treated group.

Groups	SGOT (mg/dL)	SGPT (mg/dL)	Creatinine (mg/dL)	ALP (mg/dL)
Control group	26.12 ± 2.10	32.48 ± 2.8	0.63 ± 0.03	37.95± 6.4
As pre-exposed group	72.5± 1.5***	86.12± 5.41***	1.98± 0.15***	113.41± 8.41***
Sulbactomax treated group	45.37± 3.12***	39.5± 3.33***	0.86± 0.23***	68.24 ± 6.09***

All data are expressed as mean± SD of each group. Newman keuls test was performed for statistical significance between control group vs. arsenic pre exposed group and arsenic exposed group vs. Sulbactomax treated group. Where ***p<0.001 (highly significant),**p<0.01(significant), p<0.05 (significant), ^{Ns}p>0.05 (non significant).

et al., 1998). The heme metabolic path way is highly susceptible to alterations induced by metal ions. So in the present investigation, the δ-aminolevulinic acid dehydratase enzyme activity and hemoglobin level were significantly (P<0.001) inhibited in the arsenic pre-exposed group as compared to control group after intraperitoneal administration of arsenic trioxide (AS$_2$O$_3$) for 21 days. The level of arsenic concentration was significantly increased along with significant decreased the zinc level in arsenic pre-exposed group as compared to control group. After intravenous administration of a novel fixed dose combination drug (ceftriaxone plus sulbactam with VRP1034) for 21 days treatment, the level of arsenic was decreased along with increased hemoglobin, zinc levels and δ-aminolevulinic acid dehydratase enzyme activity in the treated group.

Therefore, it is interpreted that arsenic metal interferes with heme synthesis path way that inhibits the δ-aminolevulinic acid dehydratase enzyme and these enzymes are responsible for the synthesis of heme. Due to decrease of δ ALAD enzyme activity, the heme synthesis path way is inhibited. Kannan and Flora (2004) suggested that the hemoglobin level and δ-aminolevulinic acid dehydratase enzyme activity were inhibited during arsenic exposure in rats. The principal biochemical mechanism in acute arsenic intoxication is the reversible combination of arsenic with susceptible sulfhydryl-containing enzymes.

Oxidative stress has also been identified as an important mechanism of arsenic toxicity (Flora

and Gupta, 2005; Kitchin, 2001). Arsenic induces oxidative DNA damage and increased lipid peroxidation due to excessive generation of free radicals. A number of various studies have showed that arsenic-induced the formation of reactive oxygen and nitrogen species as well as elevated DNA oxidation. Various reports have been suggested that inorganic arsenic inhibits several of the antioxidant systems in the body, such as glutathione, glutathione peroxidase, thioredoxin reductase, and superoxide dismutase (Shen et al., 2003; Shila et al., 2005; Wu et al., 2001). In our study, reduced glutathione, total thiol levels and SOD, catalase enzyme activites were significantly reduced along with significantly increased malonaldialdehyde and myloperoxidase levels (free radical mediated damage) in arsenic

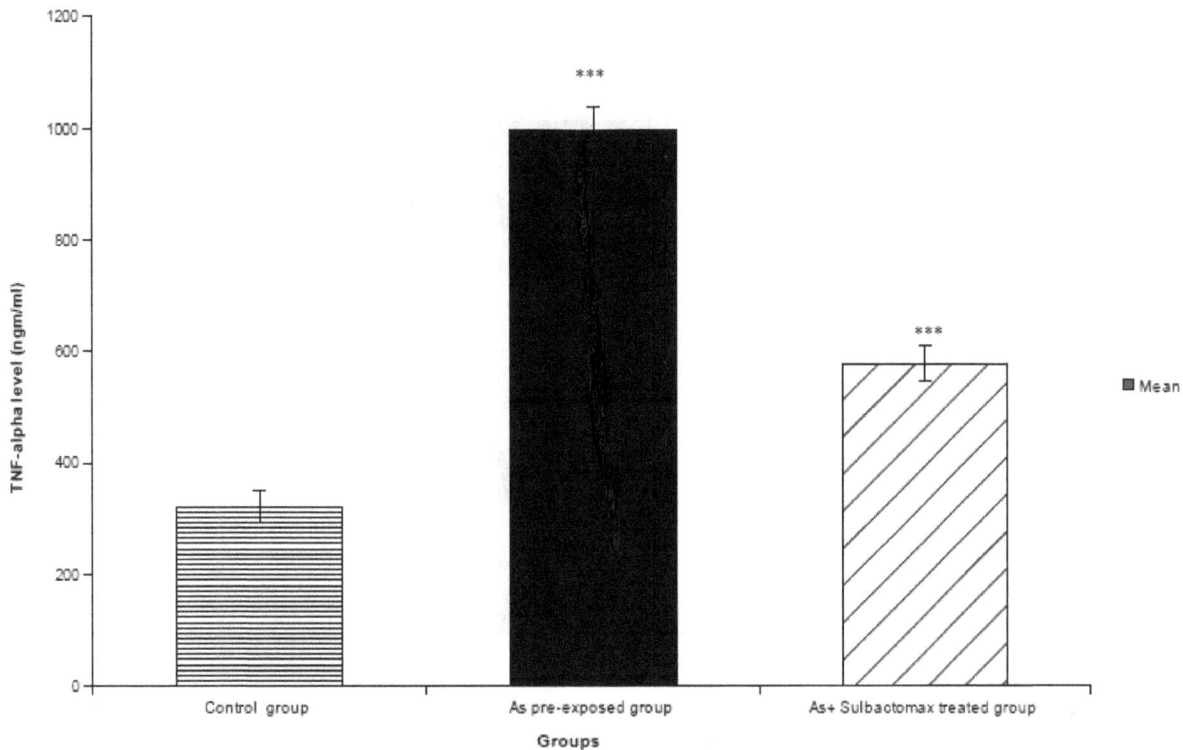

Figure 1. Status of TNF-α level in pre-exposed group and Sulbactomax treated group. Data are expressed as mean± SD of each group. Newman keuls test was performed for statistical significance between control group vs. arsenic exposed group and arsenic exposed group vs. Sulbactomax treated group. Where, ***p<0.001 (highly significant), **p<0.01(significant), *p<0.05 (significant) and ^{Ns}p>0.05 (non significant).

pre-exposed groups as compared with control group. After treatment with Sulbactomax drug, these parameters were significantly improved along with reduced free radical mediated damage levels in arsenic pre-exposed plus Sulbactomax treated group. Besides these parameters, the hepatic, renal and cytokinine (IL-6 and TNF-α) parameters were also increased in arsenic pre-exposed group as compared to control group. After treatment Sulbactomax drug for 21 days, these parameters were improved in arsenic induced plus drug treated group. Mazumdar (2005) has reported that hepatic damage occurred in experimental animals by chronic arsenic toxicity. Increased oxidative stress and cytokine response are associated with increased accumulation of collagen in the liver due to prolonged arsenic exposure. Das et al. (2005) suggested that implications of oxidative stress and hepatic cytokine (TNF-alpha and IL-6) response in the pathogenesis of hepatic collagenesis in chronic arsenic toxicity. Arsenic toxicity may cause nutritional deficit which leads to anemia and generate oxy free radicals. Thus, Sulbactomax drug may increases the antioxidant levels in the body may protect against arsenic-induced toxicity. Indeed, the administration of ascorbic acid, α-tocopherol or selenium has been shown to decrease arsenic-induced toxicity. Besides these exogenous antioxidant

compounds, various chelating agents may be useful for the prevention of arsenic toxicity. Sulbactomax drug is a novel fixed dose combination of ceftriaxone plus sulbactam along with VRP1034. VRP1034 has chelating, antioxidant properties which are protected by patent.

Various studies have been reported that ceftriaxone and sulbactam individually showed the free radical scavenging property (Lapenna et al., 1995; Gunther et al., 1993). Cephalosporins are known as thioether containing class of antibiotics which are more effective in preventing the free radical-mediated oxidation of sulfhydryl groups in the antibiotics. Beside free radical scavenging property and antimicrobial effect, ceftriaxone and sulbactam individually interact with arsenic ions and other heavy metals and form complex which chelate out from sulfhydryl group of antibiotics. Combination of ceftriaxone plus sulbactam with VRP 1034 (Sulbactomax) is the most active drug which enhanced the free radical scavenging property and also enhanced the removal of heavy metal ions. Chemical vector mediated technology was provided compatibility of cephalosporin and beta lacatmase inhibitor with VRP 1034 without interfering in the pharmacokinetic property of drug component and their role was to prevent the oxidation of methionine group, thiazolidine and dihyrothiazine present in antibiotics. There was no evidence regarding a novel

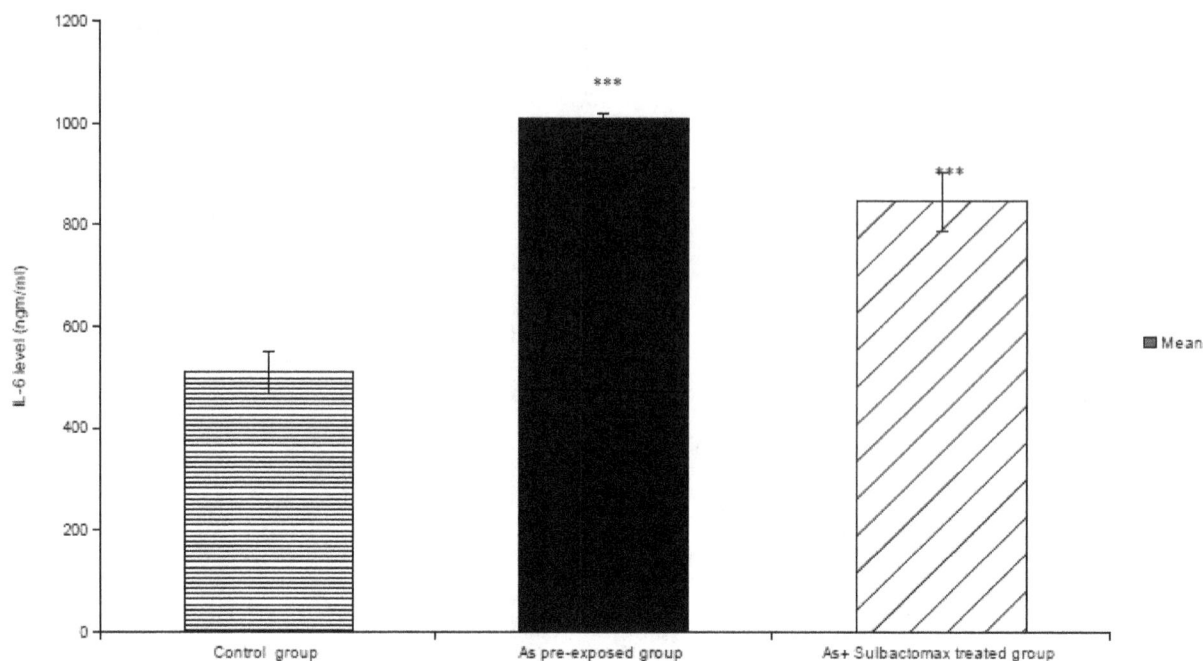

Figure 2. Status of IL-6 levels in pre-exposed group and Sulbactomax treated group. Data are expressed as mean± SD of each group. Newman keuls test was performed for statistical significance between control group vs. arsenic exposed group and arsenic exposed group vs. Sulbactomax treated group. Where ***p<0.001 (highly significant), **p<0.01(significant), *p<0.05 (significant) and ^{Ns}p>0.05 (non significant).

fixed dose combination of two antibiotic (beta lactam and betalactamase inhibitor) along with VRP1034 (Sulbactomax) has metal chelator and free radical scavenging properties. These results suggested that the novel fixed dose combination of Sulbactomax showed free radical scavenging and metal chelating properties which removes heavy metal and decreases free radical mediate damage tissue injury along with increase intracellular and extracellular antioxidant defense system and prevent hepatic and renal toxicity.

Conclusion

The conclusion of present study showed that Sulbactomax is the most and safe effective drug for the treatment of arsenic toxicity due to removal of arsenic metal along with increased δ-aminolevulinic acid dehydratase enzyme activity and also prevents free radical mediate damage tissue injury due to free radical scavenger's property and chelation ability property of drug.

ACKNOWLEDGEMENT

The authors are thankful to Management of Venus Medicine Research Centre to provide infrastructure to carry out this experiment.

REFERENCES

Aebi H (1984). Catalase. In Methods in Enzymol, Packer L (ed.). Academic Press: Orlando, FL, 125-126.

Berlin A, Schaller KH (1974). European standardized method for the determination of delta aminolevulinic acid dehydratase activity in blood. Zeit. Klin. Chem. Klin Biochem., 12: 389-390.

Das S, Santra A, Lahiri S, Mazumder DNG (2005).Implications of oxidative stress and hepatic cytokine (TNF-α and IL-6) response in the pathogenesis of hepatic collagenesis in chronic arsenic toxicity. Toxicol. Appl. Pharmacol., 204: 18-26.

Dwivedi VK, Chaudhary M, Soni A, Yadav J, Tariq A (2009). Diffusion of Sulbactomax and ceftriaxone into cerebrospinal fluid of meningitis induced rat model. Int. J. Pharmacol., 5: 307-312.

Ellman GL (1959). Tissue sulfhydrl groups. Arch. Biochem. Biophys., 82: 70-77.

Flora SJS, Bhadauria S, Pant SC, Dhaked RK (2005). Arsenic induced blood and brain oxidative stress and its response to some thiol chelators in rats. Life Sci., 77: 2324-2337.

Flora SJS, Gupta R (2005). Protective Value of Aloe vera against Some Toxic Effects of Arsenic in Rats. Phytother. Res., 19: 23-28.

Fuente HDL, Portales-pérez D, Baranda L, Barriga FD, Alanís VS, Layseca X, Amaro RG (2002).Effect of arsenic, cadmium and lead on the induction of apoptosis of normal human mononuclear cells. Clin. Exp. Immunol., 129 (1): 69-77.

Gunther MR, Mao J, Cohen MS (1993). Oxidant-scavenging activities of ampicillin and sulbactam and their effects on neutrophil functions. Antimicrob. Agents Chemother., 37(5): 950-956.

Hornhardt S, Gomolka M, Walsh L, Jung T (2006). Comparative investigations of sodium arsenite, arsenic trioxide and cadmium sulphate in combination with gamma-radiation on apoptosis, micronuclei induction and DNA damage in a human lymphoblastoid cell line. Mutat. Res., 600: 165-176.

Hu M (1994). Measurement of protein thiol groups and glutathione in plasma. Method Enzymol., 233: 380-382.

Ishak KG (1976). Mesencymal tumors o f the liver. In Hepatocellular

Carcinoma (Okuda K, Rogers RL, eds.) John Wiley, New York, pp. 247-307.

Jackson R, Grainge JW (1975). Arsenic and Cancer. Can. Med. Assoc. J., 113: 396-401.

Kannan GM, Flora SJS (2004). Chronic arsenic poisoning in the rat: treatment with combined administration of succimers and an antioxidant.Ecotoxicol.Environ. Safe, 58: 37-43.

Kitchin KT (2001). Recent advances in arsenic carcinogenesis: mode of action, animal model system and methylated arsenic metabolites. Toxicol. Appl. Pharmacol., 172: 249-261.

Kurutas EB, Arican O, Sasmaz S (2005). Superoxide dismutase and myeloperoxidase activities in polymorphonuclear leukocytes in acne vulgaris. Acta Dermatoven., 14: 39-42.

Lapenna D, Cellini L, Gioiaa S De, Mezzettia A, Ciofania G (1995) Cephalosporins are scavengers of hypochlorous acid. Biochem Pharmacol., 49(9): 1249-1254.

Mamzudar DN (2005). Effect of chronic intake of arsenic- contaminated water on liver. Toxicol. Appl. Pharmacol., 206: 169-175.

Misra HP, Fridovich I (1974). The role of superoxide anion in the auto-oxidation of epinephrine and a sample assay for Super-oxide dismutase. J. Biol. Chem., 247: 3170-3175.

Ohkawa H, Ohishi N, Yagi (1979). Assay of lipid per-oxidation in animal tissue by thio barbutric acid reaction. Anal. Biochem., 95: 351-358.

Savabieasfahani M, Lochmiller RL, Rafferty DP, Sinclair JA (1998).

Sensitivity of wild cotton rats (Sigmodon hispidus) to the immunotoxic effects of low-level arsenic exposure. Arch. Environ. Contam. Toxicol., 34: 289-296.

Shen ZY, Shen WY, Chen MH, Shen J, Zing Y (2003). Reactivre oxygen species and antioxidants in apoptosis of esopheal acncer cells induced by AS_2O_3. Int. J. Mol. Med., 11: 479-484.

Shila S, Subthara M, Devi MA Paneerselvam C (2005). Arsenic intoxication induced reduction of gluthione level and of the activity of releated enzymes in rat brain region: reversal by DL-alpha lipoic acid. Arch Toxicol., 208: 357-365.

Sikorski EE, Burns LA, McCoy KL, Stern M, Munson A (1991). Suppression of splenic accesory cell function in mice exposed to gallium arsenide. Toxicol. Appl. Pharmacol., 110: 143-156.

Sikorski EE, McCay JA, White KL, Bradley SG, Munson AE (1989). Immunotoxicity of the semiconductor gallium arsenide in female B6C3F1 mice. Fundam. Appl. Toxicol., 13: 843-158.

Singh S, Rana SVS (2007). Amelioration of arsenic toxicity by L-Ascorbic acid in laboratory rat. J. Environ. Biol., 28: 377-387.

Von OWF (1958). Poisoning. A Guide to Clinical Diagnoses and Treatment. W.B. Saunders, Philadelphia.

Weisberger JH, Williams GM (1975). Metabolism of chemical carcinogens. in Cancer (Becker FF, ed.). Plenum Press, Ney York, pp. 185-234.

Wu MM, Chiou HY, Wang TW, Hsuch YM, Wang IH, Chen CJ, Lee TC (2001). Association of blood arsenic level with increased reactive oxidant and decreased antioxidant capacity in human population northeastern Taiwan. Environ, Health. Prospect, 109: 1011-1017.

13

Captopril interferes with some serum biochemical findings

I. A. Ibrahim[1], F. S. Al-Joudi [2]*, R. Waleed Sulaiman [3] and B. Hilal AL-Saffar[4]

[1]Department of Pharmacology, Faculty of Medicine, National University of Malaysia, Kuala Lumpur, Malaysia.
[2]Faculty of Allied Health Sciences, National University of Malaysia, Kuala Lumpur, Malaysia.
[3]Department of Clinical Laboratory Science, Faculty of Pharmacy, University of Baghdad, Baghdad, Iraq.
[4]Department of Medicine, Hospital Dr. Abdul-Majeed, Karradah, Baghdad, Iraq.

Captopril is a widely used anti-hypertensive drug that acts by inhibiting angiotensin-converting enzyme. This work has been carried out to investigate the effects of captopril on some common biochemical laboratory parametres in the sera of patients receiving the drug. For this study, 40 subjects were included, all within the age range of 40 to 63 years and with newly diagnosed essential hypertension. From each patient, two samples were taken, one immediately before the start of treatment and the second one taken two weeks later. The control group comprised 30 apparently healthy volunteers of comparable ages and genders. The biochemical parameters measured in the sera were glucose, total protein (TP), urea, creatinine, total cholesterol (TC), triglycerides (TG), liver enzymes and creatine kinase (CK). Captopril exerted significant increases in the obtained readings for the concentrations of glucose, TP, urea, creatinine, TC, TG, AST and LDH. The increases in readings in the biochemical parameters may be attributable to chemical or to physical interactions. They could also be induced by physiological, enzymatic or by *in vivo* metabolic factors. By all means, these alterations that accompany captopril treatment must be taken into account by physicians and laboratory workers, to help avoid misinterpretation of laboratory data.

Key words. Captopril, biochemical laboratory tests, drug interactions.

INTRODUCTION

Drug interactions with laboratory findings have been reported previously. Examples of such interactions are the widely used aspirin and ascorbic acid, both which could interfere with serum readings to give a state of false hyperglycaemia (Young et al. 1972). Captopril, D-3-mercapto-2-methyl (propanoyl-L-proline, 1), is a selective inhibitor of angiotensin I-converting enzyme (ACE) (Cushman et al., 1978; Schmidt et al., 1986). It exerts vasodilatory effects and enhances the renal excretion of sodium (Chrysant et al., 1985; Hauger Kleven, 1985;

Hymes et al., 1983; Matcher et al., 2008; Sleight, 2001; Zanchetti et al., 2006), hence, it is mostly used in hypertension and in cardiac conditions such as post-myocardial infarction or congestive heart failure. It is also used in the preservation of kidney function in diabetic nephropathy and in cancer therapy (Lindberg et al., 2004). It has been estimated that more than 75% of the drug is rapidly absorbed in the gastro-intestinal tract and partially metabolized to produce inactive mixed disulfides with endogenous thiol compounds. Both metabolites and unchanged captopril are excreted in the urine. The half-life of captopril in the circulation is about 1.9 h in healthy volunteers (Romankiewicz et al., 1983). Although hepatotoxicity of ACE inhibitors has been rarely reported (Yeung et al., 2003), captopril toxicity, in particular, is even lower (DiBianco, 1986; Sebates et al., 2007). Nevertheless, the effects of captopril on chemical pathology data have not been disclosed. This work was performed to study the effects of captopril on some

*Corresponding author. E-mail: fajoudi@hotmail.com.

Abbreviations: TP, Total protein; **TC,** total cholesterol; **TG,** triglyceride; **AST,** aspartate aminotransferase; **ALT,** alanine aminotransferase; **LDH,** lactate dehydrogenase ; **CK,** creatine kinase.

Table 1. The *In vivo* effects of captopril on the serum biochemical parameters.

Parameters	Control	Pre-treatment	Post-treatment	M1	M2	M3
Glucose mg/dl	93.562 ± 18.402	96.80 ± 19.309	108.80 ± 33.508	NS	NS	NS
TP g/l	75.75 ± 8.434	73.80 ± 9.908	82.70 ± 6.377	NS	*	*
Urea mg/dl	32.6 ± 10.9	30.5 ± 5.9	36.9 ± 11.19	NS	NS	NS
Creatinine mg/dl	1.113 ± 0.255	1.088 ± 0.119	1.336 ±0.276	NS	*	*
TC mg/dl	191.437± 34.298	219.40 ± 51.631	226.10 ± 48.363	NS	NS	*
TG mg/dl	123.06 ± 32.947	196.20 ± 116.331	219.70 ± 153.629	NS	NS	*
AST U/l	22.937 ± 6.884	31.60 ± 9.167	31.50 ± 9.594	*	NS	*
ALT U/l	18.375 ± 6.365	14.30 ± 8.124	17.60 ± 8.487	NS	NS	NS
CK IU/l	130.50 ± 33.091	124.90 ± 26.568	156.30 ± 26.853	NS	NS	*
LDH U/l	131.75 ± 33.914	92.20 ± 33.568	101.10 ± 37.625	*	NS	*

*Significant ($p < 0.05$), NS = non significant ($p > 0.05$).
M1= control with pre-treatment, M2= pre-treatment with post-treatment.
M3= control with post-treatment

common biochemical pathology tests.

PATIENTS AND METHODS

The *in vivo* tests were conducted on the sera of forty patients newly diagnosed with essential hypertension (23 females and 17 males) selected and recruited for the study at Dr. Abdul-Majeed Hospital, Baghdad. To perform the work, ethical approval was granted by the Pharmacy School Heads of Departments meeting in October 2002 and all patients were informed of the objectives of work in advance. The patients' ages ranged from 40 to 63 years and they were all given 25 mg of captopril daily for two weeks. Pre-treatment venous blood samples were collected from each patient and post-treatment blood samples were collected two weeks after the beginning of the treatment. Serum was obtained after centrifugation of the clotted blood. Serum samples from thirty healthy subjects were used as controls. Control subjects were with no evidence of hypertension and diabetes mellitus and with ages (ranging from 35 to 67 years) and genders selected to match those of the test subjects. The biochemical parameters measured included glucose, total protein (TP), urea, creatinine, total cholesterol (TC), triglyceride (TG), aspartate aminotransferase (AST), alanine aminotransferase (ALT), lactate dehydrogenase (LDH) and creatine kinase (CK). These parameters were measured by using Randox kits (Randox Laboratories Ltd., United Kingdom) and read by spectrophotometry.

Statistical analysis

The results were expressed as the mean □ SD. A student T-test was used to examine the difference in the mean of the parameters tested. The p value of less than 0.05 was considered significant. ANOVA was used to evaluate the effects of different concentrations of the drug on the biochemical parameters.

RESULTS

The *in-vivo* effects of captopril on the biochemical parameters

Following captopril therapy, considerable increases in the readings of all biochemical parameters were obtained.

Although all test parameters of the post–treatment sera were raised compared with both the pre-treatment samples and the control samples, the rise of the TG reading was more sharp than the others. Moreover, the elevations in the concentrations of TP and of creatinine were significant in comparison with both the control and pre-treatment sera (Table 1).

The effect of captopril on enzymatic activities

The enzymatic activities were also affected by the captopril treatment. The AST levels were mildly increased in both pre- and post-treatment samples in comparison with the control subjects. The CK activity levels were increased in post-treatment samples, yet they were decreased in the pre-treatment samples. The LDH activity levels were reduced in both pre- and post-treatment sera in comparison with samples of the control subjects (Table 1).

DISCUSSION

Although the growth in therapeutic drugs has resulted in considerable progress in the treatment of diseases, side effects have been reported in many of them, including drug interactions (Young et al., 1972). Since anti-hypertension drugs are long life drugs, then, influences of drug therapy on laboratory results need monitoring to avoid more serious consequences and occasionally, erroneous therapeutic decisions (Hansten, 1979).

In a recent study, captopril was found to exert some chemical or physical interferences in measurements of some serum analytes *in vitro* and it also suggested the necessity of performing this *in vivo* work (Ibrahim and Al-Joudi, 2009).

Interference by endogenous or exogenous substances with assay techniques is a common problem in the clinical laboratory. However, increases in creatinine and urea levels with captopril therapy may be due to biological effects, as reported previously (Cirillo et al., 1988; Hirakata et al. 1984; Salway 1998; Siest et al., 1988; USP drug information for health care professional, 2001). Total protein was increased with captopril and this effect is not likely to be of metabolic origin, since it would probably take a longer time and justifiable biological stimulation to increase the protein content. So it is more likely to attribute that increase to physical or chemical interferences. The levels of TC and TG were significantly altered by captopril therapy, while captopril has been reported to improve endothelium-dependent coronary vasodilatation in patients with atherosclerosis (Fogari and Zoppi, 1999; Oskarson and Heistad, 1997), implying that TC and TG increases may be due to mobilization of fats deposited in endothelia of arteries.

The results presented in this study demonstrated clearly that some laboratory findings are altered the drug. The *in vivo* alterations in the reading of concentrations of metabolites and enzymes may be, in part, due to pharmacodynamic activities besides physical and chemical factors. However, the impact of these various factors cannot be measured separately from this current work. When drugs are detected as interfering with laboratory results, the test may have to be repeated or the laboratory personnel may need to perform special procedures to eliminate or adjust the interference. Furthermore and among other unwanted findings, the effects on the compositions of body fluids are likely to be more apparent when large doses of a drug are administered for a long-term than administration with a single dose (Ibrahim and Al-Joudi, 2009).

Although many drug-related problems come up unexpectedly and can not be predicted, a few problems are related to known pharmacological actions of the drugs and can reasonably be anticipated.

Conclusions

As drug therapy becomes more complex and as the number of individuals being treated with two or more drugs simultaneously, may increase, the ability to predict the magnitude of a specific action of any given drug diminishes. These circumstances point at the need, not only for keeping medication records for patients, but also for closer monitoring and supervision of drug therapy so that unanticipated problems can be prevented, or detected at an early stage in their development. Furthermore, alterations due to drug interference must be taken into account by physicians to avoid misinterpretation of laboratory data.

ACKNOWLEDGEMENTS

Thanks are due to all the participants in this work, including patients, Dar Al-Dawa Pharmaceuticals-Jordan, in addition to the School of Pharmacy, Baghdad University, for allowing this work to be carried out.

REFERENCES

Chrysant SG, Bal IS, Johnson B, McPherson M (1985). A comparative study of captopril and enalapril in patients with severe hypertension. J. Clin. Pharmacol. 25(2):149-151.

Cirillo VJ, Gomez HJ, Salonen J, Salonen R, Rissanen V, Bolognese JA, Nyberg R, Kristianson K. Lisinopril (1988). dose-peak effect relationship in essential hypertension. Br. J. Clin. Pharmacol. 25(5): 533-538.

Cushman DW, Cheung HS, Sobo EF, Ondetti MA (1978). Design of new antihypertensive drugs: potent and specific inhibitors of angiotensin-converting enzyme. Progr. Cardiovas. Dis. 21(3): 176-182.

DiBianco R (1986). Adverse reactions with angiotensin converting enzyme (ACE) inhibitors. Med. Toxicol. 1(2):122-41.

Fogari R, Zoppi A, Corradi L, Preti P, Mugellini A, Lusardi P (1999). Beta blockers effects on plasma lipids during prolonged treatment of hypertensive patients with hypercholesterolemia. J. Cardiovasc. Pahrmacol. 33(4): 534-539.

Hansten PD (1979). Drug Interactions. Lea & Febiger, Philadelphia.

Hauger Kleven JH (1985). Captopril in hypertensive crisis. Lancet. 28; 2(8457):732-733.

Hirakata H, Onoyama K, Iseki K, Kumagai H, Fujimi S, Omae T (1985). Worsening of anemia induced by long-term use of captopril in hemodialysis patients. Am. J. Nephrol. 4(6): 355-360.

Hymes LC, Warshow BL (1983). Captopril long term treatment of hypertension in preterm in fact and in older children. Am. J. Dis. Child. 137(3): 263-266.

Ibrahim AI, Al-Joudi FS (2009). The angiotensin-converting enzyme inhibitor, captopril, alters some biochemical laboratory parameters *in vitro*. Malaysian J. of Biochem. Mol. Biol. 17 (1): 20-22.

Lindberg H, Nielsen D, Jensen BV, Eriksen J, Skovsgaard T (2004). Angiotensin converting enzyme inhibitors for cancer treatment. Acta Oncol. 43(2):142-152.

Matchar DB, McCrory DC, Orlando LA, Patel MR, Patel UD, Patwardhan MB, Powers B, Samsa GP, Gray (2008). Systematic review: comparative effectiveness of angiotensin-converting enzyme inhibitors and angiotensin II receptor blockers for treating essential hypertension. Ann. Intern. Med. 1;148(1):16-29.

Oskarson H, Heistad D (1997). Oxidative stress produced by angiotensin II implications for hypertension and vascular injury. Circulation. 95(3): 557-559.

Romankiewicz JA, Brogden RN, Heel RC, Speight TM, Avery GS (1983). Captopril: an update review of its pharmacological properties and therapeutic efficacy in congestive heart failure. Drugs. 25(1): 6-40.

Salway JG (1998). Drug-test interactions handbook. 2nd ed., Chapman and Hall medical, London.

Schmidt M, Giesen-Crouse EM, Krieger JP, Welsch C, Imbs JL (1986). Effect of angiotensin converting enzyme inhibitors on the vasoconstrictor action of Angiotensin I on isolated rat kidney. J. Cardiovasc. Pharmacol. 8, Suppl 10:S100-105.

Sebates M, Ibanez L, Perez E, Vidal X, Buti M, Xiol X, Mas A, Guarner C, Forné M, Solà R, Castellote J, Rigau J, Laporte JR (2007). Risk of acute liver injury associated with the use of drugs. Aliment. Pharmacol. Ther. 26(11-12):1543-1544.

Siest G, Galteau MM, Malya PA, Tryding N, Delwaide P, Salway JG, Tognoni G (1988). International Federation of Clinical Chemistry, Scientific Committee: Drug interferences and drug effects in clinical chemistry. Part 3. Evaluation of biological effects of drugs. J. Clin. Chem. Clin. Biochem. 26(3):169-173.

Sleight P (2001). The role of angiotensin-converting enzyme inhibitors in the treatment of hypertension. Curr. Cardiol. Rep. 3(6): 511-518.

USP drug information for health care professional, 23rd ed. 2001, Thompson Micromedex, Taunton MA. USA.

Yeung E, Wong FS, Wanless IR, Shiota K, Guindi M, Joshi S, Gardiner G (2003). Ramipril associated hepatoroxicity. Arch. Pathol. Lab. Med. 127(11):1493-1497.

Young DS, Thomas DW, Friedman RB Pestaner LC (1972. Effects of drugs on clinical laboratory tests. Clin. Chem. 18(10): 1041-1303.

Zanchetti A, Parati G, Malacco E (2006). Zofenopril plus hydrochlorothiazide: Combination therapy for the treatment of mild to moderate hypertension. Drugs. 66 (8):1107-1115.

14

Effect of traditionally designed nutraceutical on stress induced immunoglobulin changes at Antarctica

P. Bansal[1]*, R. Sannd[2], N. Srikanth[3] and G. S. Lavekar[4]

[1]Clinical Biochemistry Laboratory, National Institute of Ayurvedic Pharmaceutical Research, Moti Bagh Road, Patiala, Punjab, India.
[2]National Institute of Ayurvedic Pharmaceutical Research, Moti Bagh Road, Patiala, Punjab, India.
[3]Central Council for Research in Ayurveda and Siddha, Ministry of Health and Family Welfare, Government of India, New Delhi, India.
[4]Central Council for Research in Ayurveda and Siddha, Department of AYUSH, Ministry of Health and Family Welfare, Government of India, New Delhi, India.

This study was conducted to establish the effect of a traditionally designed nutraceutical on stress related changes in selected immunoglobulin levels in the body. The nutraceutical was prepared from different potent herbs described in Ayurveda using standard operative procedures and were tested for heavy metal and microbial load. Initially, 21 subjects were selected in addition to 7 volunteers for control group who did not consume nutraceutical. Sampling was done at zero days and at fortnightly intervals. The levels of selected immunoglobulin IgG, IgA and IgM were estimated with turbidity metric immunoassay at different time intervals. The concentration of immunoglobulin IgA was 146±15.96 at zero day stage. The levels of these immunoglobulins were lower at all stages as compared to the concentration at zero day in trial group subjects whereas the concentration was significantly higher (t stat.>t critic. at $p<0.05$) in control group subjects. The concentration of IgG was very high to the tune of 3091±705 at zero day stage. The level of IgG was lower in trial subjects as compared to control subjects at all stages except at the 6th week stage where it was higher in trial subjects. Concentration of immunoglobulin IgM was 80.75±30.39 (t stat.>t critic. at $p<0.05$) at zero day followed by a decrease in both groups at the 2nd week, however the concentration was almost $1/3^{rd}$ in trial drug subjects as compared to the levels in control subjects followed by an abrupt increase at the 4th week. The levels increased to 106±8.94 at the 4th week stage and 115±9.35 at the 6th week stage in control subjects (even higher than at zero day) whereas the values were 46.15±11.39 and 55.38±15.34 (t stat.>t critic. at $p<0.05$) at respective stages in trial drug subjects. On the whole the pattern of fall and rise in levels of IgM were similar in the control as well as treatment group subjects at all stages. Studies revealed that the components of the nutraceutical tended to exert significant (t stat.>t critic. at $p<0.05$) anti-stress effect against stress related changes in immunoglobulin in the body due to the battery of stresses encountered at Antarctica.

Key words: *Rasayana*, immunoglobulins, *Withania somnifera*, *Tinospora cordifolia*, *Chlorophytum arundinaceum*, *Piper longum*, *Prunus amygdalus*, Antarctica, stress.

INTRODUCTION

An Antarctic expeditioner has to face and cope up with many physical and psychological stresses apart from the stress related to UV radiations, magnetic field, high wind velocity, extreme cold conditions, circadian biorhythms (Sundaresan et al., 1999) and chemicals. Men living in Antarctica also suffer significant emotional strain as a result of physical isolation and social deprivation (Roy and Deb, 1999) in addition to physiological stress imposed by few factors mentioned earlier. These have been reported to effect immunoglobulin levels also. All these changes might be due to the oxidant stress, which is enhanced by the stressors at the Antarctic region.

*Corresponding author. E-mail: bansal66@yahoo.com.

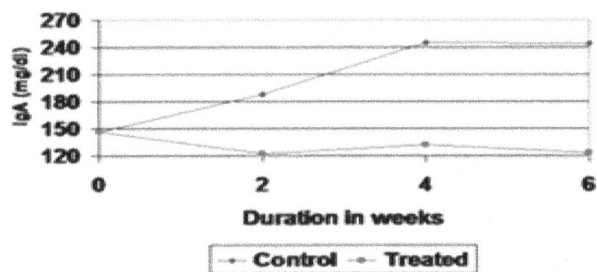

Figure 1. Effect of APY and APP on IgA.

Oxidative stress results from an imbalance in this pro-oxidant antioxidant equilibrium in favour of pro-oxidants. Oxidative stress also leads to some detrimental effects on the immune system. Attempts have been made to study the mechanisms that might be involved in man's physiological responses to polar conditions. Few of them have helped in evaluating the nutritional requirements of men working in Polar Regions.

Ayurveda, the first recorded scientific medicine in the history of the world aims to improve the quality and span of life with its major emphasis on prevention of disease and promotion of health by strengthening tissues so that they can withstand exogenous and endogenous stressors. This is achieved by modulating diet and life style as well as by the appropriate use of drugs that restore the equilibrium of the body (Dahanukar and Thatte, 1989a). Over 600 plants have been described in various Ayurvedic texts like *Charak samhita, Sushruta samhita* and various *Nighantu*. There are interesting groups of rasayana herbs that have adaptogenic properties (Dahanukar and Thatte, 1989b). According to Brekhman and Dardymov (1969), an adaptogen must produce a non-specific response that is, increase the power of resistance against multiple stressors. It should have a normalizing influence and be innocuous but should not influence normal body functions more than required. The word "rasayana" means the path that "rasa" takes (rasa-plasma; ayana-path). It is believed that in Ayurveda, the qualities of "rasa dhatu" influence the health of other "dhatu's" (tissues) of the body (Dahanukar and Thatte, 1989b). Hence any medicine that improves the quality of rasa should strengthen or promote the health of all tissues of the body. A significant part of Ayurvedic therapeutics is preventive in nature. This is the concept of "vyadhirodhak chamatav" that is, the capacity of the body to resist disease. So obviously the immune system as recognized in modern biology, which provides protection against microbes, should be a part of it.

Although there is a provision of very good nutritional diet and pills containing vitamin C and vitamin E pills to the members of the expedition, yet the scientific field lacks the data related to nutritional requirements in the Antarctic climate. At the same time Paul Coates who works in the office of dietary supplements at NIH says that "just because a food with certain compounds in it is

beneficial to health does not mean a pill containing the same compound is". Keeping the above view in mind, a regimen of various rasayana herbs including Ashwagandha (*Withania somnifera*), Guduchi (*Tinospora cordifolia*), Safed musli (*Chlorophytum Arundinaceum*), Pippali (*Piper longum*), Badam (*Prunus Amygdalus*) and some other herbs were prepared and a study was conducted to evaluate the effect of rasayana as a drink and food supplement on prevention of hazards of cold climate on the immune system in Antarctic climatic conditions.

MATERIALS AND METHODS

The study was conducted at Antarctica on Indian station "Maitri" on voluntary accepted subjects out of the 23rd Indian scientific expedition to Antarctica. There were a total of 50 team members out of which twenty one members opted for supporting this study. Initially all the members were briefed about the scope of the project and were given project proformas for their willingness to join the study as per international ethical guidelines. On day zero, blood sampling was conducted and was analysed for biochemical parameters. All estimations were done using standard methods.

Sampling was followed by feeding of subjects with the drug/food supplement "ayush poshak peye (APP)" and "ayush poshak yoga (APY)". Both drugs are coded drugs and were prepared at CCRAS head office, Delhi and supplied to the team visiting Antarctica. The standard operative procedures (SOP) for the preparation of the trial food supplements are under patent process hence can not be revealed at this stage. The trial drug was tested for heavy metals and it was free from cadmium, lead and arsenic. The food supplement was also tested for microbial load and total bacterial count was 7.3×10^4 CFU/g whereas total fungal count, enterobacteria count and *salmonella* spp. was within permissible limits.

All the subjects were provided with the food supplements daily. The "ayush poshak peye" was prepared daily at various timings depending on the convenience of members because they used to be busy with their own scientific projects. The investigators made it pertinent that all the subjects are dispensed with APP (125 ml) and APY (50 g) once daily. The subjects who could not take the drug regularly were put under dropouts. Out of 21 subjects, 8 were dropped out due to different reasons. Three of them went to ship due to some assignment and could not continue taking the drugs. Three of them had some physical problem and discontinued the drug. Two subjects who missed the food supplements continuously for more than two days were dropped out.

In addition to 21 subjects, a control group consisting of 7 subjects was maintained throughout the study period. These subjects were also team members, but were not fed with the supplement. The sampling for all the parameters was done at fortnightly intervals for up to 6 weeks. The effect of the food supplements was assessed from the levels of IgG, IgA and IgM. Immunoglobulins were estimated with turbidity metric immunoassay by using kits from Tulip Diagnostics Pvt. Ltd. A standard curve was prepared for all the immunoglobulins under study using different concentrations of the standard known controls from the same batch of the company for calibration and calculations.

RESULTS

The concentration of immunoglobulin IgA was 146±15.96 at zero day stage (Figure 1). The levels increased continuously at 2nd and 4th week stages in the control group. The level of IgA was similar at the 4th and 6th weeks in the

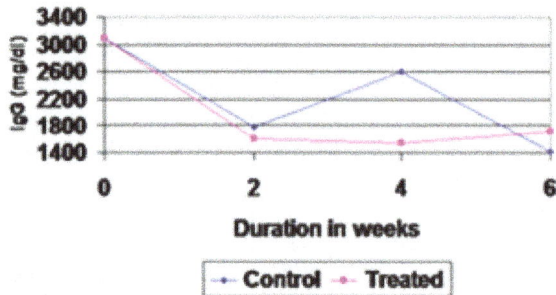

Figure 2. Effect of APY and APP on IgG.

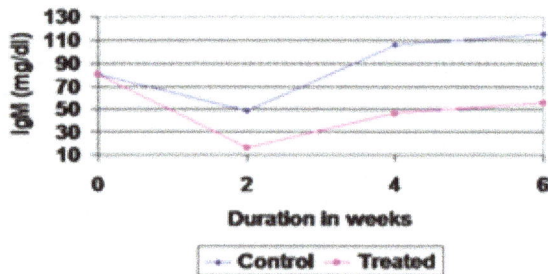

Figure 3. Effect of APY and APP on IgM.

control group. In contrast to it, the levels decreased in the trial drug group subjects at the 2nd week stage but depicted slight increase at the 4th and 6th week as compared to levels at the 2nd week stage. On the whole the levels of these immunoglobulins were lower at all stages as compared to concentrations at zero day in the trial group subjects whereas the concentration was higher in the control group subjects (t stat.>t critic. at $p<0.05$).

The concentration of IgG was very high to the tune of 3091 ± 705 at the zero day stage (Figure 2). It was lower at all stages in all the subjects irrespective of the trial drug and control as compared to levels at zero day. The level was decreased at the 2nd week, increased at the 4th week and again decreased at the 6th week stage in the control group subjects. The level of IgG was lower in trial subjects as compared to control subjects at all stages except at the 6th week stage where it was higher in trial subjects (t stat.>t critic. at $p<0.05$). The concentration of IgG decreased till the 4th week stage and slightly increased at the 6th week stage in trial drug subjects.

Concentration of immunoglobulin IgM was 80.75 ± 30.39 at the zero day stage (Figure 3). At the 2nd week stage, the level decreased in both groups, however the concentration was almost $1/3^{rd}$ in trial drug subjects as compared to the levels in control subjects. At the 4th week stage the level increased abruptly in subjects. The levels increased to 106 ± 8.94 at the 4th week stage and 115 ± 9.35 at the 6th week stage in control subjects (even higher than at zero day stage) whereas the values were 46.15 ± 11.39 and 55.38 ± 15.34 at respective stages in trial drug subjects (t stat.>t critic. at $p<0.05$). On the whole the

pattern of fall and rise in levels of IgM were similar in control as well as treatment group subjects at all the stages.

DISCUSSION

The concentration of secretory immunoglobulin IgA increased till the 4th week stage and was maintained at same level till the 6th week stage in the control group however the levels decreased or were more or less equal to zero day values in the trial drug subjects. This depicts development of tolerance in treatment group subjects in stress conditions.

Levels of IgG decreased at the 2nd week stage in both groups indicating stress related changes in the 2nd week stage of Antarctic stay. The levels in the trial group subjects were still lower as compared to the control group indicating better resistance developed in the body as an adaptive mechanism to stress. At the 4th week stage, the levels in the control group were raised significantly indicating the effect of stress on the body; however in the trial group the lower level depicted more tolerance of the body. At the 6th week stage, the levels of this immunoglobulin decreased in the control group even lower as compared to the trial drug group probably due to development of resistance of the body due to adaptation and acclimatization to the environment; however the values were almost comparable at the 2nd, 4th and 6th week stages in the trial drug group subjects.

Concentration of IgM was decreased at the 2nd week stage in both groups as compared to zero day values. This shows that the body became less prone to infections or it could be the effect of the bacteria free environment of Antarctica on the immune system. The concentration of immunoglobulin increased gradually till the 6th week stage in both groups; however the values were significantly lower in the treatment group as compared to the control group probably due to better development of resistance in the trial drug subjects against stress induced changes in immunoglobulin levels.

Amongst many medical problems immuno-suppression has been observed during Antarctica sojourn (Muller et al., 1995a; Muller et al., 1995b). There is a wide body of literature demonstrating the detrimental effects of stress on the immune system (Tingate et al., 1997; Boneau, 1990; Khansari et al., 1990). Antarctica expedition team members encounter stress, which is thought to influence immune function through the autonomic nervous system innervating lymphoid cells (Rivolier et al., 1988, Singh et al., 2004). The stress related changes in the level of these immunoglobulins in this study are in accordance with earlier Antarctic studies in which Muchmore et al. (1974) found a decline in serum IgG and IgM, whereas Tashpulatov (1974) observed a rise in serum IgG and IgM levels in Antarctic workers. However the lower level of immunoglobulins in the trial group subjects at almost all the stages in this study demonstrates that a better adaptive mechanism was adopted by these subjects in relation to stress.

A number of *Rasayana* drugs have been elaborated in *Materia Medica* of Ayurveda for the enhancement of the body's resistance. Some rasayana plants are said to prevent aging, re-establish youth, strengthen life and brain power and prevent disease (Sharma, 1983; Ghanikar, 1981) all of which imply that they increase the resistance of the body against any onslaught. The food supplement consisted of a number of herbs so its anti-stress effect on stress-induced changes in immuno-globulin levels cannot be pinpointed but can be hypo-thesized. As *Amygdalus prunus* (Badam) is a rich source of natural vitamin E and has been part of the food supplement, it would have supplied vitamin E in sufficient quantity to meet the requirements of the body (Chopra, 2004). Many of the adaptogen rasayana have already been shown to possess immuno-stimulant activity (Wagner, 1994). Whether this immuno-stimulation is res-ponsible for the adaptogenic potential is not clear till date. As mentioned earlier during stress, prostaglandin and antioxidant systems of target organs serve as natural defence mechanisms to counteract the deleterious effect of stress-induced lipid peroxidation in the target organ. In case of its depletion, stress manifestations occur; how-ever prevention of their depletion may be a mechanism for counteracting stress and the beginning of the adapta-tion process (Meerson, 1994). *Tinospora cordifolia* has also been known to show remarkable immunostimulant and immunomodulatory properties by many workers (Sainis et al., 1998, Deshmukh and Usha, 2002, Dubey et al., 2002). In a clinical study of three months duration guduchi and ashwagandha (*Withania somnifera*) were found as antioxidants. Guduchi was found to be a more effective natural antioxidant over others (De and Tripathi, 1996). Guduchi is known to be a rich source of trace elements (zinc and copper) which act as antioxidants and protects cells from the damaging effects of oxygen radi-cals generated during immune activation (Shankar and Prasad, 1998). Phenyl propene diasaccharides (Cordifo-lioside A and B) have been identified as active principles for immunostimulant action of the herb (Salil, 1997). In another study a combination of rasayana herbs including Guduchi and Ashwagandha have been shown to in-crease cellular and humoral components of immunity (Chatterjee and Das, 1996). Guduchi has been shown to enhance cellular and humoral immunity (Kapil and Sharma, 1997). Pippali has also been hypothesized to improve the immune status of patients (Abbas and Pandey, 1997). Pippali as a part of a polyherbal formula-tion has also been shown to have antistress and adapto-genic effects (Ramachandran et al., 1990).

Ashwagandha has been shown to increase non-specific general immunity in children (Parbhakar et al., 1994). It contains withanolides that have been reported to possess both immunosupression and immunostimulatory proper-ties of activating immune responses in various test

studies (Uniyal, 2002). Findings of another study indicate that increase in oxidative free radical scavenging activity of WSG may be responsible for the antistress immuno-modulatory and antiaging effects (Salil, 1997). These similar types of trends have been found in the present study. Withania as a part of a polyherbal formulation has been shown to have antistress and adaptogenic effects (Ramachandran et al., 1990).

The role of antioxidants in the immune system is multi-faceted and well known as they can serve to either suppress or enhance immune response. Depending on the desired response, different antioxidants can play diff-erent roles in balancing immune response effectively in an individual. In fact many immune cells such as phago-cytes produce free radicals in order to destroy an antigen. If an individual suffers tissue damage and at the same time due to infectious phagocytes are releasing free radi-cals to destroy an antigen, the Fenton reaction will produce a hydroxyl radical. In order to clear these oxi-dants there must be a balance for optimal response. Too many antioxidants in our diet would limit the immune cells while not enough would cause phagocytes to produce so many radicals, that they would kill themselves and again be ineffective against any antigen present. In the present study, there was stress on all the subjects and free radicals were probably generated which is also evidently reflected from the elevated serum MDA levels studied as part of the effect of the drug on the antioxidant system of the body (Bansal et al., 2007). However due to food supplements in the treatment group, the level of anti-oxidants was enough to give resistance to the body. In the treatment group probably due to more resistance, immunoglobulin levels are less; however in the control group, due to more production of free radicals, anti-oxidant level decreased and could not provide enough resistance to antigens and therefore the level of immuno-globulins were more. In the present study, the level of free radicals as demonstrated by higher MDA levels, vitamin E level and level of Immunoglobulins seem to be part of the same tale. According to a study in humans, whenever you remove antioxidants from the system, cellular integrity is hampered and the immune response is decreased and after repletion, as vitamin E levels increase, the immune response returns to have enhanced killing power. It has also been observed that vitamin E may normalize immune abnormalities in mice with murine retrovirus (Wang et al., 1994). The higher levels of vitamin E in trial group subjects also support the levels of immunoglobulin in the present study (Bansal et al., 2007).

The rasayana herbs seem to exert their effect through immunosuppressant, immunostimulant and immuno-adjuvant activities or by affecting the effector arm of the immune response. It has been found that the nervous, endocrine and immune systems are all interrelated. Imm-une products like various cytokines have been found to stimulate the hypothalamus-pituitary-adrenal axis and corticotrophin release factor (CRF), which ultimately enhances the production of adrenal corticotrophic hormone (ACTH) resulting into increased secretion of glucocorticoids which have an overall suppressive effect on the immune system. Stress also acts on the same axis and brings about changes in the immune status of the body. These rasayana drugs probably reduce stress levels by affecting antioxidant levels. So these rasayana drugs act as potent antioxidants and neuroendocrine immuno-modulators (Sehrawat et al., 2007).

Conclusion

Studies carried out on regimen prepared from different rasayana herbs described in Ayurveda revealed that the components of the drug tended to exert significant anti-stress effect against a battery of stresses encountered at Antarctica. The mechanism of action of these rasayana seems to be through its antioxidant effect. From the results obtained it may be concluded that the drug seems to have a potential adaptogenic antistress effect.

ACKNOWLEDGEMENTS

The authors acknowledge financial support from the Department of AYUSH, Ministry of Health and Family Welfare, Government of India. The authors are also grateful to NCAOR, Goa, Department of Ocean Development, for extending an opportunity to conduct this trial at Antarctica. The authors also extend their sincere thanks to all the subjects who participated in this trial without who it would have been difficult to even think of.

REFERENCES

Abbas SS, Pandey KC (1997). A double blind placebo controlled clinical trial of pippali rasayana in cases of giardiasis. Antiseptic 94(8): 250-254.

Bansal P, Sannd R, Srikanth N, Lavekar GS (2007). Antioxidant activity of coded neutraceutical rasayana products in first clinical trial at Antarctica. J. Tropical. Med. Plants. 8(2): 178-182.

Brekhman I, Dardymov IV (1969). New substances of plant origin which increase non-specific resistance. Ann. Rev. Pharmacol. 9: 419-430.

Boneau RH, Kielcoltglase JK, Glaser R (1990). Stress-induced Modulation of the Immune Response. Ann. NY Acad. Sci. pp. 253-269.

Chatterjee S, Das SN (1996). Immunopotentiating effect of proimmu, a polyherbal formulation. Indian J. Pharmacol. 28(1): 58.

Chopra HK (2004). Almonds a panacea. Longevity and Health (April) p. 17.

Dahanukar SA, Thatte UM (1989a). Therapeutic approaches in Ayurveda Revisited, Popular Prakashan, Mumbai, pp. 74-130.

Dahanukar SA, Thatte UM (1989b). Therapeutic approaches in Ayurveda Revisited, Poppular Prakashan, Mumbai, pp. 109-110.

De RK, Tripathi PC (1996). Role of certain indigenous drugs antioxidants in aging. International seminar on free radical mediated diseases and Ayurveda, IMS BHU, Varanasi, 2-4 Sept. 1996.

Deshmukh A, Usha D (1994). In vitro effect of Tinospora cordifolia on PMN function. Update Ayurveda-94, Bombay, India 24th-26th feb. pp. 63.

Dubey GP, Aggarwal A, Juneja G, Dixit SP (2002). Immunosenescence in elderly- prevention and management strategies. In Proceedings of Conference on Rasayana 25th-26th march. pp. 170-179.

Ghanikar BG (1981). Sushrut Samhita Sutra Sthana, Motilal Banarsi Das, Varanasi, 1/7: 3.

Kapil A, Sharma S (1997). Immunopotentiating compound from Tinospora cordifolia. J. Ethnopharmacol. 58:89.

Khansari DN, Murgo AJ, Faith R (1990). Effect of stress on immune system. Immunology Today 11(5): 170-175.

Meerson FZ (1984). Antioxidant factors in the organism a system of natural prevention of stress induced and ischaemic injuries. In Adaptation, stress and prophylaxis. Springer verlag, Berlin. pp. 242-294.

Morley JE (1981). The Endocrinology of the Opiates and the Opioid Peptides. Metabolism 30: 195-209.

Muchmore HG, Tatem BA, Worley RA, Shurley JT (1974). In E. Scott, O.G. Edholm, and E.K.E., Gunderson (Eds) Immunoglobulins during south polar isolation. pp. 135-140.

Muller HK, Lugg DJ, Ursin H, Quinn D, Donovan K (1995a). Immune response during an Antarctic summer. Pathology. 27:186-190.

Muller HK, Lugg, DJ, Quinn D (1995b). Cell mediated immunity in Antarctic wintering personnel. Immunol. Cell. Biol. 73 (4): 316-320.

Parbhakar C, Koshy V, Menon S, Viadyanathan V (1994). Effect of withania somnifera, asparagus racemosus and lakshadi thalim in improving the physical and mental health of school children. Medicinal .Nutri. Res. Communications 2:15-18.

Ramachandran U, Divekar HM, Grover SK, Srivastava KK (1990) New experimental model for the evaluation of adaptogenic products. J. Ethnopharmacol 29: 275-281.

Rivolier J, Goldsmith R, Lugg DJ, Taylor AJW (1988), Man in the Antarctic. Taylor and Grancis, London, New York Philadelphia, pp. 121.

Roy DD, Deb NC (1999). Role stress profiles of scientists and defense personnel in 15th Antarctic Expedition. 15th IEA scientific report 13 pp. 371-375.

Sharma P (1983). Charak Samhita Chikitsa Sthan, Chowkhamba Orientalis, Varanasi.; Ch 6(7-8):3-4.

Sainis KB, Ramakrishana R, Sumariwala PF, Sipahimilani AT, Chintalwar GJ, Banerji A (1998). Immunomodulatory effects of Tinospora cordifolia. Update Ayurveda, 98, Mumbai, India. L, 3.1, 11-14, Feb, 1998.

Salil KB (1997). Antioxidant activity of glycowithanolides from withania somnifera. Ind. J. Experimental. Biol. 35(3): 236-239.

Sehrawat D, Kaur J, Bansal P, Bikshapathi T (2007). Immunopotentiating and immunoprophylactic activities and probable mechanism of action of rasayana drugs in Indian system of medicine. In: Abstract book 33rd Indian Immunology Society Conference on Molecular and Clinical Immunology in Health and Disease on 28-31, Jan 2007, AIIMS, New Delhi.

Shankar AH, Prasad AS (1998). Zinc and immune function the biological basis of an altered resistance to infection. Amer. J. Clin. Nutr. 68 (2):447.

Singh AK, Kaushik V, Mathew TL (2004). Effect of cold, isolation and ultraviolet radiation stress on human immune system. 19th IEA scientific report 17: 239-246.

Sundaresan G, Laxmikanth A, Sachdeva U (1999). Exposure to Antarctica summer affects the immunoglobulin rhythm but not the oral temperature, 15th IEA scientific report. pp. 353-359.

Tashpulatov RIU (1974). Microbial flora and human immunity in Antarctica. Antarktika; Doklady Komissi, 13: 183-189.

Tingate TR, Lugg DJ, Mullar HK, Strowe RP, Pierson DL (1997). Antarctic isolation Immune and viral studies. Immunol. Cell. Biol. 75: 275-283.

Uniyal MR (2002). Effective ayurvedic medicinal plants used in rasayana therapy. In proceedings of Ayurvedic conference on Rasayana 25-26 March, RAV, Delhi. pp. 180-187.

Wagner H, Norr H, Winterhoff H (1994). Plant adaptogens. Phytomedicine. 1: 63-76.

Wang Y, Huang DS, Liang B, Watson RR, 1994 Nutritional status and immune responses in mice with murine AIDS are normalized by vitamin E supplementation. J. Nutr. 124: 2024-2032.

15

In vitro biodegradation of keratin by dermatophytes and some soil keratinophiles

Mukesh Sharma*, Meenakshi Sharma and Vijay Mohan Rao

Microbiology Laboratory, Department of Botany, University of Rajasthan, Jaipur-302004, India.

The present investigation was aimed to evaluate the *in vitro* biodegradation of keratin by clinical isolates of dermatophytes and soil fungi. Ten fungal species, out of which, six (*Chrysosporium indicum, Trichophyton mentagrophytes, Scopulariopsis* sp., *Aspergillus terreus, Microsporum gypseum* and *Fusarium oxysporum*) were isolated from soil and four clinical (*Trichophyton rubrum, Trichophyton verrucosum, Trichophyton tonsurans* and *Microsporum fulvum*) were obtained from human skin. The isolates were tested for their keratin degradation ability on human and animal (cow and buffalo) hair baits. The rate of keratin degradation was expressed as weight loss over three weeks of incubation. Human hair had the highest rate of keratin degradation (56.66%) by colonization of *C. indicum.* whereas *M. gypseum* and *T. verrucosum* were highly degraded (49.34%) to animal hairs. There was a significant difference (p < 0.05) in keratin substrate degradation rates by the examined fungi. Human hair served as an excellent source for the biodegradation of keratin by the isolated test fungi as compared to animal hair. Releasing protein showed maceration of the keratin substrates by the test fungi. The present study reveals that, the isolated test fungi play a significant impact on biodegradation of keratin substrates for betterment of environmental hazards.

Key words: Fungal species, keratin substrates, keratin degradation, released protein, environmental hazards.

INTRODUCTION

Keratin is a major component of hair, feathers and wool and is the most complex of the cytoskeletal intermediate filament proteins of epithelial cells (Latkowski and Freedberg, 1999). The durability of keratins is a direct consequence of their complex architecture. In addition to keratin, keratinaceous materials such as skin, hair, nails, hoofs and horns contain a large proportion of non-keratin proteins. A large number of fungi, including yeasts, dermatophytes and other moulds, grow on human skin, hair and nails. The term 'keratinolytic' is used for fungi exhibiting the enzymatic ability to attack and utilize keratin. Degradation of keratin by microorganisms is performed by specific proteases that is, keratinases (Onifade et al., 1998; Wang and Shih, 1999; Gradisiar et al., 2000; Sandali and Brandelli, 2000; Kim et al., 2001; Allpress et al., 2002; Longshaw et al., 2002; Yamamura

et al., 2002; Gessesse et al., 2003; Singh, 2003).

Proteolytic enzymes like trypsin, pepsin and papain are largely produced in the presence of keratinous substrates in the form of hair, feather, wool, nail, horn etc. during their degradation, (Gupta and Ramnani, 2006). The complex mechanism of keratinolysis involves cooperative action of sulfitolytic and proteolytic systems. Keratinases are robust enzymes with a wide temperature and pH activity range and are largely serine or metallo proteases. A distinctive feature of keratin is its relatively high sulfur content due to the presence of sulfur containing amino acids namely cystine, cysteine and methionine. The disulphide bonds considered to be responsible for the stability of keratin and its resistance to enzymatic degradation (Kunert, 1989, 1995). Dermatophytes are often present in skin and invade the keratin tissues. Surveys of kerationophilic fungi from different habitats have indicated that, several species of dermatophytes and non-dermatophytic fungi inhabit soil (Kushwaha, 1983), air (Marchiso et al., 1994) and sewage sludge (Muhsin and Hadi, 2000). The ability of various fungal

---

*Corresponding author. E-mail: mukeshsharma_uniraj@yahoo. co.in.

species to degrade keratin *in vitro* from substrates such as hair (Marchiso et al., 1994; Bahuguna and Kushwaha, 1989; Deshmukh and Agrawal, 1985; Malviya et al., 1993), wool (Safranek and Goos, 1982; Al-Musalam and Radwan, 1990) and feathers (Kushwaha, 1983; El-Naghy et al., 1998; Kaul and Sumbali, 1999) have been examined.

Since soil often receives high amounts of keratin substrates due to human activities, it was of utmost interest to examine the degradation capability of the dominant fungal species in polluted soil for two common keratinaceous substrates. In the present paper, an in vitro degradation of keratin substrates by ten species of fungi isolated from soil and superficial layer of skin of patients infected with dermatomycoses has been studied and discussed.

MATERIALS AND METHODS

Ten fungal species were tested for present investigation in which six species namely *Chrysosporium indicum*, *Trichophyton mentagrophytes*, *Scopulariopsis* sp., *Aspergillus terreus*, *Microsporum gypseum* and *Fusarium oxysporum* were collected from soil and four species namely *Trichophyton rubrum*, *Trichophyton verrucosum*, *Trichophyton tonsurans* and *Microsporum fulvum* were isolated from superficial skin of human dermatomycoses patients in Jaipur (India) during 2006. The method of soil fungal isolation, identification and purification has been described previously (Vanbreuseghem, 1952; Muhsin and Hadi, 2000). The clinical fungi were isolated from dermatomycoses patients in S.M.S. Hospital, Jaipur under the able guidance of skin specialists. The fungal species were grown on Sabouraud Dextrose Agar (SDA) for two weeks. A spore suspension of each species was prepared following the method of Kunert (1989) with slight modification. In this method, a loopful spore material of each fungus was mixed separately with 1 ml of sterilized distilled water under sterilized condition. The keratin degradation by each fungal species was expressed as percentage weight loss. A total of 50 mg keratin such as human and animal hair were cut into fragments (2 cm long), washed and surface sterilized with ethanol (3%).

A buffer solution (pH 6.5) was prepared using 0.04 g of KH_2PO_4 in 100 ml distilled water in sterilized conical flasks (100 ml volume). Each flask contained 20 ml buffer solution and was inoculated with 0.2 ml of a spore suspension of each fungal species to which 50 mg of a selected substrate was added. Three replicate flasks of each fungus per substrate were prepared and incubated at 27°C in shaker incubator at 80 rpm. A control, having the keratin substrates without fungal suspension, was also run along with the test flasks. The fungal cultures were filtered after 1, 2 and 3 weeks of incubation. The keratin fragments were collected on Whatman filter paper No. 1, washed gently to remove fungal hyphae, dried at 75°C for 48 h and weighed. Changes in the pH of the culture medium with incubation time were also determined using pH meter. The protein released into the culture medium due to the keratin substrate degradation was determined (Lowry et al., 1951). During this method different dilutions of BSA solutions were prepared by mixing stock BSA solution (1 mg/ ml) and water in the test tube. The final volume in each of the test tubes is 5 ml. The BSA range is 0.05 to 1 mg/ ml. From these different dilutions, pipette out 0.2 ml protein solution to different test tubes and add 2 ml of alkaline copper sulphate reagent (analytical reagent). Mix the solutions well. This solution was incubated at room temperature for 10 min. Then add 0.2 ml of reagent Folin Ciocalteau solution (reagent solutions) to each tube and incubate for 30 min. Zero the colorimeter with blank

and take the optical density (measure the absorbance) at 660 nm.

Statistical analysis of data

The data collected was analyzed using one-way analysis of variance. The effects were considered significant when p value of ANOVA F-test was < 0.05.

RESULTS

The degradation rate of the human and animal hair substrates by the test fungi are presented in Figure 1. A significant difference ($p < 0.05$) in degradation rates of each keratin substrate by the fungal species was observed (Table 1). Human hair was highly colonized and degraded by *C. indicum* and *M. gypseum* at 56.66 and 49.34% of weight loss, respectively and less degraded by *Scopulariopsis* sp. at 17.34% of weight loss, over 3 weeks of incubation (Figure 1). *M. gypseum* (49.34%) and *T. verrucosum* (49.34%) showed the greatest degradation of animal hair (Table 1). These keratinous substrates exhibited low degradation by *F. oxysporum* (16.66%). Generally, there was an increase in the degradation rate of the keratin substrates, in terms of weight loss, by the fungi during incubation.

The amount of the protein released into the culture medium varied with the fungi and with different sources of keratin (Table 2). The highest amount of protein released in the cultures containing human hair was by *M. gypseum* (68 µg/ml) followed by *M. fulvum* (63 µg/ml). The lowest amount of protein released in culture medium reported by *Scopulariopsis* sp. (17 µg/ml). In cultures containing animal hair the protein released were highest in *T. verrucosum* (66 µg/ml) followed by *T. mentagrophytes* (62 µg/ml) and lowest rate of protein released was in *Scopulariopsis* sp. (13 µg/ml). Changes in pH of the culture medium due to the substrates degradation process were observed (Figures 2a and b). The pH of the control cultures did not change during the incubation periods. Keratin substrate degradation was accompanied by alkalination of the culture medium with elevation of pH from 6.5 to 8.4, when human hair was keratin substrates (Figure 2a) and 6.5 to 8.1 when animal hair was keratinous materials (Figure 2b). Present study showed erosion or perforation of the keratin substrates by the different fungal species.

DISCUSSION

The keratin substrates revealed different rates of keratin degradation with the fungi under investigation. It is generally believed that keratin degradation is due to the enzymatic action of the fungi, which is indicated by the substrate weight loss, as well as by the release of soluble products into the culture medium (Deshmukh and Agrawal, 1985). However, the distinctive feature of

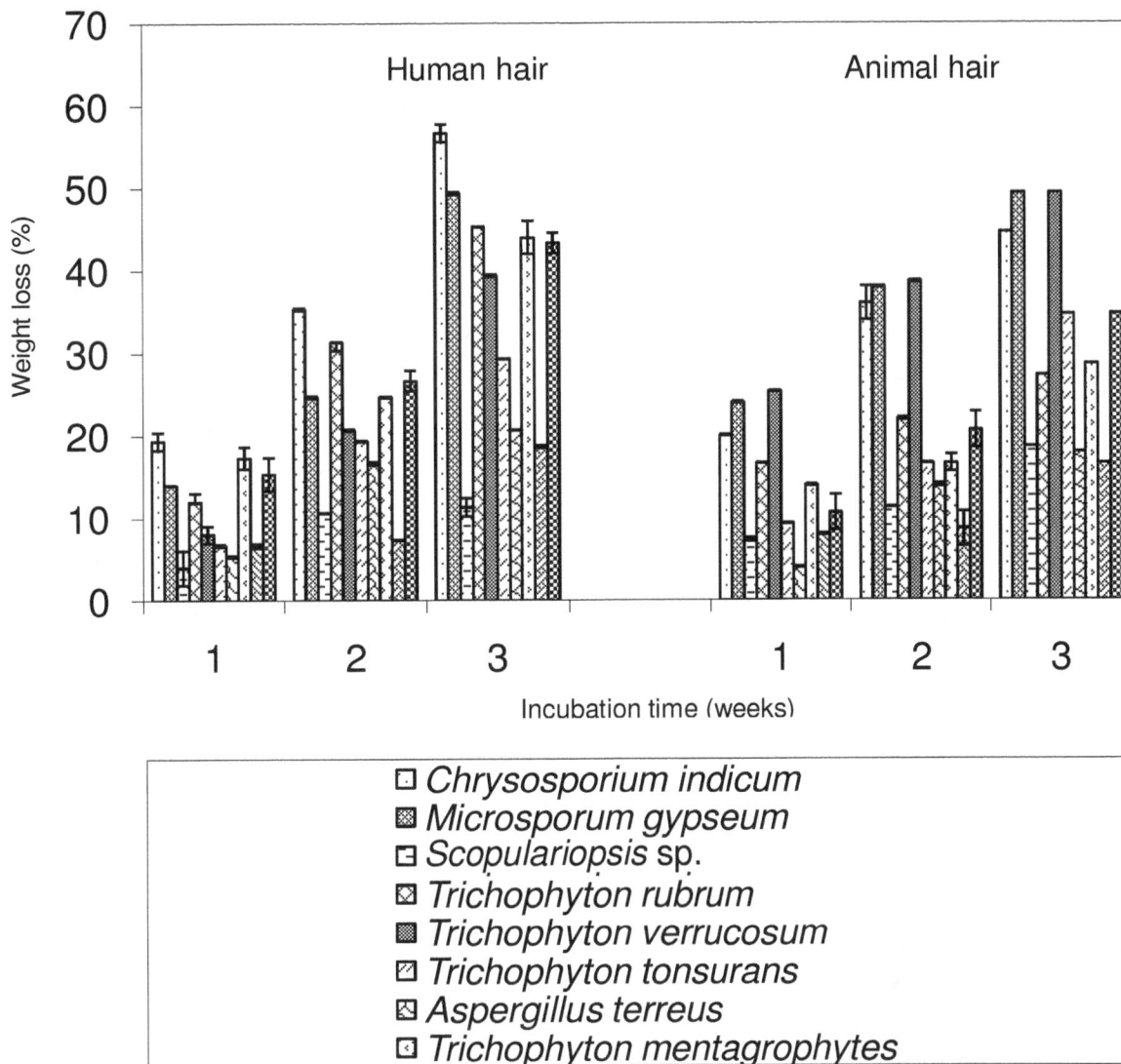

Figure 1. Degradation rate percentages of keratin substrates by fungal species during the incubation periods.

Table 1. Weight remained of keratin substrates by tested fungi in culture medium during incubation time (Initial weight = 50 mg).

Fungal species	Weight remained of keratin substrates(mg)					
	Human hair			Animal hair		
	1st week	2nd week	3rd week	1st week	2nd week	3rd week
Chrysosporium indicum	40.33±0.04	32.33±0.13	21.67±0.05	40.00±0.05	32.00±0.04	27.67±0.08
Trichophyton mentagrophytes	41.33±0.15	37.67±0.41	28.00±0.09	43.00±0.13	41.67±0.05	35.67±0.12
Microsporum gypseum	43.00±0.01	37.67±0.09	25.33±0.03	38.00±0.17	31.00±0.01	25.33±0.11
Trichophyton rubrum	44.00±0.31	34.33±0.03	27.33±0.18	41.67±0.09	39.00±0.09	36.33±0.17
Trichophyton verrucosum	46.00±0.09	39.67±0.35	30.33±0.13	37.33±0.04	30.67±0.01	25.33±0.07
Trichophyton tonsurans	46.67±0.05	40.33±0.14	35.33±0.27	45.33±0.07	41.67±0.04	32.67±0.12
Microsporum fulvum	42.33±0.02	36.67±0.19	28.33±0.16	44.67±0.01	39.67±0.13	32.67±0.09
Fusarium oxysporum	46.67±0.15	46.33±0.01	40.67±0.09	46.00±0.24	45.67±0.09	41.67±0.15
Aspergillus terreus	47.33±0.09	41.67±0.16	39.67±0.01	48.00±0.05	43.00±0.16	41.00±0.16
Scopulariopsis sp.	48.00±0.03	44.67±0.11	41.33±0.09	46.33±0.03	44.33±0.13	40.67±0.05

Values are means (n=3) ± SE, the results were considered significant when p < 0.05.

Table 2. Protein release by tested fungi in culture medium baited with two keratin substrates.

Fungal species	Protein release (µg/ml)	
	Human hair	Animal hair
Chrysosporium indicum	57	54
Trichophyton mentagrophytes	52	59
Microsporum gypseum	68	62
Trichophyton rubrum	49	38
Trichophyton verrucosum	53	66
Trichophyton tonsurans	33	45
Aspergillus terreus	34	31
Scopulariopsis sp.	17	13
Fusarium oxysporum	29	23
Microsporum fulvum	63	51

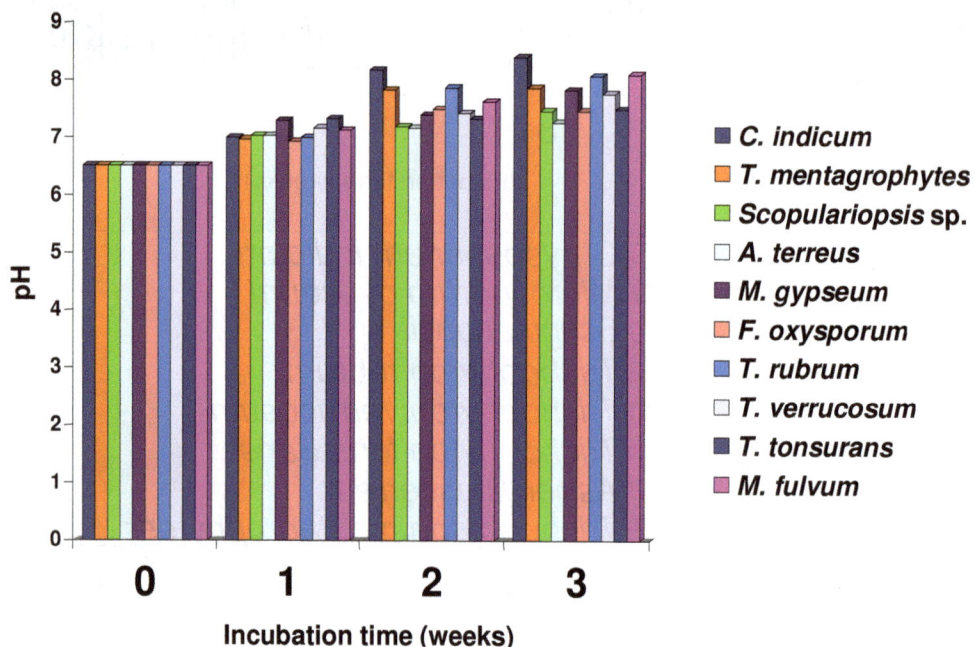

Figure 2a. pH changes of the culture medium due to keratin degradation of human hair by the fungi during the incubation periods.

keratin is its high cysteine content, that makes it more resistant to enzymatic digestion (Kaul and Sumbali, 1999). The enzyme was found to be a halophilic serine proteinase with unique substrate specificity (Namwong et al., 2006). The amount of keratinolytic enzyme in the culture fluid was dependent on the initial pH of the culture medium (Santos et al., 1996). The keratinolytic nature of the examined fungi towards different substrates may be related to their ability to produce cysteine as reported in other studies (Kunert, 1995; Kaul and Sumbal, 1999). The keratinolytic activity of dermatophytes has long attracted the attention of mycologists, biochemists and physicians. Some groups have emphasized the specific

properties of fungal enzymes able to digest keratin, a substrate extremely resistant to the action of physical and chemical agents (Deshmukh and Agrawal, 1982; Kaaman and Forslin, 1985; Takatori et al., 1983). Others have isolated and characterized keratinolytic enzymes of various species of dermatophytes (Asahi et al., 1985; Sanyal et al., 1985; Takiuchi et al., 1982) and their role in virulence and the pathogenesis of mycoses has been postulated (Davies and Zaini, 1984; Eleuterio et al., 1973).

The present study has revealed that *C. indicum* has been the most active keratinolytic fungus on human hair. This species released the highest protein in medium after

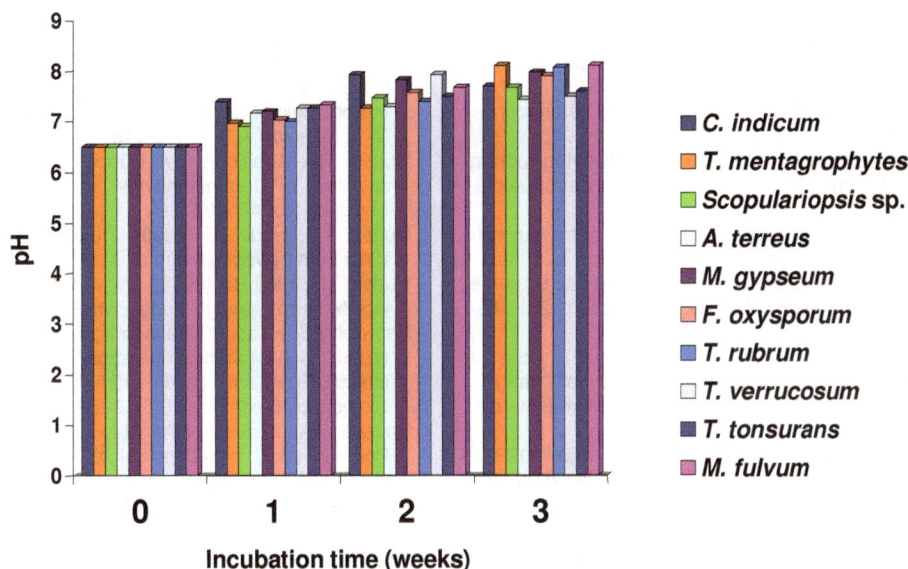

Figure 2b. pH changes of the culture medium due to keratin degradation of animal hair by the fungi during the incubation periods.

M. gypseum and M. fulvum respectively. Rajak et al. (1991) also studied the keratin degradation of human scalp hair by C. indicum, isolated from the soils of a gelatin factory in Jabalpur, India. Marchiso et al. (1991) reported the C. indicum was the most active keratinolytic fungi. In our study, M. gypseum released highest amount of protein in culture medium. This is in agreement with the study of Muhsin and Hadi (2001) who also reported that M. gypseum released high protein in culture medium. Scopulariopsis sp. and F. oxysporum showed the lowest ability to degrade human and animal hair and both released low levels of protein in culture. Marchiso and Fusconi (2001) studied the morphological evidence for keratinolytic activity of Scopulariopsis spp. isolates from nail lesions and the air. Although, T. mentagrophytes is known as a dermatophyte; however, it has been recorded as a human pathogen (Latkowski and Freedberg, 1999) and is frequently isolated from soil. It has also been reported that, this species is a high producer of keratinase (Muhsin et al., 1997) and proteinase (Aubaid and Muhsin, 1998). Wawrzkiewicz et al. (1991) studied in vitro keratinolytic activity of some dermatophytes and found that T. verrucosum degrade guinea pigs hair only and release enzymes mainly to the medium. In this study, M. gypseum and T. verrucosum both showed high ability to degrade animal hair in which T. verrucosum released high amount of protein into the culture medium followed by M. gypseum. The amount of keratinolytic enzyme production depended on substrates concentrations (Park and Son, 2007).

Consequently, pH of the media changed towards alkalinity. It has been stated that keratinolytic fungi often alkalinize culture media (Kaul and Sumbali, 1999). Our observation is in conformity with earlier observations

(Hasija et al., 1990). It is apparent that, there are numerous nondermatophytic filamentous keratinophilic fungi belonging to diverse taxonomic groups. The ability of these fungi to invade and parasitize tissues is associated with and depends upon use and breakdown of keratin. A great variety of non-dermatophytic filamentous fungi can utilize keratin for their growth and are strongly keratinolytic with no concrete evidence of their pathogenic role; their mere isolation in culture from lesions on skin or other sites should not ascribe them any etiological significance. Thus, it can be concluded that fungi isolated by us, which are keratinophilic in nature play an important role not only in pathogenicity but also in biodegradation of keratin substrates.

ACKNOWLEDGEMENTS

The authors are grateful to the Head, Department of Botany, University of Rajasthan, Jaipur, India for providing research facilities and to Dr. V. N. Saxena, S.M.S. Medical College and Hospital, Jaipur, India for providing the help in isolation of dermatophytic fungi from dermatomycoses patients.

REFERENCES

Allpress JD, Mountain G, Gowland PC (2002). Production, purification and characterization of an extracellular keratinase from Lysobacter NCIMB 9497. Lett. Appl. Microbiol., 34: 337-342.

Al-Musalam AA, Radwan SS (1990). Wool-colonizing microorganisms capable of utilizing wool-lipids and fatty acids as sole sources of carbon and energy. J. Appl. Bacteriol., 69: 806-813.

Asahi M, Lindquist R, Fukuyama K, Apodaca G, Epstein WL, Mckerrow JH (1985). Purification and characterization of major extracellular proteinases from Trichophyton rubrum. Biochem. J., 232: 139-144.

Aubaid AH, Muhsin TM (1998). Partial purification and kinetic studies of exocellular proteinase from Trichophyton mentagrophytes var. erinacei. Mycoses, 41: 163-168.

Bahuguna S, Kushwaha RKS (1989). Hair perforation by keratinophilic fungi. Mycoses, 32: 340-343.

Davies RR, Zaini F (1984). Enzymatic activities of Trichophyton rubrum and the chemotaxis of polymorphonuclear leucocytes. Sabouraudia, 22: 235-141.

Deshmukh SK, Agrawal SC (1982). In vitro degradation of human hair by some keratinophilic fungi. Mykosen, 25: 454-458.

Deshmukh SK, Agrawal SC (1985). Degradation of human hair by some dermatophytes and other keratinolphilic fungi. Myksosen, 28: 463-466.

Eleuterio MK, Grappel SF, Caustic CA, Blank F (1973). Role of keratinases in dermatophytosis. III. Demonstration of delayed hypersensitivity to keratinases by the capillary tube migration test. Dermatologica, l47: 255-260.

El-Naghy MA, El-Katany E, Fadl-Allah EM, Nazeer WW (1998). Degradation of chicken feathers by Chrysosporium georgiae. Mycopathologia, 143: 77-84.

Gessesse A, Hatti-Kaul R, Gashe BA, Mattiasson B (2003). Novel alkaline proteases from alkaliphilic bacteria grown on chicken feather. Enzyme Microb. Technol., 32: 519-524.

Gradisiar H, Kern S, Friedrich J (2000). Keratinase of Doratomyces microsporus. Appl. Microbiol. Biotechnol., 53: 196-200.

Gupta R, Ramnani P (2006). Microbial keratinases and their prospective applications: An overview. Appl. Microbiol. Biotechnol., 70(1): 21-33.

Hasija SK, Malviya H, Rajak RC (1990). Keratinolytic ability of some fungi isolated from gelatin factory campus, Jabalpur (MP). Proc. Nat. Acad. Sci. India, 3: 305-309.

Kaaman T, Forslind B (1985). Ultrastructural studies on experimental hair infections in vitro caused by Trichophyton mentagrophytes and T. rubrum. Act. Dermatologica. Venereologica (Stock), 65: 536-569.

Kaul S, Sumbali G (1999). Production of extracellular keratinase by keratinophilic fungal species inhabiting feathers of living poultry birds (Gallus domesticus): A comparison. Mycopathologia, 146: 19-24.

Kim JM, Lim WJ, Suh HJ (2001). Feather-degrading Bacillus species from poultry waste. Process. Biochem., 37: 287-291.

Kunert J (1989). Biochemical mechanism of keratin degradation by the actinomycete Streptomyces fradiae and the fungus Microsprum gypseum: A comparison. J. Basic Microbiol., 9: 597-604.

Kunert J (1995). Biochemical mechanism of keratin degradation by keratinolytic fungi. In: Govil, C.M., Bilgrami, K.S., Singh, D. (eds). Platinum jubilee volume of Indian Botanical Society. Agrawal Printers Meerut, India, pp. 89-98.

Kushwaha RKS (1983). The in vitro degradation of peacock feathers by some fungi. Mykosen, 26: 324-326.

Latkowski JM, Freedberg IM (1999). Epidermal cell kinetics, epidermal differentiation, and keratinization. In: Freedberg, I.M., Eisen, A.Z., Wolff, K. (eds). Fitzpatrick's Dermatology in General Medicine. McGraw-Hill, New York, pp. 133-144.

Longshaw CM, Wright JD, Farrell AM, Holland KT (2002). Kytococcus sedentarius, the organism associated with pitted keratolysis, produces two keratin-degrading enzymes. J. Appl. Microbiol., 93: 810-816.

Lowry OH, Rosebrough NJ, Farr AL, Randall RJ (1951). Protein measurement with the folin phenol reagent. J. Biol. Chem., 193: 265-275.

Malviya HK, Rajak RC, Hasija SK. (1993). In vitro degradation of hair keratin by Graphium penicilloides; Evidence for sulphytolysis and peptidolysis. Crypto. Bot., 3: 197-201.

Marchiso VF, Fusconi A (2001). Morphological evidence for keratinolytic activity of Scopulariopsis spp. isolates from nail lesions and the air. Med. Mycol., 39 (3): 287-294.

Marchiso VF, Curetti D, Cassinelli C, Bordese C (1991). Keratinolytic and keratinophilic fungi in the soils of Papua New Guinea. Mycopathologia, 115(2): 113-119.

Marchiso VF, Fusconi A, Rigo S (1994). Keratinolysis and its morphological expression in hair digestion by airborne fungi. Mycopathologia, 127: 103-115.

Muhsin TM, Aubaid AH, Al-Duboon AH (1997). Extracellular enzyme activities of deramatophytes and yeast isolates on solid media. Mycoses, 40: 465-469.

Muhsin TM, Hadi RB (2000). Dermatophytes and other keratinolytic fungi from sewage sludge at Basrah, Iraq. Mycoses (in press).

Muhsin TM, Hadi RB (2001). Degradation of keratin substrates by fungi isolated from sewage sludge. Mycopathologia, 154: 185-189.

Namwong S, Hiraga K, Takada K, Tsunemi M, Tanasupawat S, Oda K (2006). A halophilic serine proteinase from Halobacillus sp. SR5-3 isolated from fish sauce: purification and characterization. Biosci. Biotechnol. Biochem., 70(6): 395-401.

Onifade AA, Al-Sane NA, Al-Mussallam AA, Al-Zarbam S (1998). Potential for biotechnological application of keratin-degrading microorganisms and their enzymes for nutritional improvement of feathers and other keratins as livestock feed resources. Biores. Technol., 6: 1-11.

Park GT, Son HJ (2007). Keratinolytic activity of Bacillus megaterium F7-1, a feather-degrading mesophilic bacterium. Microbiol. Res., 23, [Epub ahead of print].

Rajak RC, Parwekar S, Malviya H, Hasija SK (1991). Keratin degradation by fungi isolated from the grounds of a gelatin factory in Jabalpur, India. Mycopathologia, 114(2): 83-87.

Safranek WW, Goos RD (1982). Degradation of wool by saprophytic fungi. Can. J. Microbiol., 28: 137-140.

Sandali S, Brandelli A (2000). Feather keratin hydrolysis by a Vibrio sp. strain kr2. J. Appl. Microbiol., 89: 735-743.

Santos RMDB, Firmino AA, De Sa CM, Felix CR (1996). Keratinolytic activity of Aspergillus fumigatus Fresenius. Curr. Microbiol., 33(6): 364-370.

Sanyal AK, Das SK, Banerjee AB (1985). Purification and partial characterization of an excellular proteinase from Trichophyton rubrum. Sabouraudia, 23: 165-178.

Singh CJ (2003). Optimization of an extracellular protease of Chrysosporium keratinophilum and its potential in bioremediation of keratinic wastes. Mycopathologia, 156: 151-156.

Takatori K, Udagawa S, Kurata H, Hasagawa A (1983). Microscopic observation of human hairs infected with Microsporum ferrugineum. Mycopathologia, 81: 129-133.

Takiuchi I, Higuchi D, Sei Y, Koga M (1982). Isolation of an extracellular proteinase (keratinase) from Microsporum canis. Sabouraudia, 20: 281-288.

Vanbreuseghem R (1952). Biological technique for the isolation of dermatophytes from the soil. Ann. Soc. Belge. Med. Trop., 32: 173-178.

Wang JJ, Shih JCH (1999). Fermentation production of keratinase from Bacillus licheniformis PWD-1 and a recombinant B. subtilis FDB-29. J. Ind. Microbiol. Biotechnol., 22: 608-616.

Wawrzkiewicz K, Wolski T, Lobarzewski J (1991). Screening the keratinolytic activity of dermatophytes in vitro. Mycopathologia, 114(1): 1-8.

Yamamura S, Morita Y, Hasan Q, Yokoyama K, Tamiya E (2002). Keratin degradation: A cooperative action of two enzymes from Stenotrophomonas sp. Biochem. Biophys. Res. Commun., 294: 1138-1143.

16

Influence of road transportation during hot summer conditions on oxidative status biomarkers in Iranian dromedary camels (*Camelus dromedarius*)

Saeed Nazifi[1]*, Mahdi Saeb[2], Hasan Baghshani[2] and Saeedeh Saeb[2]

[1]Department of Clinical Studies, School of Veterinary Medicine, Shiraz University, Shiraz, Iran.
[2]Department of Basic Sciences, School of Veterinary Medicine, Shiraz University, Shiraz, Iran.

Transportation causes stress in livestock that may alter numerous physiological variables with a negative impact on production and health. The objective of the current study was to investigate the effects of road transport on oxidative stress biomarkers in camels. Ten Iranian dromedary camels were selected and subjected to a journey of approximately 300 km in a truck by road in August 2008. Blood samples were collected immediately before loading at 8:30 A.M., after 1 h transportation, at 9:30 A.M., and at the end of the journey after unloading at 1:30 PM. Final blood sample was taken 24 h after arrival. Plasma concentrations of malondialdehyde and α-tocopherol, erythrocyte superoxide dismutase and whole blood glutathione peroxidase activities were measured using validated methods. The mean concentration of MDA (1.87 ± 0.26 nmol/mL) and glutathione peroxidase activity (297.86 ± 25.68 U/g Hb) in basal pre-transport conditions show significant increase 24 h after arrival. The mean concentration of α-tocopherol (5.22 ± 0.74 µmol/L) and superoxide dismutase activity (1742.5 ± 74.36 U/g Hb) in basal pre-transport conditions had no significant change during and after transportation. Results suggest that transport stress causes an oxidative challenge in dromedary camels and represent novel biomarkers for stress-associated disease susceptibility and welfare assessment. However, further research efforts should be directed towards understanding the role of particular antioxidants and oxidants on the stressful conditions.

Key words: Dromedary camel, road transportation, oxidative status, malondialdehyde, α-tocopherol, glutathione peroxidase, superoxide dismutase.

INTRODUCTION

The dromedary camel is one of the most important domestic animals in the arid and semi arid regions as it is equipped to produce high quality food at comparatively low costs under extremely harsh environments (Yagil, 1982; Yousif and Babiker, 1989). It can survive well on sandy terrain with poor vegetation and may chiefly consume feeds unutilized by other domestic species (Shalah, 1983). Camel husbandry in Iran is almost localized in southern provinces, while its slaughter and meat consumption is done in most parts of this country. Sale and slaughter are the most usual reasons for transporting camels in Iran.

Farm livestock experience a variety of stressors that can modify normal behavior and growth, leading to losses in perf

formance. Transport is a critical phase in animal production and utilization and often considered as one of the main causes of stress raising considerable interest, both in economic and animal welfare terms (Mormede et al., 1982). During transport, animals are exposed to a variety of potential stressors such as motion of the vehicle, noise, vibrations, centrifugal forces, rapidly changing light conditions, heat, cold, poor air quality, deck height, mixing of unfamiliar groups, poor road conditions and the possible lack of water and feed (Parrott et al., 1998 ; Broom, 2000; Hurtung, 2003). Animal health can be impaired by various pre-transport and transport conditions and may cause injury, reduce performance and can promote the development of diseases in animals (Hurtung, 2003). In addition, the time between removing animals from farms to the slaughter house has a major impact on meat quality (Kadim et al., 2006). Stress is directly related to lipid oxida-

*Corresponding author. E-mail: nazifi@shirazu.ac.ir.

tion in muscle (McClelland, 2004) and the ratio and the proportions of oxidation in meat are also influenced by pre-slaughter (example, stress) and post-slaughter events (Linares et al., 2007). Signs of transportation stress have been demonstrated in different animal species by, for example, increased adrenal cortical activity (Ruiz-de-la-Torre et al., 2001), decreased humoral immunity (Machenzie et al., 1997), increased morbidity and mortality due to infectious diseases in the few weeks following transportation (Hurtung, 2003; Chirase et al., 2004). Increased susceptibility for infections such as pneumonia following transportation is well documented in animal species such as cattle (Knowles 1999; Hurtung 2003; Chirase et al., 2004), sheep (Hurtung, 2003), goat (Kannan et al., 2000), horse (Smith et al., 1996; Knowles, 1999) and camel (Wernery and Kaaden, 2002). Although there are many factors affecting animal susceptibility during these times, vulnerability may be traced back to biochemical processes within the body.

Oxidative stress resulting from increased production of free radicals and reactive oxygen species (ROS), and/or a decrease in antioxidant defense, leads to damage of biological macromolecules and disruption of normal metabolism and physiology (Trevisan et al., 2001). When reactive forms of oxygen are produced faster than they can be safely neutralized by antioxidant mechanisms, oxidative stress results. These conditions can contribute and/or lead to the onset of health disorders in animals (Miller et al., 1993; Chirase et al., 2004). Oxidative stress has been implicated in the pathophysiology of transport-related maladies (McBride et al., 2001; Chirase et al., 2004; Pregel et al., 2005; Urban-Chmiel, 2006; Wernicki et al., 2006; Burke et al., 2009). Diminished antioxidant defenses or excess oxidative species resulting from transport stress may be deleterious to tissues, and may be linked to the manifestation of disease. Stress of any origin is capable of depleting the body's antioxidant resources (Sconberg et al., 1993). Serum concentrations of the antioxidant vitamin E are reduced in transported steers, and exposure to simulated dust storm after being transported further decreases these concentrations (Chirase et al., 2001). Vitamin E addition to receiving diets seemed beneficial for increasing average daily gain and decreasing shipping fever mortality (Galyean et al., 1999), highlighting the importance of antioxidants to health. Physical and psychological stressors have an influence on the increase of catabolic reactions, which may cause an increase of ROS production (Wernicki et al., 2006). Concentration of ROS is increased in white blood cells isolated from calves after shipping, likely as a result of enhanced respiratory burst (Wernicki et al., 2006). Chirase et al. (2004) observed decreased serum total antioxidant capacity in transported calves. Similarly, Pregel et al. (2005) surmised that total antioxidant status of serum was a useful tool for measuring stress in transported calves. Moreover, heat stress has been recognized as one of the most common problems encountered during road transportation of livestock (Mitchell and Kettlewell, 1998) and con-

tributes to transportation-induced stress during summer months (Stull and Rodiek, 2000).

It is well established that different animal species and even animals of the same species but different genetic background respond differently to the same stressor (Hall et al., 1998). No reference is available about oxidative changes under stressful conditions in dromedary camels. As camels are normally raised and thrive under harsh environmental and dietary conditions, we thought it worthwhile to assess if these desert animals would respond differently to the stressful stimulus of transportation. The objective of the study reported here was to determine the effect of transportation under hot summer conditions on oxidative stress biomarkers in Iranian dromedary camels. Plasma malondialdehyde (MDA) concentration was measured because MDA is one of the end-products of lipid peroxidation, and the extent of lipid peroxidation is most frequently measured by estimating MDA levels (Lata et al., 2004). Estimating the activities of enzymatic antioxidants, such as superoxide dismutase (SOD) and glutathione peroxidase (GSH-Px), is another means of evaluating oxidative stress (Kleczhowski et al., 2003). α-tocopherol measurement also was done because of its antioxidant properties.

MATERIALS AND METHODS

Animals and transportation

The 10 clinically healthy Iranian dromedary camels (Camelus dromedarius), 5 males and 5 females, ranging in age from 3 - 4 years and weighing about 300 kg were selected for the study. The camels had been reared at the Camel Research Institute in Yazd province of Iran, which is supervised by an experienced veterinarian. Preliminary procedures (handling, physical restraint, loading, and unloading) were undertaken by the same staff and blood sampling was always carried out by the same operator. The journey took place on one day in August 2008. Transportation of the camels was conducted between Camel Research Institute in Baphgh to Yazd and back to Institute, on smooth roads in an open truck which took 5 h (about 300 Km). Stocking density was about 1 m^2 per animal. Environmental temperature and relative humidity during the journey was 32 - 36°C and 17 - 25%, respectively. The camels had similar feeding and watering conditions ad libitum before and after the journey. During the journey there was no feed, water or unloading for rest.

Blood sampling and processing

Blood samples were collected immediately before loading, at 08:30 am (T1), after 1 h transport, at 09:30 am (T2) and immediately after transport termination and unloading, on arrival at Institute (T3) at 01:30 pm. Final blood sample was taken 24 h after arrival (T4). Blood samples were collected by jugular venepuncture into vacuum containers containing EDTA (Becton Dickinson, NJ, USA). The blood tubes were placed on ice until laboratory arrival (<2 h). The samples were centrifuged at 750 g for 20 min, and then the plasma was pipetted into different aliquots and stored at -70°C until analysis for plasma content of MDA and α-tocopherol. Haemoglobin concentration was measured by Cyanmethaemoglobin method. SOD activity was measured by a modified method of iodophenyl nitrophenol phenyltetrazolium chloride (RANSOD kit, Randox Com United Kingdom). This method employs xanthine and xanthine oxidase

Table 1. Mean (± SE) of circulating oxidative status indices in female and male dromedary camels($n=5$ in each gender) before transport (T1), 1 h after transport initiation (T2), on the end of transportation (T3) and 24 h after arrival (T4).

Parameter*		T1	T2	T3	T4
Plasma malondialdehyde	Female	1.67 ± 0.24	1.62 ± 0.36	1.95 ± 0.11	2.12 ± 0.27
(nmol/mL)	Male	2.07 ± 0.49	1.72 ± 0.51	2.16 ± 0..39	2.35 ± 0.28
	Total	1.87 ± 0.26 [a]	1.67 ± 0.29 [a]	2.06 ± 0.19 [a,b]	2.23 ± 0.21 [b]
α-tocopherol	Female	5.51 ± 1.28	4.43 ± 0.95	4.32 ± 0.74	5.18 ± 0.98
(µmol/L)	Male	4.94 ± 0.92	5.2 ± 0.83	4.11 ± 0.97	3.56 ± 1.09
	Total	5.22 ± 0.74	4.8 ± 0.6	4.22 ± 0.57	4.37 ± 0.74
Glutathione peroxidase	Female	317.52 ± 38.71	242.6 ± 33.68	279.86 ± 27.42	343.84 ± 28.96
(U/g Hb)	Male	278.2 ± 35.73	297.24 ± 29.11	301.44 ± 21.16	322.14 ± 25.44
	Total	297.86 ± 25.68 [a]	269.92 ± 22.88 [a]	290.65 ± 16.72 [a]	332.99 ± 18.53 [b]
Superoxide dismutase	Female	1837 ± 82	1436 ± 90	1950 ± 40	1823 ± 116
(U/g Hb)	Male	1648 ± 117	1740 ± 146	1810 ± 173	1814 ± 97
	Total	1742 ± 74	1588 ± 96	1880 ± 87	1819 ± 71

*No significant difference was observed in any parameter between sexes at all sampling times. Time □ gender interaction was not significant for all parameters ($P < 0.05$). [a,b]Mean (± SE) in each row with no common superscript differ significantly ($P < 0.05$).

(XOD) to generate superoxide radicals which react with 2-(4- iodo-phenyl)-3-(4-nitrophend)-5-phenyltetrazolium chloride (INT) to form a red formazan dye. The superoxide dismutase activity was then measured by the degree of inhibition of this reaction. One unit of SOD was considered a 50% inhibition of reduction of INT under the condition of the assay. Glutathione peroxidase activity (GSH-Px) was measured by the method of Paglia and Valentine (1967) (RAN-SEL kit, Randox Com, United Kingdom). GSH-Px catalyses the oxidation of glutathione (GSH) by cumene hydroperoxide. In the presence of glutathione reductase (GR) and NADPH, the oxidised glutathione (GSSG) is immediately converted to the reduced form with a concomitant oxidation of NADPH to $NADP^+$. The decrease in absorbance was measured at 340 nm by UV-visible spectrophotometer (Pharmacia LKB Biochrom, England). Total MDA concentration of the plasma was measured by reverse phase high-pressure liquid chromatography with UV detection at 310 nm after derivatisation with 2, 4-dinitrophenylhydrazine as described by Pilz et al. (2000). α-tocopherol content of plasma was measured in deproteinized, hexane-extracted plasma using reverse phase high-pressure liquid chromatography with UV detection at 295 nm (Solichova et al., 2003).

Statistical analysis

The data are presented as mean ± standard error (SE). A two way (time, gender) repeated measures analysis of variance was applied for statistical analysis. The level of significance was set at $P < 0.05$. Significances between means were assessed using the least-significant-difference procedure. All calculations were performed using SPSS/PC software.

RESULTS

The mean ± SE of the measured parameters of oxidative status in Iranian dromedary camels subjected to road transportation are presented in Table 1. Mean concentration of MDA increased significantly 24 h after arrival. No significant change was observed in plasma concentrations of α-tocopherol during and after transportation. Erythrocyte glutathione peroxidase activity showed a significant increase 24 h after arrival. The increase in superoxide dismutase activity on arrival and 24 h later was not signifi-

cant. No significant difference was observed in any parameter between sexes at all sampling times. Time × gender interaction was not significant for all studied parameters.

DISCUSSION

The transportation stress can be viewed as a combination of a number of concurrent stressors that are likely to operate during and around transportation, and which exert both "physical", "physiological" and "psychological" stressful stimuli (Knowles, 1998). Oxidative stress may exacerbate psychological and physiological demands brought about by stressful conditions. Oxidative stress, involving an imbalance between the production of free radicals and the capability of an organism to absorb their excess, has been proposed to play a role in the pathogenesis of several infectious diseases of domestic animals (Miller et al., 1993; Lykkesfeldt and Svendsen, 20-07). Oxidative stress is extremely dangerous because it does not exhibit any symptoms and is recognizable with great difficulty by means of common methods of analysis. Oxidative stress promotes the insurgence of serious pathologies as a result of the degenerative damage of cellular structures (Freidovich, 1999; Matsuo and Kaneko, 20-00; McCord, 2000). Measure of oxidative stress allows es-timation of the real status of physiological defense and prevention of the appearance of correlated pathologies (Piccione et al., 2007). The inevitability of livestock transport makes stress associated with transportation an appropriate field of focus. Physiological stress due to transportation or inappropriate housing elevates oxidative stress in calves as measured by plasma ascorbate levels or serum total antioxidant capacity (Cummins and Brunner, 1991; Tyler and Cummins 2003; Chirase et al. 2004; Pregel et al. 2005; Wernicki et al., 2006).

Lipid peroxidaion is a general mechanism whereby free

radicals induce tissue damages, and implicated under several diverse pathological conditions (Halliwell and Chirico 1993). Malondialdehyde (MDA) has been widely applied as the most common biomarker for the assessment of lipoperoxidation in biological and medical sciences (Bird and Draper, 1984; Suttnar et al., 2001). MDA concentration in the present study at basal pre-transport state was lower than values (30.44 ± 2.89 µmol/L) obtained by Mohamed (2008) in healthy dromedary camels. The mentioned difference may be due to different methodologies utilized in two trials. In the present experiment significant increase in plasma MDA concentrations was observed 24 h after transport termination. Our finding is in conformity with the reported results in cattle (Chirase et al. 2004; Wernicki et al. 2006). Chirase et al. (2004) found that serum MDA concentrations in calves tripled after transportation. Wernicki et al. (2006) reported large increases in plasma thiobarbituric acid reactive substances on the first 3 days after transportation, with a gradual decline on the sixth day, and a return to baseline levels on the ninth day post-transport. Oxidative stress associated with cattle transport has also been evidenced by excessive accumulation of leukocyte lipid oxidation products (Urban-Chmiel, 2006). In horses, after the transportation and training exercises a significant increase of lipid peroxidetion was observed (Ishida et al. 1999). Burke et al. (2009) fail to detect any increase in plasma MDA 7 days after transport in beef calves and reported a decline in MDA concentration 7 days after transport. The week of recuperation time given to steers in that experiment may have been sufficient for recovery from oxidative insult and return to normal redox balance.

Higher concentrations of MDA in dromedaries after transportation under hot temperatures may be explained by higher levels of glucocorticoids and adrenaline-induced pathways of aerobic energy production associated with stress, which generate reactive oxygen metabolites and thus lipid peroxidation (Freeman and Crapo, 1982; Nockels et al., 1996). Glucocorticoids, as the final effecttors of the hypothalamic–pituitary–adrenal (HPA) axis, participate in the control of whole body homeostasis and the organism's response to stress. In mammals, lipid peroxidation was induced while the nonenzymatic antioxidant capacity was decreased, and enzymatic antioxidant systems were suppressed in the liver (Ohtsuka et al., 1998), erythrocytes (Orzechowski et al., 2000), skeletal muscle and lymphoid organs (Pereira et al., 1999) after administration of glucocorticoid hormones. Furthermore, presence of heat stress conditions in the present study, in addition to other stressful stimuli, can induce the metabolic changes that are involved in the induction of oxidative stress. Heat stress can enhance the formation of ROS and induce oxidative stress in cells (Flanagan et al., 1998; Lord-Fontaine and Averill-Bates, 2002) and intact animals (Hall et al., 1994; Harmon et al., 1997; Bernabucci et al., 2002; Lin et al., 2006).

Antioxidants are implied in the inactivation or transformation of oxidants, which can either be transformed by anti-

oxidant enzymes into less reactive forms or which can react with antioxidant molecules that are chemically stable.α-tocopherol is considered to be the most biologically active of several forms of vitamin E. It is an effective lipophilic antioxidant which protects lipid membranes against peroxidation (Solichova et al., 2003). Cattle that were stressed or were given a stress treatment of ACTH and epinephrine injections were found to have reduced α-tocopherol concentrations in their plasma, neutrophils, and red blood cells (Sconberg et al., 1993). After stress cattle may have reductions in α-tocopherol concentrations in certain tissues. Supplemental vitamin E may be required after stress to restore α-tocopherol in tissues (Nockels et al., 1996). The importance of antioxidant status during shipping and receiving can be further substantiated by work demonstrating decreased average daily gain and increased bovine respiratory disease in conjunction with decreased post-transport concentrations of serum vitamins A and E (Chirase et al., 2001). Based on our results no significant difference was observed in plasma α-tocopherol concentration during and after transportation in dromedary camel. However, caution should be observed in interpreting vitamin E status based solely on plasma α-tocopherol values, because of plasma vitamin E values may not accurately reflect body status of vitamin E (Sconberg et al., 1993; Nockels et al., 1996).

Antioxidant enzyme levels are sensitive markers of oxidative stress. Both increased and decreased antioxidant enzyme levels have been reported in different conditions as a consequence of enhanced ROS production either by up-regulation of enzyme activity or utilization of the antioxidant enzymes to counter the ROS. Superoxide dismutase (SOD) is a metalloenzyme which catalyses the dismutation of $O_2^{.-}$ into O_2 and H_2O_2. It is important in the antioxidative defense mechanism and protects against lipid peroxidation (Halliwell and Chirico, 1993; Miller et al. 1993). Erythrocytes, due to their role and their propensity to generate radical species, may be considered as sensitive and intermediate cells in oxidative reactions (Sato et al., 1998). Various kinds of stressors increase lipid peroxidation levels and therefore SOD activity (Gaal et al., 1993; Lata et al., 2004). The higher erythrocyte SOD activity attributed to elevated temperatures during summer months has been reported in cattle (Bernabucci et al., 2002). At the present study no significant increase was observed in erythrocyte SOD activity due to road transporttation stress. This could be due to the fact that we measure erythrocyte SOD activity, and perhaps whole blood SOD activity might be a better measure.

The seleno-enzyme glutathione peroxidase, as one of the primary antioxidant enzymes, contributes to the oxidative defense of animal tissues by catalyzing the reduction of hydrogen and lipid peroxides (Arthur 2000). GSH-Px functions in cellular oxidation-reduction reactions to protect the cell membrane from oxidative damage caused by free radicals (Flohe et al., 1973). Whole blood G-SH-Px activity at basal conditions reported here is in line with the data reported for the dromedary camel by Corbera et al. (2001). GSH-Px activity show significant increase 24 h after arrival in comparison with previous activity concentrations and this is different from results of Burke et al. (2009) that reported GSH-Px activity does not fluctuate appreciably in calf leukocytes due to two-stage weaning and subsequent transport. This difference might be due to different sampling schedule because in the mentioned trial sampling was done 7 days post-transport and this time might be sufficient for recovery from oxidative damage as Wernicki et al. (2006) showed a post-transport decline in lipid oxidation on the sixth day. We speculate that the significant increase in GSH-Px activity might be an indirect compensatory response of cells to increased oxidant challenge due to stressful stimuli of transportation under hot temperatures. Increased GSH-Px activity attributed to elevated temperatures during summer has been reported in cattle (Bernabucci et al., 2002). In humans, GSH-Px activity increases after long-term exercise training regimens, a response which may be an adaptive mechanism against physical stressors (Tessier et al., 1995; Tauler et al., 2006). Stabel et al. (1989) reported that signs of morbidity coincided with elevated plasma GSH-Px in weaned and transported calves challenged with *Manhemia hemolytica*.

In line with the data reported for other species mentioned above, our findings show significant increased concentrations of plasma MDA due to transportation stress in dromedaries which, in parallel with increased whole blood GSH-Px activity supported the hypothesis that transport stress represents an oxidative challenge for livestock. These findings suggest novel biomarkers for stress-associated disease susceptibility and welfare assessment. Better understanding of the mechanisms of ROS production and further investigation of their effects on the stress-related maladies are essential to obtain new insight into this issue and eventually develop new diagnostic, prophylactic and therapeutic strategies. However, further research efforts should be directed towards understanding the role of particular antioxidants and oxidants on the stressful conditions because of this fact that susceptibility to disease is enhanced when physiological stress increases oxidative stress. In the other hand, because of the potential effects of lipid peroxidation on tissue types, it would be interesting and economically relevant to find out whether plasma MDA concentrations are related to carcass characteristics of meat animals.

ACKNOWLEDGMENT

The authors gratefully acknowledge the staff of the Camel Research Institute for their kind helps in this study.

REFERENCES

Arthur JR (2000). The glutathione peroxidases. Cell Mole Life Sci. 57: 1825–1835.

Bernabucci U, Ronchi B, Lacetera N, Nardone A (2002). Markers of oxidative status in plasma and erythrocytes of transition dairy cows during hot season. J. Dairy Sci. 85: 2173–2179.

Bird RP, Draper HH (1984). Comparative studies on different methods of malonaldehyde determination. Methods Enzymol. 105: 299-305.

Broom DM (2000). Welfare assessment and welfare problem areas during handling and transport In: Grandin T editor. Livestock Handling and Transport. 2nd edn. CAB International 43-61.

Burke NC, Scaglia G, Boland HT, Swecker Jr. WS (2009). Influence of two-stage weaning with subsequent transport on body weight, plasma lipid peroxidation, plasma selenium, and on leukocyte glutathione peroxidase and glutathione reductase activity in beef calves. Vet. Immunol. Immunopathol. 127: 365–370.

Chirase NK, Greene W, Purdy CW, Loan RW, Auvermann BW, Parke DB, Walborg EF, Stevenson DE, Xu Y, Klaunig JE (2004). Effect of transport stress on respiratory disease, serum antioxidant status, and serum concentrations of lipid peroxidation biomarkers in beef cattle. Am. J. Vet. Res. 65: 860–864.

Chirase NK, Greene LW, Purdy CW, Loan RW, Briggs RE, McDowell LR (2001) Effect of environmental stressors on ADG, serum retinol and alpha-tocopherol concentrations, and incidence of bovine respiratory disease of feeder steers. J. Anim. Sci. (Suppl.1) 79:188.

Corbera JA, Gutierrez C, Morales M, Montel A, Ontoya JA (2001). Assessment of blood glutathione peroxidase activity in the dromedary camel. Vet .Res. 32: 185–191.

Cummins KA, Brunner CJ (1991). Effect of calf housing on plasma ascorbate and endocrine and immune function. J. Dairy Sci. 74:1582–1588.

Flanagan SW, Moseley PL, Buettner GR (1998). Increased flux of free radicals in cells subjected to hyperthermia detection by electron paramagnetic resonance spin trapping. FEBS Lett. 431:285–286.

Flohe L, Gunzler WA, Schock HH (1973). Glutathione peroxidase: a selenoenzyme. FEBS Lett. 32:132–134.

Freeman BA, Crapo JD (1982). Biology of disease-free radicals and tissue injury Lab. Inves. 47: 412-426.

Freidovich I (1999). Fundamental aspects of reactive oxygen species, or what's the matter with oxygen? Ann. N.Y. Acad. Sci. 893: 13-20.

Gaal T, Mezes M, Miskucza O, Ribiczey-Szabo P (1993). Effect of fasting on blood lipid peroxidation parameters of sheep. Res .Vet .Sci 55: 104–107.

Galyean ML, Perino LJ, Duff GC (1999) Interaction of cattle health/immunity and nutrition. J. Anim. Sci. 77: 1120-1134.

Hall SJG, Broom DM, Kiddy GNS (1998) Effect of transportation on plasma cortisol and packed cell volume in different genotypes of sheep. Small Rum Res. 29: 233–237.

Hall DM, Buettner GR, Matthes RD, Gisolfi CV (1994) Hyperthermia stimulates nitric oxide formation: electron paramagnetic resonance detection of NO-heme in blood. J. Appl. Physiol. 77: 548–553.

Halliwell B, Chirico S (1993) Lipid peroxidation: its mechanism, measurement, and significance. Am. J. Clin. Nut. (Suppl. 1) 57:715S – 725S.

Harmon RJ, Lu M, Trammel DS, Smith BA (1997) Influence of heat stress and calving on antioxidant activity in bovine blood. J Dairy Sci. (Suppl. 1) 80: 264.

Hurtung J (2003). Effects of transport on health of farm animals. Vet. Res. Comm. 27: 525-527.

Ishida N, Hobo S, Takahashi T, Nanbo Y, Sato F, Hasegawa T, Mukoyama H(1999). Chronological changes in superoxide-scavenging ability and lipid peroxide concentration of equine serum due to stress from exercise and transport. Equine Vet. J. 30: 430-433.

Kadim IT, Mahgoub O, Al-Kindi A, Al-Marzooqi W, Al-Saqri NM (2006). Effects of transportation at high ambient temperatures on physiological responses, carcass and meat quality characteristics of three breeds of Omani goats. Meat Sci. 73:626–634.

Kannan G, Terrill TH, Kouakou B, Gazal OS, Gelaye S, Amoah EA, Samake S (2000). Transportation of goats: effect on physiological stress responses and live weight loss. J. Anim. Sci. 78: 1450–1457.

Kleczhowski M, Klucinski W, Sikora J, Jdanocviez M, Dziekan P (2003). Role of antioxidants in the protection against oxidative stress in cattle: non-enzymatic mechanisms (part 2). Polish J. Vet. Sci. 6: 301-308.

Knowles TG (1999). A review of the road transport of cattle. Vet. Rec. 144: 197–201.

Knowles TG (1998). A review of the road transport of slaughter sheep.

Vet. Rec. 143: 212–219.

Lata H, Ahuja GK, Narang APS, Walia L (2004). Effect of immobilization stress on lipid peroxidation and lipid profile in rabbits. J. Clin. Biochem. 19: 1–4.

Lin H, Decuypere E, Buyse J (2006). Acute heat stress induces oxidative stress in broiler chickens. Comp. Biochem. Physiol. A 144: 11–17.

Linares MB, Berruga MI, Bornez R, Vergara H (2007). Lipid oxidation in lamb meat: Effect of the weight, handling previous slaughter and modified atmospheres. Meat Sci. 76: 715–720.

Lord-Fontaine S, Averill-Bates DA (2002) Heat shock inactivity cellular antioxidant defenses against hydrogen peroxide: protection by glucose. Free Radical Biol. Med. 32: 752–765.

Lykkesfeldt J, Svendsen O (2007) Oxidants and antioxidants in disease: oxidative stress in farm animals. Vet. J. 173: 502–511.

Machenzie AM, Drennan M, Rowan TG, Dixon JB, Carter SD (1997) Effect of transportation and weaning on humoral immune responses of calves. Res. Vet. Sci. 63:227–230.

Matsuo M, Kaneko T, In: Z Radak (Ed.) (2000). The chemistry of reactive oxygen species and related free radicals. Leeds Human Kinetics 1–34.

McBride KW, Greene LW, Chirase NK, Kegley EB, Cole NA (2001). The effects of ethoxyquin on performance and antioxidant status of feedlot steers. J Anim. Sci. 79: 285.

McClelland GB (2004). Fat to the fire: The regulation of lipid oxidation with exercise and environmental stress. Comp. Biochem. Physiol. B 139: 443–460.

McCord JM (2000). He evolution of free radicals and oxidative stress. Am. J Med. 108: 652-657.

Miller JK, Brzezinska-Slebodzinska E, Madsen FC (1993). Oxidative stress, antioxidants, and animal function. J. Dairy Sci. 76:2812–2823.

Mitchell MA, Kettlewell PJ (1998). Physiological and welfare of broiler chickens in transit; solutions not problems. Poultry Sci. 77: 1803–1814.

Mohamed HE (2008). Antioxidant status and the degree of oxidative stress in dromedary (Camelus dromedarius) with or without endometritis. Vet. Res. 2: 1-2.

Mormede P, Soissons J, Bluthe RM , Rault J, Legarff G, Levieux D, Dantzer R (1982). Effect of transportation on blood serum composition, disease incidence, and production traits in young calves. Influence of the journey duration. 13:369–384.

Nockels CF, Odde KG, Craig AM (1996). Vitamin E supplementation and stress affect tissue alpha-tocopherol content of beef. J. Anim. Sci. 74: 672-677.

Ohtsuka A, Kojima H, Ohtani T, Hayashi K (1998). Vitamin E reduces glucocorticoid-induced oxidative stress in rat skeletal muscle. J. Nut. Sci. Vitaminol. 44: 779– 786.

Orzechowski O, Ostaszewski P, Brodnicka A, Wilczak J, Jank M, Balasinska B, Grzelkowska K, Ploszaj T, Olczak J, Mrowczynska A (2000). Excess of glucocorticoids impairs whole-body antioxidant status in young rats. Relation to the effect of dexamethasone in soleus muscle and spleen. Hormon. Metab. Res. 32: 174–180.

Paglia DE, Valentine WN (1967). Studies on the quantitative and qualitative characterization of erythrocyte glutathione peroxidase. J. Lab. Clin. Med. 70: 158–169.

Parrott RF, Hall SJG, Lloyd DM (1998). Heart rate and stress hormone responses of sheep to road transport following two different loading procedures. Anim. Welf 7: 257-267.

Pereira B, Bechara EJ, Mendonca JR, Curi R (1999). Superoxide dismutase, catalase and glutathione peroxidase activities in the lymphoid organs and skeletal muscles of rats treated with dexamethasone. Cell Biochem. Function 17:15– 19.

Piccione G, Borruso M, Giannetto C, Morgante M, Giudice E (2007). Assessment of oxidative stress in dry and lactating cows. Acta Agric. Scand A 57: 101-104.

Pilz J, Meineke I, Gleiter CH (2000). Measurement of free and bound malondialdehyde in plasma by high-performance liquid chromatography as the 2, 4-dinitrophenylhydrazine derivative. J. Chromat. B 742:315–325.

Pregel P, Bollo E, Cannizzo FT, Biolatti B, Contato E, Biolatti PG (2005) Antioxidant capacity as a reliable marker of stress in dairy calves transported by road. Vet. Rec. 156:53–54.

Ruiz-de-la-Torre JL, Velarde A, Diestre A, Gispert M, Hall SJ, Broom DM, Manteca X (2001). Effect of vehicle movements during transport on the stress response and meat quality of sheep. Vet. Rech. 148:227–229.

Sato Y, Kanazawa S, Sato K, Suzuki Y (1998) Mechanism of free radical induced haemolysis of human erythrocytes: II. Hemolysis by lipid soluble radical initiator. Biol. Pharmacol. Bull. 21: 250–256.

Sconberg S, Nockels CF, Bennet DW, Bruyninclese W, Blancquaret AMB , Craig A (1993). Effect of shipping handling, adrenocortiocotropic hormone and epinephrine on α-tocopherol content of bovine blood. Am. J. Vet. Res. 54: 1287–1293.

Shalah MR (1983). The role of camels in overcoming world meat shortage. Egyptian J. Vet. Sci. 20:101–110.

Smith BL, Jones JH, Hornof WJ, Miles JA, Longworth KE, Willits NH (1996) Effect of road transport on indices of stress in horses. Equine Vet. J. 28: 446-454.

Solichova D, Korecka L, Svobodova I, Musil F, Blaha V, Zdansky P, Zadak Z (2003). Development and validation of HPLC method for the determination of α-tocopherol in human erythrocytes for clinical applications. Ann. Biann. Chem. 376: 444–447.

Stabel JR, Spears JW, Brown TT, Brake J (1989). Selenium effects on glutathione peroxidase and the immune response of stressed calves challenged with Pasteurella hemolytica. J. Anim. Sci. 67: 557–564.

Stull CL, Rodiek AV (2000). Physiological responses of horses to 24 hours of transportation using a commercial van during summer conditions. J. Anim. Sci. 78: 1458-1466.

Suttnar J, Masova L, Dyr E (2001). Influence of citrate and EDTA anticoagulants on plasma malondialdehyde concentrations estimated by high-performance liquid chromatography. J. Chromat. B 751:193–197.

Tauler P, Aguilo A, Gimeno I, Fuentespina E, Tur JA, Pons A (2006) Response of blood cell antioxidant enzyme defenses to antioxidant diet supplementation and to intense exercise. Euro. J. Nut. 45: 187–195.

Tessier F, Margaritis I, Richard M, Moynot C, Marconnet P (1995). Selenium and training effects on the glutathione system and aerobic performance. Med. Sci. Spor. Exe. 27: 390–396.

Trevisan M, Browne R, Ram M, Muti P, Freudenheim J, Carosella AN, Armstrong D (2001). Correlates of markers of oxidative status in the general population. Am. J. Epid. 154: 348–356.

Tyler PJ, Cummins KA (2003). Effect of dietary ascorbyl-2-phosphate on immune function after transport to a feeding facility. J. Dairy Sci. 86: 622–629.

Urban-Chmiel R(2006). The influence of transport stress on oxidative stress parameters in bovine leukocytes. Slovakia. Vet. Res. 43:243–246.

Wernery U, Kaaden OR (2002). Infectious Diseases in Camel Ides, 2nd edn, Blackwell Wissenschafts-Verlag Berlin 97.

Wernicki A, Urban-Chmiel R, Kankofer M, Mikucki P, Puchalski A, Tokarzewski S, (2006). Evaluation of plasma cortisol and TBARS levels in calves after short-term transportation. Rev. Med. Vet. 157: 30–34.

Yagil R (1982). Camels and Camel Milk. FAO Animal Production and Health. Publications Division, Food and Agriculture Organization of the United Nations. Via delle Terme di Caracalla, 00100 Rome, Italy, 26.

Yousif OK, Babiker SA (1989). The desert camel as meat animals. Meat Sci. 26: 245–254.

17

Effect of experimental *Schistosomiasis mansoni* infection on serum levels of iron, zinc and copper in the olive baboon (*Papio anubis*)

Mungatana N. W. K.[1,2], Ngure R. M.[2], Shitandi A. A.[2*], Mungatana C. K.[2] and Yole D. S.[1]

[1]Institute of Primate Research, P. O. Box 24481, Karen, Nairobi, Kenya.
[2]Department of Biochemistry, Egerton University, P. O. Box 536, Njoro, Kenya.

Schistosoma mansoniasis **is a disease of grave concern due to its high morbidity and mortality in parts of the world. This study aimed at providing insight into the pathogenesis of *S. mansoniasis* as an aid in the development of effective control methods. Iron, zinc and copper concentrations were spectrophotometrically measured in sequential serum specimens obtained from baboons throughout the course of acute *S. mansoni* infection, following curative treatment with praziquantel and following post-treatment challenge with a second cercarial infection. The initial infection resulted in a two-fold increase in copper concentrations by Day 102 post-infection. Iron concentrations fell to almost half of pre-infection concentrations by Day 123 post-infection, while those of zinc fell to a third of pre-infection concentrations by Day 81 post-infection. These changes were seen to recover several weeks following treatment, though pre-infection concentrations were never achieved. Haptoglobin, a sensitive biomarker in the acute phase response of *S. mansoni*, was also measured at all sampling points. Haptoglobin changes were in concordance with those of the cations. The findings demonstrate that iron, zinc and copper are reactants in the acute phase response of *S. mansoni* in the nonhuman primate model, *Papio anubis*. Furthermore, these reactants are modulated in challenge infections and may be important in the immunopathology of the disease.**

Key words: Acute phase response, *Schistosomiasis mansoni*, serum iron, serum zinc, serum copper.

INTRODUCTION

During the acute phase response (APR), which takes place at the very beginning of the inflammatory process, there are changes in concentrations of a large number of serum proteins and cations (Hirvonnen, 2000). These changes are attributed to the host response to tissue injury and are therefore useful indicators for the course and severity of a disease (Alsemgeest, 1994; Mikhail and Mansour, 1982).

The APR is initiated at the site of tissue injury by mononuclear cells, which release a broad spectrum of pro-inflammatory mediators including the pro-inflammatory cytokines TNF-α, IL-1, IL-6 and IFN-γ. These mediators initiate changes in the homeostatic control of the injured animal. However, measurements of cytokines are difficult as the cytokines are very transient in the blood and prone to many interfering factors (Murata et al., 2004). It is therefore more useful to measure the end-point biomarkers, the acute phase proteins, as a gauge of the inflammatory process. Haptoglobin has been demonstrated a sensitive acute phase protein in *S. mansoniasis* in the baboon (Mungatana et al., 2007) and in other hepatosplenic schistosomiasis (El-Sahly et al., 1985).

Iron, zinc and copper ions have also been shown to vary in APRs of certain parasitic diseases (Mikhail and Mansour, 1982; Mwangi et al., 1995; Dede et al., 2008). Zinc and iron concentrations decline substantially, whereas serum copper concentrations may increase (Hayes, 1994; Dede et al., 2008). These changes reflect changes in cation binding of serum proteins (Hoppe et al., 2008), and more importantly, alterations in cellular uptake mechanisms (Anderson et al., 2000; Hirvonen, 2000).

S. mansoniasis is a disease of grave concern due to its high morbidity and mortality in certain parts of the world.

---

*Corresponding author. E-mail: ashitandi@yahoo.com.

Studies that would give insight into the pathogenesis of the disease may aid in development of effective control methods. Since controlled and effective studies of the disease in humans are not possible, the baboon provides an excellent primate model for the disease (Farah et al., 2000). To the best of our knowledge, this work is the first to profile changes that occur in serum cation levels in the APR of primate infection with *S. mansoni*. Furthermore, the changes produced in the serum levels of these cations were down modulated in challenge infection suggesting that these cations may be involved in the immunopathology of *S. mansoniasis*.

MATERIALS AND METHODS

Parasites

Biomphalaria pfeifferi snails were collected from Kangundo Division, Machakos Disrict, Kenya. They were screened by exposure to strong light to ensure that they did not have any schistosomes. The snails were housed in plastic trays at temperature between 25-28°C for 12 h of light /12 h of darkness in a snail room at the Institute for Primate Research, Nairobi. They were fed on lettuce throughout the experimental period. The snails were then infected individually with 3-6 miracidia, artificially hatched from eggs of *S. mansoni* harvested from infected baboon feaces. The infected snails were maintained in the same conditions for four weeks, after which they were put in the dark until they were required for cercarial shedding. Cercariae for infection were then obtained by exposing the infected snails to artificial light (100 watt lamp) for 1-3 h.

Hosts

Ten Kenyan olive baboons, *Papio anubis*, weighing about 6.5 kg and caught in a high altitude, non-schistosomiasis-endemic region, were used for the study. The animals were handled humanely and ethically through out the experimental period under the guidance of the Institute of Primate Research's ethics committee. The animals were quarantined for 90 days, prior to initiation of experiments, during which time they were screened for common bacterial, viral and parasitic infections. They were also tuberculin tested according to the standard procedures at the Institute of Primate Research, Nairobi, Kenya. They were tested for prior schistosomiasis infection by the Kato technique (Katz et al., 1972) and miracidia-hatching test (Yole et al., 1996).

As a further safeguard against the possibility of prior exposure to schistosomiasis, serum was obtained from each animal and tested for specific *S. mansoni* soluble worm antigen preparation (SWAP) IgG using enzyme linked immunosorbent assay (ELISA), as described by Nyindo et al. (1999). Only animals with antibody concentrations below 22 μg/ml $\pm$ 3 standard deviations of the value, obtained from a laboratory colony-born baboon and negative on the Kato test were used in the study. The animals were individually caged and fed on baboon pellets twice a day. They were also fed on fruits and vegetables twice a week, while water was provided *ad libitum*.

Infection schedules comprised a large cercarial dose, of about 1000 cercariae, given once as a single-dose infection. All infections were done percutaneously following anaesthetisation by the pouch method (Farah et al., 2000). Five of the baboons were infected and the remaining five were used as uninfected controls. The controls were sampled as the infected baboons.

The baboons were all sampled for blood on Day 0 (pre-infection) and on Days 27, 54, 81 and 88 post-infection. Additionally, from Day 27 post-infection, all animals were examined weekly by miracidia hatching test for presence of viable schistosome eggs in the stool. The infected animals were then all treated with a curative dose of praziquantel on Day 88 post-infection. This treatment was repeated 14 days later and was accompanied by sampling. All the five baboons were subsequently sampled on Days 123, 136, 164, 221 and 228 post-infection. On Day 228, the baboons were infected with a post-treatment challenge of 1,000 cercariae. The infection was done as previously described. The baboons were then sampled on Days 242, 256, 270, 284 and 298 post-infection.

Blood collection

The baboons were anaesthetized with ketamine/xylaxine and 10 ml of blood was collected from the inguinal vein of each animal, allowed to clot and centrifuged (Farah et al., 2000). The serum thus separated was then stored at -70°C ready for analysis.

Gross pathology

The baboons were euthanised at the end of the experiments by intravenous administration of heparinized sodium pentobarbitol. The viscera were exposed after median incision and the extent of tissue inflammation, the density of granulomas, the extent of heaptic fibrosis and size of mesenteric lymph nodes was assessed to subjectively determine the gross pathology (Farah et al., 2000).

Determination of serum cations

Iron, zinc and copper concentration determinations were spectrophotometrically determined as described by Passey et al. (1985). The serum was digested with spectrosol grade concentrated nitric acid, diluted with de-ionized water and analyzed by aspiration into an atomic absorption spectrophotometer. Iron concentrations were measured at a wavelength of 248.3 nm and a current of 8 mAmps, within a linear range of up to 5 parts per million (ppm). Zinc concentrations were measured at wavelength of 213.9 nm, current 3 mAmps, within a linear range of up to 0.4 ppm. Copper concentrations were determined at a wavelength of 324.7 nm, current 5 mAmps, within a linear range of up to 4 ppm.

Haptoglobin determination

Haptoglobin (Hp) was measured using the method described by Makimura and Suzuki (1982) with modifications by Conner et al. (1988). The assay uses purified bovine Hp as standard. This test is based on the ability of Hp to bind to haemoglobin (Hb) and retain peroxidase activity at acidic pH, whereas free Hb loses its peroxidase activity. On addition of the substrate, peroxidase activity. Resulted in a proportionate color change, which was read off on an ELISA plate reader at 450 nm.

Statistical analysis

Statistical analysis of the data was performed using SPSS and Excel software programmes. Excel was used to compare trends in the analytes measured in infected animals and in uninfected controls. The data were considered statistically significant at p <0.05 as determined by one-way ANOVA.

RESULTS

Changes in the mean serum iron concentrations (±SEM) in *S.*

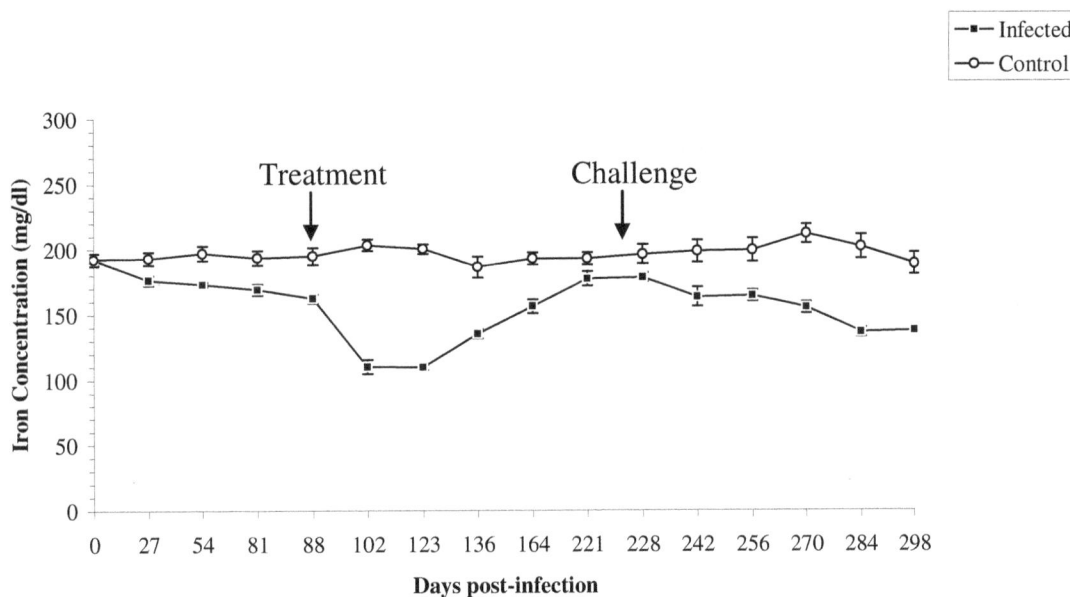

Figure 1. Mean serum iron concentrations (± S.E.M.) in μg/dl in *S. mansoni* infected baboons and in uninfected controls (n=5). Infected animals were subjected to an initial infection at Day 0, treated curatively with praziquantel at Day 88 and given a challenge infection at Day 228.

mansoni infected baboons, treated with praziquantel at Day 88 and challenged at Day 228, and their uninfected controls, are depicted in Figure 1. Following infection, mean iron concentration of the infected animals decreased gradually from pre-infection concentrations 192.0 ± 4.8 μg/dl, to reach 162.0 ±3.4 μg/dl at Day 88 post-infection, when the baboons were treated. Following treatment, the mean levels continued to decrease until Day 102, when the mean iron concentration had decreased to 109.6 ± 2.2 μg/dl. These levels were maintained until Day 123 post-infection. After this, the iron levels started to increase, reaching a peak mean concentration of 178.0 ± 3.4 μg/dl at Day 221 post-infection. At Day 228 post-infection when the baboons were challenged with *S. mansoni* cercariae, peak on concentrations were maintained. Following challenge, the serum iron levels began a gradual decrease to reach mean concentrations of 137.0 ± 2.4 μg/dl by Day 298, when the experiment was terminated. Mean serum concentrations in the infected animals showed significant differences from those of the uninfected controls at all post-infection sampling points (p<0.05).

Changes in the mean serum zinc concentrations (±SEM) in *S. mansoni* infected baboons, treated with praziquantel at Day 88 and challenged at Day 228, and uninfected controls, are depicted in Figure 2. Zinc concentrations of infected animals drastically fell from a mean pre-infection concentration of 230.0 ± 3.5 to 149.0 ± 3.2 μg/dl at Day 27 post-infection. The zinc levels continued to decline to reach 74.0 ± 2.7 μg/dl at Day 81 post-infection. These concentrations were maintained until Day 88 post-infection, when the baboons were treated.

Following treatment, mean zinc concentrations remained at about treatment levels until Day 123 post-infection. After that, they gradually increased to182.0 ± 3.6 μg/dl by Day 228, when the animals were challenged with a second cercarial infection. Following challenge, the zinc levels showed a slight, gradual, fluctuated decrease to reach a concentration of 177.0 ± 8.4 μg/dl at Day 298 post-infection, when the experiment was terminated. Mean serum zinc concentrations of infected animals differed significantly from those of uninfected controls at all post-infection sampling points (p<0.05).

Changes in the mean serum copper concentrations (±SEM) in *S. mansoni*_infected baboons, treated with praziquantel at Day 88 and challenged at Day 228, and uninfected controls, are depicted in Figure 3. Following infection, mean serum copper concentrations showed a gradual increase from Day 27 post-infection, with the increase taking a more drastic trend from Day 54 post-infection to reach levels of 128.0 ± 5.0 μg/dl at Day 88 post-infection, when the baboons were treated. Following treatment, the copper levels continued to increase to reach a peak of 143 ± 5.8 μg/dl at Day 102 post-infection. After this, mean copper concentrations drastically declined to reach near control levels of 83 ± 1.7 μg/dl by Day 228, when the baboons were challenged by a second cercarial dose. Following challenge, the mean copper levels started to gradually increase, reaching levels of 96.0 ± 2.8 μg/dl by Day 298 post-infection when the experiment was terminated. From Day 54 post-infection, mean copper concentrations of the infected animals were statistically different from those of the uninfected controls at p< 0.05, except for days 27, 228, 242, 256 and 270.

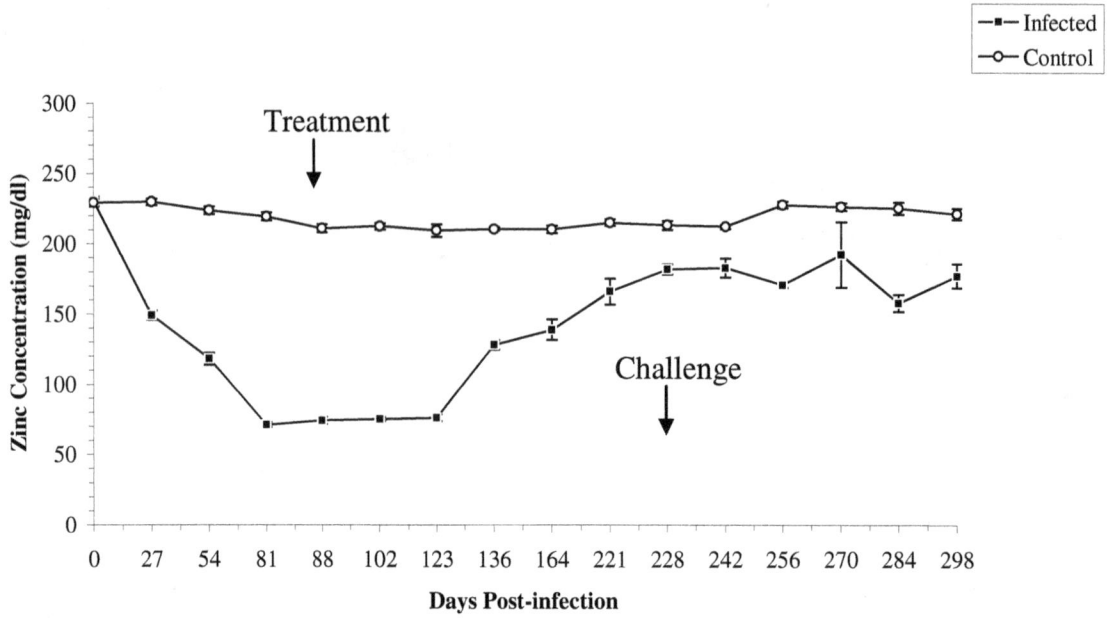

Figure 2. Mean serum zinc concentrations (± S.E.M.) in µg/dl in S. mansoni infected baboons and in uninfected controls (n=5). Infected animals were subjected to an initial infection at Day 0, treated curatively with praziquantel at Day 88 and given a challenge infection at Day 228

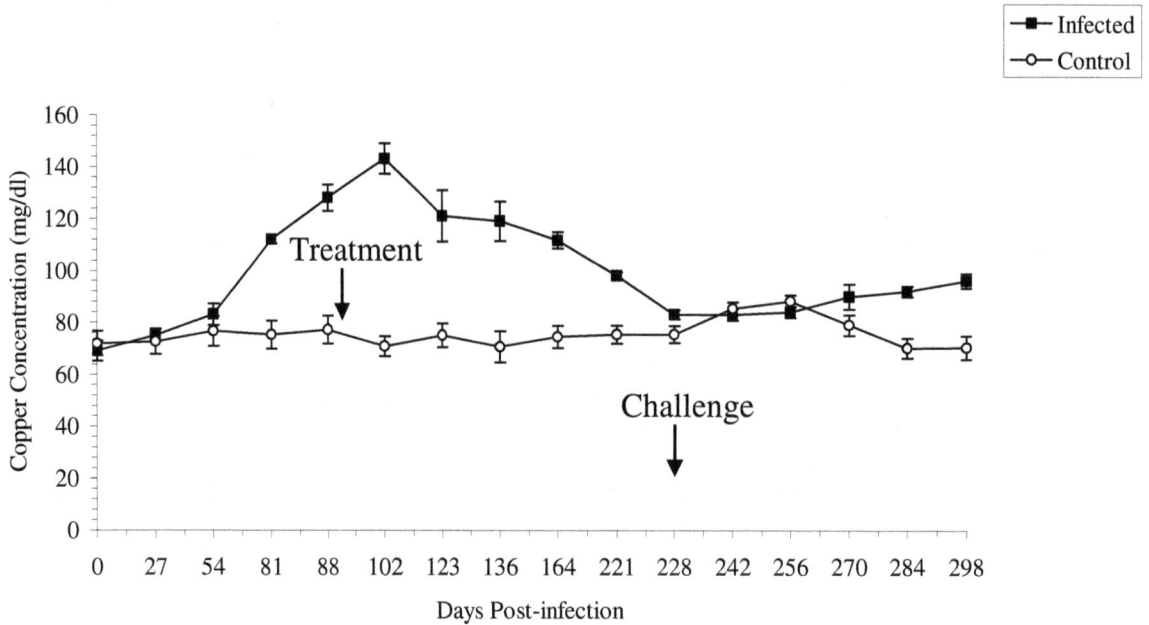

Figure 3. Mean serum copper concentrations (± S.E.M.) in µg/dl in *S. mansoni* infected baboons and in uninfected controls (n=5). Infected animals were subjected to an initial infection at Day 0, treated curatively with praziquantel at Day 88 and given a challenge infection at Day 228.

Changes in the mean serum haptoglobin concentrations (±SEM) in *S. mansoni* infected baboons, treated with praziquantel at Day 88 and challenged at Day 228, and the uninfected controls, are depicted Figure 4. The mean haptoglobin concentrations showed a four-fold increase from pre-infection concentrations of 1.68 ± 0.23 to 8.2 ± 0.81 g/l by Day 88 post-infection when treatment was instituted. Following treatment, the levels slightly increased to 8.4 g/l ± 0.86 g/l on Day 102. After this, they gradually decreased to near pre-infection levels of 2.3 ±

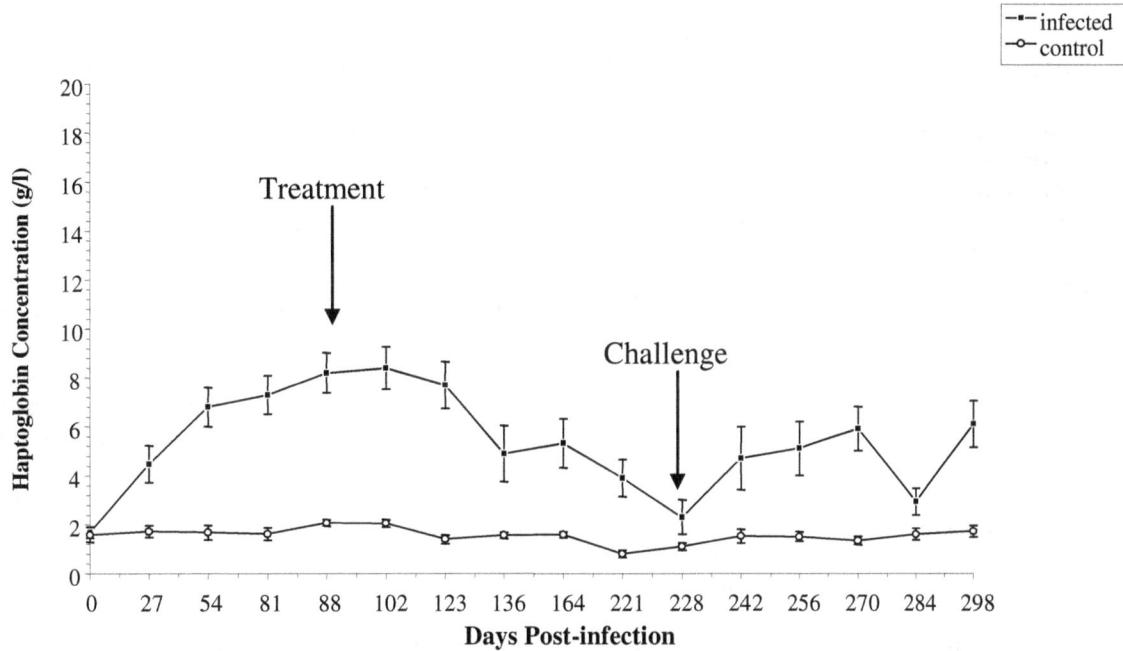

Figure 4. Mean serum haptoglobin concentrations (± S.E.M.) in g/l in *S. mansoni* infected baboons and in uninfected controls (n=5). Infected animals were subjected to an initial infection at Day 0, treated curatively with praziquantel at Day 88 and given a challenge infection at Day 228.

0.71 g/l on Day 228, when the animals were challenged with *S. mansoni* cercariae. On challenge with the parasites the haptoglobin levels showed a slow but gradual decrease to reach a concentration of 6.1 ± 0.95 g/l on Day 298, when the experiment was terminated. The haptoglobin concentrations following infection, and thereafter following challenge, were statistically different from pre-infection concentrations, until at Day 221 post-infection. Mean haptoglobin concentrations of the infected baboons were also significantly different ($P < 0.05$) from those of uninfected controls.

All infected animals tested positive for infection by the miracidia hatching assay from the sixth week post-infection. Also from the sixth week post-infection, all animals exhibited loss of appetite and passed loose stools. At termination of the experiment, the animals appeared to have lost body condition and developed ascitis, hepatomegally, splenomegally and slightly enlarged mesenteric lymph nodes. The livers of the animals demonstrated moderate density of granulomas and tissue fibrosis.

DISCUSSION

Changes in concentrations of serum cations are recognized as one of the first, and most sensitive, indicators of a metabolic change in animals as a result of parasitic infection (Beisel, 1977). In the present experiment, initial infection of the baboons resulted in decreases in serum iron and zinc concentrations, while copper concentrations were seen to increase. Following treatment and subse-

quent recovery of the animals, iron and zinc levels rose while those of copper fell, but pre-infection concentrations were never achieved. When the animals were challenged with a second cercarial infection, iron and zinc levels were seen to drop and copper levels increased. These changes were however more gradual and less pronounced than those following the initial infection.

The response sensitivity of iron was found to be relatively moderate with the greatest decrease of 43.2% observed on Day 102 of the experiment. These results agree with a study by Laudage and Schirp (1996), who found that schistosomiasis, was a rare cause of iron-deficiency anaemia. Moreover, Serum iron has been found to decline substantially in the APR of a number of infections (Lohuis et al., 1988a and 1988b; Mwangi et al., 1995). Decline of serum iron during the APR is a reflection of diminished brush border iron uptake and increase in iron binding proteins such as haptoglobin (Anderson et al., 2000), hepcidin (Hoppe et al., 2008), transferrin and ferritin (Sheikh et al., 2007). These proteins bind iron and preserve it, in an effort to prevent iron deficiency anaemia (Hirvonen, 2000). Indeed, upregulation of expression of genes responsible for synthesis of these iron-binding proteins has been demonstrated in acute phase response (Sheikh et al., 2007). A concurrent study by Mungatana et al. (2007) done on similarly infected baboons found evidence that haptoglobin is indeed a sensitive reactant in the APR of *S. mansoni*. Reduction of iron might be of importance in the resistance of infections which causative organisms require iron for growth (Hirvonen, 2000).

During the APR of many illnesses, there is a suddenly developing hypozincemia as a result of a sequestration of zinc within hepatocytes, thymocytes and marrow cells. Zinc sequestration then usually persists throughout the duration of the illness (Lao et al., 1987; Rhodes and Kluh, 1993). This pattern was observed in the present study when zinc dropped fairly rapidly following infection with S. mansoni. This drop was sustained throughout the illness before curative treatment, and a decrease as low as 69.1% was achieved on Day 81 of the experiment. The sequestration of zinc is the secondary consequence of the induced expression, caused by pro-inflammatory cytokines, of metallothionein 1 and 2 genes within these responding cells. Newly synthesized metallothionein proteins generate binding sites for zinc, and lead to the hypozincemia, which characterizes acute phase reactions (Lao et al., 1987; Rhodes and Kluh, 1993;, Taylor, 1996). In addition, zinc is a component of many enzymes, including thymulin, and other protein structures such as the zinc fingers in transcription activation factors.

Dramatic loss of zinc may therefore occur during catabolic processes, especially following severe infections. Acute zinc deficiency may then be precipitated as anabolic processes requiring zinc supervene (Vallee and Falchuk, 1993).

Copper was observed to increase up to 61% on Day 42 post- infection. In a previous similar mice experiment (Mungatana et al., 2006), copper responses were observed to be poor, with increases of only up to 16% on Day 42 post- infection. The differences in responses between mice and baboon models could be due to variations in plasma half-lives of copper between the two species. The major circulating form of copper is the blue glycolprotein, ceruloplasmin, which is synthesized in the liver. Each molecule of ceruloplasmin contains 6-8 atoms of copper. The functions of this acute phase protein are still unclear, but it is important in iron metabolism as a ferroxidase, and may have a role in regulating copper transport (Danks, 1995). It is an acute phase reactant and can increase greatly in response to infection, injury, chronic inflammatory conditions or steroid hormones (Hirvonen, 2000). Serum copper and ceruloplasmin are both increased in these circumstances as ceruloplasmin normally carries about 95% of the circulating copper (Taylor, 1996).

Depressed serum zinc and elevated serum copper levels have been demonstrated in the acute phase responses of various parasitic infections (Dede et al., 2008). A study by Mikhail and Mansour (1982) found that plasma copper in patients with active S. mansoni infection was significantly elevated and zinc was significantly depressed, and the degree of this correlated with the associated hepato-splenic complications of the disease. At clinical recovery, levels of copper and zinc were significantly improved in all the patients from pre-treatment values, but were within the normal range only in patients without complications. This indicated that complications associated with schistosomiasis could delay normalization of copper and zinc levels beyond the clinical cure stage. The mice study by Mungatana et al. (2006) found that some residual tissue pathology persists for some time even after curative treatment. These observations were echoed in this study when complications, including hepatomegally, splenomegally and ascites, were noted in the infected animals at perfusion. This may have hindered resolution of the acute phase response, and attainment of pre-infection concentrations of the acute phase reactants.

The pathophysiological mechanisms involved in the changes in cation concentrations are not well understood. The anorexia observed in most infected hosts could also be a contributory factor to the changes seen in serum levels of the cations. Moreover, a phenomenon termed "nutritional immunity" was proposed in which the host prevents the multiplication of bacteria in body fluids by preventing their acquisition of nutritionally important trace minerals (Weinberg, 1978). Such a mechanism has been well characterized for iron, which is tightly chelated in extracellular fluids by transferrin and lactoferrin (Weinberg, 1978; Mwangi et al., 1995). However, "nutritional immunity" involving zinc or copper has not yet been demonstrated. Adequate zinc levels are however known to be necessary for almost all manifestations of an animal's immune response, including macrophage phagocytosis and intracellular killing, the early activation stages of antibody formation, and sequestration of zinc in the APR may then be a means of the body to conserve the vital cation. In addition, macrophage activation may also be enhanced by low serum zinc values (Taylor, 1996). Copper also is an essential trace element as copper-containing metallo-enzymes are important in iron and catecholamine metabolism, haemoglobin, elastin and collagen synthesis and free radical scavenging (Cousin, 1985; Danks, 1995).

After treatment, all the cations monitored showed positive trends back towards their normal pre-infection concentrations, in response to tissue recovery. When the baboons were challenged with a second cercarial dose 228 days after the first infection, the post-challenge serum concentrations of copper and iron attained significant differences ($p < 0.05$) from pre-infection concentrations. The post-challenge concentrations of copper, though elevated, did not significantly differ from post-infection/pre-challenge concentrations. The post-challenge changes were notably less marked than those following the initial infection. This may be explained by the fact that the liver had not fully regained its original capacity for synthesis, following fibrosis caused by the first infection, by the time of the challenge infection. Moreover, hepatic injury has been shown to be moderated by repeated exposure to S. mansoni cercariae (Farah et al., 2000), thus producing a moderated APR. Roth et al. (1994) also found that repeated challenge of a host results in attenuation of the release of cytokines involved in the APR, which results in downmodulation of the APR. This is an important observation as most human infections in endemic areas occurs repeatedly.

Conclusion

The results of the study indicate that during the acute phase of *Schistosoma mansoni* infection of baboon models, there is a characteristic increase in serum copper concentrations and a decrease in serum iron and zinc concentrations. These changes were seen to reduce gradually following curative treatment. However, the concentrations did not attain normal haematological values even after curative treatment, probably because tissue trauma and inflammation had not completely ceased or due to down regulation of pro-inflammatory cytokines such as TNF-α, IL-1, IL-6 and IFN-γ.

ACKNOWLEDGMENTS

The authors are grateful to the staff at the Institute of Primate Research, Karen, and the Biochemistry Department of Egerton University for all their technical assistance.

REFERENCES

Alsemgeest M (1994). Blood concentrations of acute phase proteins in cattle as markers for disease. Phd. Thesis, Faculty of Verterinary Medicine, Utrecht University, pp 39-54.

Anderson GJ, Frazer D, Wilkins S, McKiie T, Becker M, Murphy TL (2000). Impaired intestinal iron transport during the acute phase reaction. Queens Instit. Med. Res. 14: 01-02.

Beisel WR (1977). Magnitude of host nutritional response to infection. Am. J. Clin. Nutr. 30:1236-1247.

Cousin RJ 1(985). Absorption, transport, and hepatic metabolism of copper and zinc. Am. J. Physiol. 65:238-309.

Danks DM (1995). Disorders of copper transport. The Metabolic Basis of Inherited Disease. 7th edn, NewYork, McGraw-Hill, pp 2211-2235.

Dede S, Deger Y, Deger S, Tanritanir P (2008). Plasma levels of zinc, copper, copper/zinc ratio, and activity of carbonic anhydrase in equine piroplasmosis. Bio.l Trace Element Res. 25(1):41-45.

Farah IO, Nyindo M, King CL, Hau J (2000). Hepatic granulomatous response to Schistosoma mansoni eggs in BALB/c mice and olive baboons. J. Comprehensive Pathol. 12: 7-14.

El-Sahly AM, El-Tablawy ZM, Ahdel-Rahman HM (1985). Serum haptoglobin levels in hepatosplenic schistosomiasis. J. Egyp. Med, Assoc. 6: 8151-159.

Hayes, MA (1994). Functions of cytokines and acute phase proteins in inflammation. Lumsden J. Health (ed) 6th Congress of the ISACB proceedings, Guelph, Canada. 194: 1-7.

Hirvonen J (2000). Hirvonen's Phd thesis on acute phase response in dairy cattle. Helsingin Yliopiston Verkkojulkaisut. University of Helsinki, Faculty of Veterinary Medicine, pp 1-300.

Hoppe M, Lonnerdal B, Hossain B, Olsson S, Nilsson F, Lundberg PA, RoDjer S, Holthen L (2008). Hepcidin, Interleukin-6 and hematological iron markers in males before and after Heart Surg. (2008). J. Nutri. I Biochem. (Epub ahead of print).

Katz ZN, Chaves A, Pellegrino J (1972). A simple device for quantitative thick smear technique in Schistosomiasis mansoni. Res. Inst. Trop. Med. Sao Paulo. 14: 397-400.

Lao TH, Chin RH, Swaminathan R, Panesar NS Cockram CS (1987). Erythrocyte zinc in differential diagnosis of hyperthyroidism in pregnancy: a preliminary report. British Med. J. 294: 1064-1065.

Laudag G, Schirp J (1996). Schistosomiasis – a rare cause of iron deficiency anaemia. Leber Magen Darm. 6:216-218.

Lohuis A, Verheijden H, Burvenic C, Miert S (1988a). Pathophysiological effects of endotoxins in ruminants: Changes in body temperature and reticulo-rumen motility, and the effects of repeated administration. Vet. Q. 10: 109-116.

Lohuis JA, Verheijden H, Burvenic C, Miert S (1988b). Pathophysiological effects of endotoxins in ruminants: metabolic aspects. Vet. Q. 10: 117-125.

Mikhail M, Mansour M (1982). Complications of human schistosomiasis and their effect on levels of plasma copper, zinc and serum Vitamin A. Clin. Nutr. 36: 289-296.

Mungatana NWK, Ngure R, Kariuki SM, Yole DS. (2006). Assessment of the acute phase response in experimental infection of mice with Schistosoma mansoni. The Int J. Trop. Med. 3;1.

Mungatana NWK, Ngure RM, Yole DS (2007). Acute Phase response of albumin and haptoglobin in experimental infection of the olive baboon, Papio Anubis, with Schistosoma mansoni. Scandinavian J. Lab. Anim. Sci. 34 (2): 119-126.

Murata H, Shimada N, Yoshioka N (2004). Current research on acute phase proteins in veterinary diagnosis: an overview. Vet. J. 168: 3-5.

Mwangi SM, McOdimba F, Logan-Henfrey L (1995). The effect of Trypanosoma brucei brucei infection on rabbit plasma iron and zinc concentrations. Acta Tropica. 59: 283-291.

Nyindo MT, Kariuki P, Mola P, Farah I, Elson L, Blanton R, King C (1999). Role of adult worm antigen-specific IgE in acquired immunity to Schistosoma mansoni infection in baboons. Infect. Immunol. 67: 636-638.

Passey RB, Malif KC, Fuller R (1985). Quantitation of zinc in nitric acid-digested serum by atomic absorption spectrophotometry. Ann. .Biochem. 151: 462-465.

Rhode D, Kluh A (1993). Zinc fingers. Sci. Am. J. 28: 1-2.

Roth J, McClellan JL, Kluger MJ, Zeisberger E (1994). Attenuation of fever and release of cytokines after repeated injection of lipopolysaccharide in guinea-pigs. J. Physiol. 477: 177-185.

Sheikh N, Dudas J, Ramadori G (2007). Changes of gene expression of iron regulatory proteins during turpentine oil-induced acute phase response in the rat. Lab. Invest. 87: 713-724.

Sturrock RF, Butterworth AE, Houba V (1976). Schistosoma mansoni in the baboon (Papio anubis): Parasitological responses of Kenyan baboons to different exposures of a local parasite strain. Parasitology 73: 239-253.

Taylor A (1996). Detection and monitoring for disorders of essential trace elements. Ann. .Clin. Biochem. 33:486-510.

Vallee BL, Falchuk KH (1993). The biochemical basis of zinc physiology. Physiol. Rev. 73:79-118.

Weinberg ED (1978). Iron and infection. Microbiol. Rev. 42:45-66.

Yole DS, Pemberton R, Reid GDF, Wilson RA (1996). Protective immunity to Schistosoma mansoni induced in the olive baboon, Papio anubis by the irradiated cercariae vaccine. Parasitology. 112: 37-46.

18

Effect of ferric oxide nanoparticles on microtubules organization

Ali Khaleghian[1,2], Gholam Hossein Riazi[2], Shahin Ahmadian[2], Mahmoud Ghafari[2], Marzieh Rezaie[1], Akira Takahashi[3], Yutaka Nakaya[3] and Hossein Nazari[1,3]*

[1]Department of Biochemistry and Hematology, Semnan University of Medical Sciences, Semnan, Iran.
[2]Department of Biochemistry and Biophysics, University of Tehran, 13145-1384, Iran.
[3]Department of Nutrition and Metabolism, University of Tokushima, Tokushima, Japan.

Nanoparticles (NPs) are widely used in several manufactured products. The small size of nanoparticles facilitates their uptake into cells as well as transcytosis across epithelial cells into blood and lymph circulation to reach different sites, such as the central nervous system. Studies have shown different risks of Fe_2O_3NPs in the neuronal system and other organs. They are membrane-bound layer aggregates or single particles that could not enter only cells, but also in mitochondria and nuclei. Therefore, these particles can interact with cytoplasmic proteins such as microtubules (MTs). MTs are cytoskeleton proteins that are essential in eukaryotic cells for a variety of functions, such as cellular transport, cell motility and mitosis. MTs play an important role in neurons and to act as a substances transport such as neurotransmitters. Single Fe_2O_3NPs in cytoplasm can interact with these proteins and affect their crucial functions in different tissues. In this study, we showed the effects of Fe_2O_3NPs on MTs organization and structure using ultraviolet spectrophotometer and fluorometry. The fluorescent spectroscopy showed a significant tubulin conformational change in the presence of Fe_2O_3NPs and the ultraviolet spectroscopy results showed that Fe_2O_3NPs causes MTs depolymerization and decrease turbidity intensity as well as increase spectra emission. The aim of this study was to find the potential risks that Fe_2O_3NPs pose to human organs and cells for cancer treatment.

Key words: Ferric oxide, tubulin, microtubule, protein interaction, nanoparticle.

INTRODUCTION

Ferric oxide nanoparticles (Fe_2O_3NPs) are used in numerous manufactured products, including cosmetics, sunscreen, toothpaste and medicines. However, there is insufficient knowledge about the potential risks they posses (Baalousha et al., 2008). The effect of nanoparticles (NPs) on the human body is increasing concern as these particles are found in an ever-expanding variety of goods. Humans are exposed to NPs through inhalation, ingestion, dermal contact and injection (Baalousha, 2008; Frazer, 2001). The small size of NPs facilitates their uptake into cells as well as transcytosis across epithelial cells into blood and lymph circulation to reach sensitive target sites where they then persist (Lomer et al., 2004). Previous studies have observed the translocation of NPs along axons and dendrites of neurons as well as access to the central nervous system and ganglia (Stearns et al., 2001). Therefore, the study of the potential risks presented by NPs is immediately needed.

The lungs are continuously exposed to environmental particles (Rutishauser et al., 2007). Several studies have reported phagocytosis of particles by lung epithelial cells (Stearns et al., 2001). Fe_2O_3NPs do not clear from the cells; rather, they persist there, and their concentration is increases. Analytical transmission electron microscopy of different cell culture types showed single Fe_2O_3 particles, small aggregates free and membrane-bound layer aggregates in cytoplasm after exposure to Fe_2O_3NPs. In another study, Fe_2O_3 (20 - 30 nm) was detected as free single particles in cytoplasm. Exposure cells to Fe_2O_3NPs leads to their accumulation in different organs where

*Corresponding author. E-mail: hossen253@yahoo.co.uk.

there are some vital proteins, such as microtubule, that can interact with them (Oberdorster et al., 2004). Although many studies have been done on Fe_2O_3 toxicity in animal models and cell cultures (Oberdorster, 2004; Muhlfeld, 2007), there is little data on the interaction of Fe_2O_3NPs with sub cellular structures.

MTs are cytoskeletal proteins that are essential in eukaryotic cells for a variety of functions, such as cellular transport, cell motility and mitosis. They are crucial in the development and maintenance of cell shape, the transport of vesicles, organelles and other components throughout cells, cell signaling, cell division and mitosis. MTs in neurons are used to transport substances such as neurotransmitters. MTs are formed from polymers of tubulin heterodimers (α- and β-tubulin) that polymerize from end to end (Lowe, 2001; Desai, 1997). Agents that interfere with MTs assembly also interfere with their dynamics and function, and inhibit all MTs functions such as cell division and neurotransmitter transportation (Downing, 2000; Jordan, 2004). Evidence of oxidative stress responses after NPs endocytosis indicates that research on possible cellular interactions between such particles and vital proteins, such as microtubules (MTs), is urgently needed. By studying NPs' biological and toxicological effects, its interactions with sub cellular structures, such as MTs, may be clarified.

To investigate the effects of Fe_2O_3NPs on intracellular interactions, we studied the effects of Fe_2O_3NPs on tubulin organization and MTs protein function by ultraviolet (UV) spectrophotometer and fluorescent spectroscopy. To understand the mechanism of these reactions, repolymerization tests were carried out. Fe_2O_3NPs were added after tubulin polymerization to discover whether Fe2O3NPs disrupted the structure of MTs. Fluorescent spectroscopy was used to study conformational changes in tubulin that altered its function.

MATERIALS AND METHODS

Materials

Ultra fine Fe_2O_3, with particles averaging 100 nm, was a gift from Dr. mohmadi (engineering faculty, University of Tehran, Tehran, Iran). The particles were suspended in piperazine-1,4-bis (2-ethanesulfonic acid) (PIPES) buffer (Merck, Darmstadt, Germany) to obtain an 8-mg/ml final concentration. Before use, KOH was added to adjust the pH of the colloidal solution to 6.9. Particles were then sonicated with a Bandelin sonicator (Bandelin, Berlin, Germany) for 3 min and immediately added to the protein solution. EGTA, guanosine-5'-triphosphate (GTP), ATP, glycerol and $MgSO_4$ were acquired from Sigma (Dorset, England). Phosphocellulose P11 was obtained from Whatman (Florham Park, USA). All other chemicals (NaCl, KOH and ANS, Merck) were from analytical grade and used without further purification. All solutions were prepared with double distilled water and were kept at 4°C before use.

Purification of tubulin

After homogenization in PEM buffer (100 mM PIPES, pH 6.9, 1 mM EGTA, 2 mM $MgSO_4$) and 1 mM MgATP, followed by two cycles of temperature dependent assembly and disassembly, MTs proteins were prepared from sheep brains. PMG (100 mM PIPES, pH 6.9, 2 mM $MgSO_4$, 1 mM EGTA and 3.4 M glycerol) was used as polymerization buffer. Microtubule-associated proteins tubulin was prepared by chromatography on phosphocellulose P11 with a slight modification of the method used by Weingarten et al. (1974). Eluted tubulin fractions were stored at 70°C for further study. The protein concentration was determined using the Bradford reagent (Bio-Rad, Hercules, USA) with bovine serum albumin as standard (Marshal and William, 1993).

UV spectroscopy

Turbidimetric assay of MTs and tubulin was carried out by incubating the protein in PIPES buffer (with the final concentration of 2 mg/ml) in cuvettes at 37°C in a thermostatically controlled UV spectrophotometer (Varian, Melbourne, Australia). Turbidity change was measured at 350 nm. To examine the effect of NPs on polymerization, the MTS and tubulin proteins were pre-incubated with NPs at 4°C for 30 min, and polymerization was initiated with the addition of 1 mM GTP. The mixture was warmed to 37°C.

Fluorescence spectroscopy

All fluorescence experiments were carried out using a Varian eclipse spectrofluorometer equipped with a computer to add and subtract spectra. Denaturation of tubulin was measured as tryptophan emission after excitation at 295 nm. Interactions in the presence of increasing concentrations of Fe_2O_3NPs were carried out at 25°C. To test conformational changes, 8-anilino-1-naphthalenesulfonic acid (ANS) was used to detect whether the Fe_2O_3-treated tubulin had an exposed hydrophobic surface area. The excitation wavelength was 380 nm, and emission was monitored between 450 - 550 nm. All measurements used 2 µM tubulin and all experiments were carried out at 25°C.

Electronic microscopy

Exposure of tubulin samples to 37°C for MTs polymerization was studied by electronic Microscopy. Tubulin polymerization was occurring in suitable temperature as a control. Inhibitory effect of Fe_2O_3 on MTs polymerization process was concentration dependent manner.

RESULTS

Inhibition of MTs polymerization by Fe_2O_3NPs

The effect of Fe_2O_3NPs on tubulin polymerization was measured as shown in Figure 1. MTs assembly was clearly inhibited by different concentrations of Fe_2O_3NPs in 2 mg/ml, MTs solution comparison with control sample. Specifically, Fe_2O_3NPs inhibited both the rate and extent of MTs assembly, and it had influence on the rate of MTs nucleation by increasing the time.

MTs repolymerization assay

MTs organization (assembled and disassembled) was

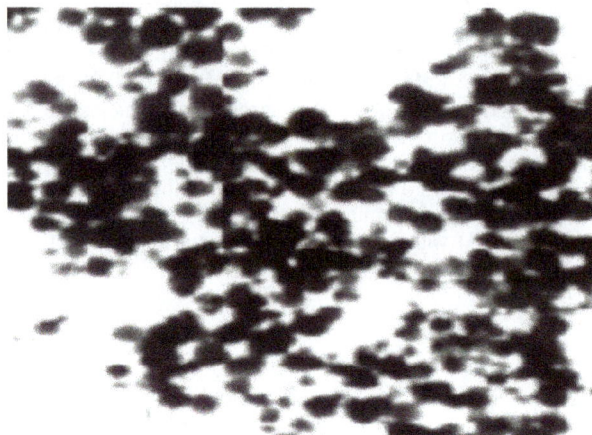

Figure 1. Structure of γ-Nano Fe2O3 powder, 99% magnetic. APS: 20 - 30 nm, SSA: >50 m2/g - Color: red brown, Morphology: nearly spherical, Bulk Density: 1.20 g/cm3, True density: 5.24 g/cm3,
APS = Average Particle Size, SSA = Special surface area.

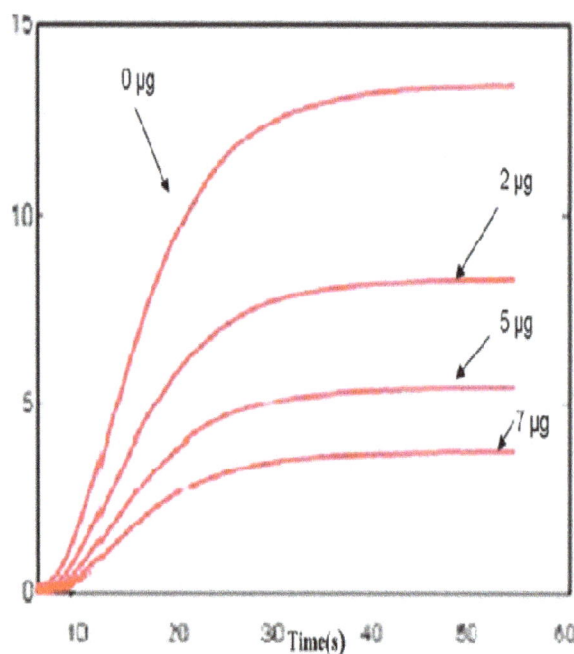

Figure 2. Effect of Fe_2O_3 on microtubule organization. Proteins (2 mg/ml) were preincubated at 4 °C in piperazine-1,4-bis (2-ethanesulfonic acid) buffer with different concentration of Fe_2O_3. Polymerization was initiated by adding 1 mM guanosine-5'-triphosphate. The turbidity was monitored at 350 nm at 37 °C. 0 μg/ml Fe_2O_3; 2 μg/ml Fe_2O_3; 5 μg/ml Fe_2O_3; 7 μg/ml Fe_2O_3.

cooling to 4 °C, re-warming the solution to 37 °C which induced assembly, as a control. Repolymerization assays showed that polymerized MTs made depolymers again after 30 min incubation in 4 °C (Figure 2).

Effect of Fe_2O_3NPs on MTs dynamics at steady state

By adding Fe_2O_3 to a MTs solution at equilibrium point (Figure 3, vertical arrow), a decrease in turbidity was observed and the reaction reached a new equilibrium. The new equilibrium was the same as the equilibrium observed for MTs treated with Fe_2O_3NPs at zero time (Figure 1).

Intrinsic fluorescence spectra

To obtain structural information at the tertiary level, intrinsic (tryptophan) fluorescence spectrum of tubulin in the presence of different concentrations of Fe_2O_3NPs was measured. Fluorescence analysis indicated that the interaction of Fe_2O_3NPs with tubulin resulted in fluorescence quenching of surface-exposed tryptophans in tubulin. Figure 4 shows that the fluorescence intensity decreases with increasing amount of Fe_2O_3NPs.

Increasing of tubulin-bis-ANS fluorescence by Fe_2O_3NPs

There are several low affinity sites and one high affinity site for the polar molecule ANS in tubulin. Tubulin-ANS complex has a strong fluorescence and is extremely environmentally sensitive. Therefore, it is a useful tool for probing the conformational state of the tubulin dimer. Tubulin-ANS fluorescence has been used to determine the nature of interactions. Tubulin (2 μM) was incubated in the presence of various concentrations of Fe_2O_3NPs for 10 min at 4 °C. ANS (50 μM final concentration) was added to the tubulin Fe_2O_3NPs solution and incubated again for 7 min. Figure 5 show that Fe_2O_3NPs made a concentration-dependent increase in tubulin-ANS fluorescence. Furthermore, incubation of tubulin with ANS before the addition of Fe_2O_3 generated similar results (data not shown).

Effect of Fe_2O_3 on MTs organization

In agreement the behavior of Fe_2O_3 on MTs organization will down electronics microscopy experiments. Exposure of tubulin samples to 37 °C for MTs polymerization was inhibited by Fe_2O_3. Tubulin assembling occurs in suitable temperature as a control. Inhibitory effect of Fe_2O_3 on MTs polymerization process was concentration dependent manner (Figure 5).

Discussion

Although Fe_2O_3NPs are widely used in various commercial products, there is insufficient knowledge about their side effects (Brannon-Peppas, 2004; Brunner,

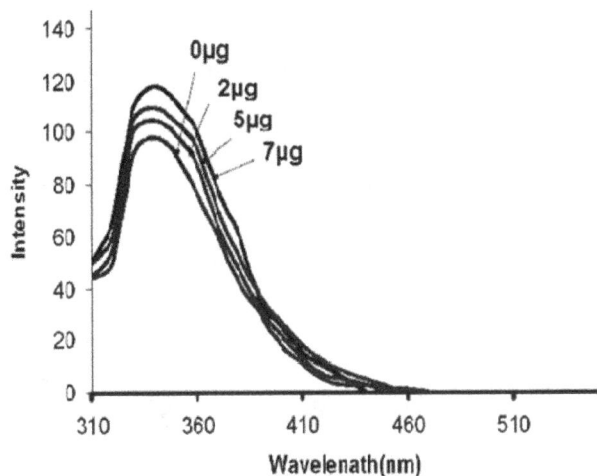

Figure 3. Emission spectra (excited at 295 nm) of 2 μM tubulin in the presence of increasing concentrations of Fe2O3. The solution conditions were 1 M piperazine-1,4-bis (2-ethanesulfonic acid), pH 6.9, containing 1 mM EGTA and 2 mM MgSO4.The excitation and emission band passes were 5 nm. 1) 0 μg/ml Fe_2O_3; 2 μg/ml Fe_2O_3; 5 μg/ml Fe_2O_3; 7 μg/ml Fe_2O_3.

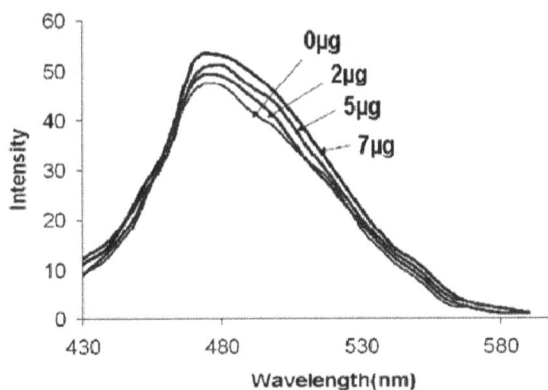

Figure 4. Emission spectra (excited at 380 nm) of 2 μM tubulin in the presence of increasing concentrations of Fe_2O_3.
Tubulin was mixed with 0 μg/ml Fe_2O_3 (curve 1), 2 μg/ml Fe_2O_3 (curve 2), 5 μg/ml Fe_2O_3 (curve 3) and 7 μg/ml Fe_2O_3 (curve 4) for 10 min at 4 °C.The 8-anilino- 1-naphthalenesulfonic acid (50 μM final concentration) was added and after 7 min fluorescence was measured. The solution conditions were 1 M piperazine-1,4-bis (2-ethanesulfonic acid), pH 6.9, containing 1 mM EGTA and 2 mM MgSO4. The excitation and emission band passes were 5 nm.

2006). Studies have shown that some neurons exposed to Fe_2O_3 initiate a cellular process that can ultimately lead to cell death, and according to Brannon et al, ultrafine Fe_2O_3 induces apoptosis (Brannon et al., 2004). There are few reports about the cytotoxic and genotoxic effects of Fe_2O_3NPs. Some studies have shown an interaction

between Fe_2O_3NPs and some proteins, such as human plasma fibrinogen, but the toxic effect of Fe_2O_3NPs on MTs protein has not yet been elucidate (Fortina, 2004; Ferrari, 2005).

In this study, 0̄.50 μg/ml Fe_2O_3NPs was used. We showed that Fe_2O_3NPs inhibited tubulin polymerization and decrease MTs nucleation rate. The inhibition of tubulin polymerization depended on the concentration of Fe_2O_3NPs (Figure 1). Assembled MTs were disassembled by cooling and then reassembled by re-warming; repolymerization was observed (Figure. 2). These results show that Fe_2O_3NPs inhibited MTs repolymerization. Also, Fe_2O_3 did not induce formation of MTs aggregate even at high concentrations, while some agents at high concentrations induced irregular aggregates of MTs at that cooling state and did not appear to depolymerize these aggregates (Valiron and Caudron, 2001).

Fe_2O_3NPs decreased tubulin polymerization, and data showed that Fe_2O_3NPs also induce MTs depolymerization and change the MTs steady state equilibrium to a new equilibrium (Figure 3). Results show approximately 35 mg/ml IC_{50} for Fe_2O_3NPs. In the presence of 20 mg/ml Fe_2O_3NPs, the normal activity was 66%, and in 50 mg/ml Fe_2O_3NPs, activity decreased to 33%. Results indicated that Fe_2O_3 affected both soluble tubulin and tubulin in MTs structure, suggesting that tubulin conformational change led to decreased tubulin polymerization ability.

Using intrinsic fluorescent spectroscopy, we identified changes in protein conformation involved in protein function alteration. We used excitation wavelength at 295 nm so changing in emission wavelength showed tryptophan environment changes (Bhattacharya et al., 1996). Fe_2O_3NPs modified the polarity in the vicinity of tryptophan residues, and we therefore observed fluorescence quenching and the maximum blue shift of the emission wavelength (Figure 4). Fluorescence experiments with ANS demonstrated that Fe_2O_3NPs induce increases in fluorescence emission (Figure 5). The tubulin ANS complex's increase in fluorescence may result from exposing some of intrinstic tubulin's hydro-phobic pockets for ANS binding (Sarkar et al., 1995). Alternatively, binding may induce a conformational change in tubulin leading to increased ANS binding or tubulin-ANS fluorescence.

These results indicated that Fe_2O_3NPs induce conformational changes in tubulin that cause changes in tryptophan position, moving them towards GTP binding sites in protein structures. GTP has fluorescent quenching ability (Solomaha and Palfrey, 2005), so intrinsic fluorescence was decreased and blue shift was observed. Conformational changes in protein allowed some hydrophobic pockets to be reached, increasing tubulin-ANS fluorescence. Both GTP and its binding site in tubulin have a crucial role in tubulin polymerization. GTP should be hydrolyzed to guanosine-5′-diphosphate when tubulin wants to polymerize to MTs form (Desai and Mitchison, 1997). Fluorescence data showed conformational changes in

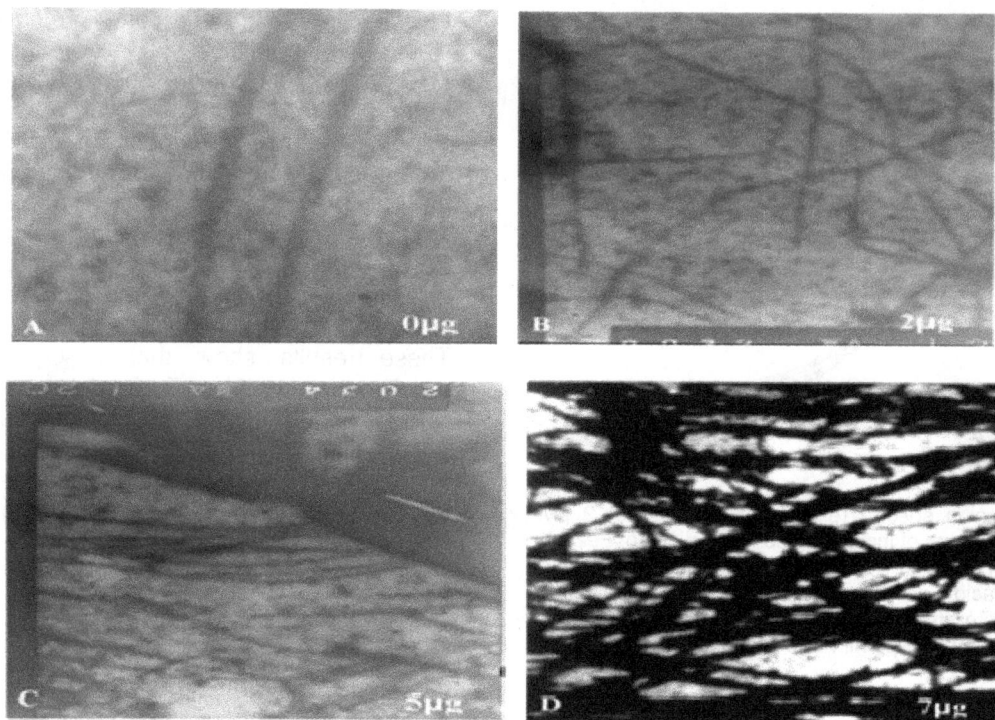

Figure 5. Effect of Fe2O3 on MTs organization.
Exposure of Tubulin samples to 37°C for MTs polymerization was studied by electronic Microscopy.
(A) Tubulin assembling occurs in suitable temperature as a control. Inhibitory effect of Fe2O3 on MTs
polymerization process was concentration dependent manner. Tubulin was mixed with 2 µg/ml Fe_2O_3
(B) 5 µg/ml Fe_2O_3 (C) 7 µg/ml Fe_2O_3 and (D) for 10 min.

GTP binding sites, and turbidimetric assays demonstrated that these changes result in the suppression of tubulin polymerization (Sparreboom et al., 2005).

In conclusion, our research has shown that Fe_2O_3NPs have an inhibitory effect on tubulin polymerization. Experiments demonstrated that 2 mg/ml tubulin protein activity fell to 50% in the presence of 35 µg/ml Fe_2O_3NPs. Furthermore, the same concentrations of Fe_2O_3NPs that disrupted MTs performance inhibited tubulin polymerization. Fe_2O_3NPs interact with both tubulin and MTs protein, which change their folding and results in function alteration. The GTP binding site in tubulin is also affected by Fe_2O_3NPs, leading to tubulin function changes. Finally In this study we find the potential risks that Fe_2O_3NPs pose to human organs and cells for cancer treatment. Ultimately, long-term exposure to NPs can be dangerous, and companies should further research on the hazards associated with NPs.

ACKNOWLEDGEMENTS

We would like to thank Dr. Mohhamadi (engineering faculty, University of Tehran for preparing Fe_2O_3NPs. This work was supported by a grant from the Institute of Biochemistry and Biophysics, Tehran University, Tehran, Iran.

REFERENCES

Baalousha M, Maniculea A, Cumberland S, Lead J, Kendall K (2008). Aggregation and Surface Properties of Iron Oxide Nanoparticles. Environ.l Toxico. Chem.

Frazer L (2001). Titanium dioxide: environmental white knight. Environ Health Perspect. (109): 174-177.

Lomer MC, Hutchinson C, Volkert S, Greenfield SM, Catterall A, Thompson RP, Powell JJ(2004). Dietary sources of inorganic microparticles and their intake in healthy subjects and patients with Crohn's disease. Br. J. Nutr. 92: 947-955.

Stearns RC, Paulauskis JD, Godleski JJ(2001). Endocytosis of ultrafine particles by A549 cells. Am. J. Respir. Cell. Mol. Biol. 24: 108-115.

Rothen-Rutishauser B, Muhlfeld C, Blank F, Musso C, Gehr P (2007). Translocation of particles and inflammatory responses after exposure to fine particles and nanoparticles in an epithelial airway model. J. Part. Fibre. Toxicol. P.49.

Oberdorster G, Sharp Z, Atudorei V, Elder A, Gelein R, Kreyling W, Cox C (2004). Translocation of inhaled ultrafine particle to the brain. J. Inhal. Toxicol. 16: 37-45.

Muhlfeld C, Geiser M, Kapp N, Gehr P, Rothen-Rutishauser B (2007). Evidence for clearance through microvasculature. J. Part. Fibre. Toxicol. P. 47.

Lowe JLi H, Dowing H, Nogales E (2001). Refined structure of tubulin at 3.5 Å resolution. J. Mol. Biol. (313): 1045-1057.

Desai A, Mitchison TJ (1997). Microtubule polymerization dynamics. J. Annu. Rev. Cell. Dev. Biol. (13): 83-117.

Downing KH (2000). Structural basis for the interaction of tubulin with proteins and drugs that affect microtubule dynamics. J. Annu. Rev. Cell. Dev. Biol. 16: 89-111.

Jordan MA, Wilson L (2004). Microtubules as a target for anticancer drugs. J. Nature. Rev. (4): 253-265.

Weingarten MD, Suter MM, Littman DR, Kirschner MW (1974).

Properties of the depolymerization products of microtubules from mammalian brain. J. Biochem. 27: 5529-5537.

Marshal T, William KM (1993). Bradford protein assay and the transition from an insoluble to soluble dye complex: Effect of sodium dodecyl sulphate and other additives. J. Biochem. Biophys. Methods. (26): 237-240.

Brannon-Peppas L, Blanchette JO (2004). Nanoparticle and targeted systems for cancer therapy. J. Adv. Drug Deliv. Rev. (56): 1649-1659.

Brunner TJ, Wick P, Manser P, Spohn P, Grass RN, Limbach LK, Bruinink A, Stark WJ (2006). Comparison to Asbestos, Silica, and the Effect of Particle Solubility. J. Environ. Sci . Tech. (40): 4374-4381.

Fortina P, Kricka LJ, Surrey S (2005). the promise and reality of new approaches to molecular recognition. J. Trends Biotechnol. 23: 168–173.

Ferrari M (2005). Cancer nanotechnology: opportunities and challenges. Nat Rev Cancer. (5): 161–71.

Valiron O, Caudron N, Job D (2001). Microtubule dynamics. J. Cell. Mol. Life. Sci. (58): 2069-2084.

Bhattacharya A, Bhattacharyya B, Roy S (1996). Fluorescence energy transfer measurement of distances between ligand binding sites of tubulin and its implication for protein-protein interaction. J. Protein. Sci. (5): 2029-2036

Sarkar N, Mukhopadhyay K, Parrack PK, Bhattacharyya B (1995). Aging of tubulin monomers using 5,5´-bis(8-anilino-1-naphthalenesulfonate) as a probe. J. Biochem. (34): 13367-13373.

Solomaha E, Palfrey HC (2005). Conformational changes in dynamin on GTP binding and oligomerization reported by intrinsic and extrinsic fluorescence. J. Biochem. 391: 601-611.

Sparreboom A, Scripture CD, Trieu V (2005). Comparative preclinical and clinical pharmacokinetics of a cremophor-free, nanoparticle albumin-bound paclitaxel (ABI-007) and paclitaxel formulated in Cremophor (Taxol). J. Clin. Cancer. Res. 11: 4136–4143.

19

Standardized alcoholic extract of *Phyllanthus fraternus* exerts potential action against disturbed biochemical parameters in diabetic animals

Munish Garg[1]*, Chanchal Garg[2], V. J. Dhar[3] and A. N. Kalia[3]

[1]Department of Pharmaceutical Sciences, Maharshi Dayanand University, Rohtak-124001, Haryana, India.
[2]Department of Pharmaceutical Chemistry, Faculty of Pharmacy, Jamia Hamdard, New Delhi-110062, India.
[3]Department of Pharmacognosy, I. S. F. College of Pharmacy, Moga-142001, Punjab, India.

Alcoholic extract of *Phyllanthus fraternus* Webster whole plant (PFAE) prepared by successive solvent treatment was administered at a dose of 500 mg/kg body weight once in a day for 21 days to the alloxan induced diabetic albino rats. Certain biochemical parameters that is lipid profile (total cholesterol, high density lipoprotein, triacylglycerols), kidney functions (urea, creatinine) and liver functions (alkaline phosphate, alanine aminotransferase, aspartate aminotransferase) were evaluated and compared with normal and standard drug tolbutamide (200 mg/kg body weight) administered group. As a result, drug treatment has significantly improved the disturbed biochemical parameters at variable degrees when compared with standard drug. The phytochemical studies conducted for standardization of the extract showed the presence of tannins and flavonoids as major phytoconstituents. The total phenolics content was found to be 37.51 mg/g of drug extract. Quantitative estimation carried out on two major flavonoids by HPTLC confirmed a concentration of 1.706% w/w rutin and 5.614% w/w of quercetin present in the alcoholic extract. In conclusion, owing to the positive potential activity against disturbed biochemical parameters associated with diabetes, *P. fraternus* can be used effectively in the management of this deadly disease.

Key words: *Phyllanthus fraternus,* renal functions, liver functions tests, lipid profile, total phenolics.

INTRODUCTION

Diabetes mellitus is a chronic metabolic disorder with vascular components that is characterized by disturbances in carbohydrates, lipids and protein metabolism (Pickup and Williams, 2003). Studies have shown that good metabolic control is beneficial in slowing the progression of these complications in diabetes (Floretto et al., 1998; Renu et al., 2004). Several herbal drugs in different formulations have been experimented in search of an effective treatment for diabetes and certain claims of cure are on record (Jung et al., 2003). The plants of genus *Phyllanthus* (Euphorbiaceae) are widely distributed and long been used in traditional medicines and due to the presence of potential phytoconstituents, it has led to some promising findings in several disorders

(Kirtikar and Basu, 1987; Calixto et al., 1998). A few species of this genus have also been reported to possess antidiabetic activity in addition to our earlier reported antidiabetic activity of the same plant (Higashino et al., 1992; Garg et al., 2008). Based on the above, an attempt was made to screen the efficiency of PFAE on associated biochemical parameters in diabetic albino rats.

MATERIALS AND METHODS

Plant material

The plant material of *Phyllanthus fraternus* Webster (whole plant) was collected from medicinal garden of University Institute of Pharmaceutical Sciences, Panjab University, Chandigarh, Authenticated and voucher specimen was stored in the department (Number ISF/Ph/VS-103). The material was then shade dried, powdered and stored at 25°C. Alcoholic extract of the whole plant were prepared by soxhlet extraction using petroleum ether as solvent by soxhlation. Extract was concentrated by rotaevaporator,

*Corresponding author. E-mail: mgarg2006@gmail.com.

vacuum dried and stored. In animal studies, the extract was triturated with freshly prepared 0.3% w/v carboxyl methyl cellulose (CMC) solution to obtain a suspension of concentration 0.3 g/ml for oral administration to the animals.

Phytochemical analysis

Preliminary phytochemical screening of the alcoholic extract was carried out using standard method (Peach and Trecy, 1956). The total phenolics content were estimated by Folin-Ciocalteu reagent method. Total phenolic contents present in the plant extract were calculated as gallic acid equivalents (GAE) by applying the formula, $C = c * V/m$. Where, C-total content of phenolic compounds (mg/gm) plant extract in GAE, c-the concentration of gallic acid established from the calibration curve (mg/ml), V- volume of extract (ml), m- weight of pure plant extract (gm).

For fingerprinting studies, 10 mg/ml solutions of extracts in respective solvents were applied in triplicate on HPTLC plate (Silica gel 60 F 254 Aluminum sheets, 10 X 10 cm) and developed in solvent system benzene: ethyl acetate: formic acid (80:20:5) up to 90 mm in twin trough chamber. The developed chromatogram was scanned at 254 nm wavelength for detection of active compounds in the absorbance mode using CAMAG scanner III. Quantitative estimation of rutin and quercetin were carried out by HPTLC using solvent systems ethyl acetate: methanol: water: formic acid (100:13.5:10:2.5) and benzene: ethyl acetate: formic acid (40:10:2.5) respectively, by calibration curve method and compared with standard compounds rutin and quercetin.

Test animals

Albino rats of either sex (5 - 6 weeks) weighing 150 - 200 g were obtained from animal house, I. S. F. college of pharmacy, moga, punjab, kept in teflon cages and maintained under controlled conditions (22 - 28°C temp, 60 - 70% relative humidity) at 12 h dark/light cycle, fed with standard rat pellet diet (Hindustan Lever, India) and given water *ad libitum*. All drugs and chemicals were of analytical grade.

Diabetes was induced in animals by injecting freshly prepared alloxan monohydrate in sterile normal saline at a dose of 150 mg/kg body weight, intraperitoneally (Aruna et al., 1999). To prevent fatal hypoglycemia due to massive pancreatic insulin release, rats were treated with 20% glucose solution intraperitoneally after 6 h followed by supply of 5% glucose solution bottles in their cages for next 24 h (Barry et al., 1997). The animals shown blood glucose level >200 mg/dl after 72 h were considered as diabetic. Experiments performed were complied with committee for the purpose of control and supervision of experiments on animals (CPCSEA) New Delhi, India (registration no: 816/04/C/CPCSEA). The experimental protocol was duly approved by the institutional ethical committee.

Experimental design

In this experiment, a total of 24 rats (18 diabetic surviving rats, 6 normal rats) were used. The rats were divided into four groups (A-D) of six rats each. Group A and B served as normal and diabetic control while group C and D served as PFAE treated (500 mg/kg body weight) and standard drug tolbutamide (200 mg/kg body weight) treated groups, respectively. The dosage of the drug extract administered to the animals decided as 500 mg/kg body weight b.w. by oral administration on the basis of previously conducted studies (Khanna et al., 2002; Adeneye et al., 2006). The animals were administered drug for 21 days once daily by oral route using an intra gastric tube and weighed after every three days. On the last

day before sacrificing the animals, the blood was withdrawn by cardiac puncture and placed into sterile container plastic tubes. Serum was separated by centrifugation (4000 rpm, 10 min) and transferred to Eppendorf tubes. All serum samples were stored at -80°C (deep freezer) until analysis. Total cholesterol, high density lipoprotein (HDL), serum triacylglycerols, urea, creatinine, serum alkaline phosphate (ALP), aspartate aminotransferase (AST) and alanine aminotransferase (ALT) estimations were carried out as per methods described (Allain et al., 1974; Jacob and Denmark 1960; Talke and Schubert, 1965; Bartel et al., 1972; Varley, 1975). Data was statistically analyzed with one way ANOVA and Students't' test for paired observations. All data expressed as mean $\pm$ standard error.

RESULTS

In the phytochemical analysis, PFAE have shown presence of tannins and flavonoids as major phytoconstituents. Total phenolic contents were calculated as 37.51 mg/g present in the extract. HPTLC fingerprinting studies revealed six numbers of spots and showed the presence of 1.706% w/w of rutin and 5.614% w/w of quercetin contents, respectively. Changes in body weight in treated and untreated animals are shown in Figure 1. Significant reduction in body weight was observed in diabetic animals. Administration of PFAE improved the body weight and the values were quite comparable with the effect of standard drug. In biochemical studies, the results expressed in Table 1 reveal that PFAE administration has significantly reduced the elevated levels of serum cholesterol and triacylglycerols in alloxan-induced albino rats.

The HDL levels which were reduced in diabetic animals were significantly (P < 0.001) elevated by the drug treatment to the extent of normal animals and even better than standard drug. Serum urea and creatinine were observed (Table 2) as badly affected in diabetic rats and their values were quite elevated. However, PFAE administration for 21 days to the alloxan-induced diabetic animals has reduced the elevated levels significantly at variable degrees. The response in serum creatinine was better than the serum urea contents. The effect of PFAE administration on liver function tests are expressed in Table 3 which reveals that an elevation of alkaline Phosphate (ALP), Aspartate aminotransferase (AST) and (Alanine aminotranferase) ALT were observed in diabetic rats. The drug treatment has significantly controlled the elevated levels not to the extent of standard drug. However, a potential positive effect was observed on all the liver parameters as a whole.

DISCUSSION

The plants are considered as biosynthetic laboratory for a multitude of compounds that exert physiological effects. Secondary metabolites are the compounds which are responsible for imparting therapeutic effects. The plants of genus *Phyllanthus* have been reported to contain

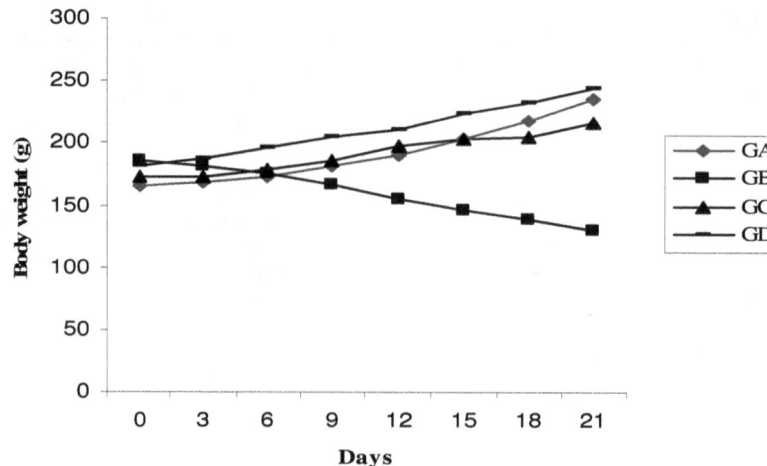

Figure 1. Effect of PFAE on body weight in diabetic albino rats after 21 days of treatment. GA = normal control, GB=diabetic control, GC = alcoholic extract treated and GD = standard drug tolbutamide treated group.

Table 1. Effect of PFAE on lipid profile in diabetic albino rats after 21 days of treatment (mg/dl).

Group	Serum cholesterol (mg/dl)	Serum triacylglycerols (mg/dl)	HDL (mg/dl)
Normal control	152.3 ± 3.76	61.3 ± 0.96	45.8 ± 0.70
Diabetic control	254.8 ± 9.62	102.0 ± 1.03	33.8 ±0.60
Alcoholic extract (500 mg/kg)	166.0 ± 8.01[c]	70.6 ± 0.92[c]	43.0 ± 0.68[c]
Tolbutamide (200 mg/kg)	155.5 ± 7.29[c]	66.8 ± 1.25[c]	45.5 ± 0.76[b]
F Value (ANOVA) (3,18)	26.59	147.21	51.22

Data are expressed as Mean ± SEM (n = 6). Statistical significance in comparison to control, [a] = $P < 0.5$, [b] = $P < 0.01$, [c] = $P < 0.001$, student's t-test.

Table 2. Effect of PFAE on renal profile in diabetic albino rats after 21 days of treatment.

Group	Serum urea (mg/dl)	Serum creatinine (mg/dl)
Normal control	25.5 ± 1.46	0.49 ± 0.036
Diabetic control	58.7 ± 3.11	1.2 ± 0.056
Alcoholic extract (500 mg/kg)	30.6 ± 2.20[c]	0.59 ± 0.026[c]
Tolbutamide (200 mg/kg)	28.7 ± 2.21[c]	0.62 ± 0.064[c]
F Value (ANOVA) (3,18)	24.05	21.81

Data are expressed as Mean ± SEM (n = 6). Statistical significance in comparison to control, [a] = $P < 0.5$; [b] = $P < 0.01$; [c] = $P < 0.001$, student's t-test.

Table 3. Effect of PFAE on liver functions tests in diabetic albino rats after 21 days of treatment.

Group	ALP (µ/L)	AST (µ/L)	ALT (µ/L)
Normal control	122.7 ± 4.25	72.8 ± 3.05	40.0 ± 2.12
Diabetic control	321.1 ±7.66	144.8 ± 1.99	96.5 ± 3.15
Alcoholic extract (500 mg/kg)	159.17±12.70[c]	104.3 ± 2.75[c]	76.7± 3.08[b]
Tolbutamide (200 mg/kg)	131.50±4.98[c]	81.7 ± 1.71[c]	63.83 ± 3.1[c]
F Value (ANOVA) (3,18)	98.22	133.79	45.83

Data are expressed as Mean ± SEM (n = 6). Statistical significance in comparison to control, [a] = P<0.5; [b] = $P < 0.01$; [c] = P<0.001, student's t-test. ALP = Serum alkaline Phosphate, AST = Aspartate aminotransferase, ALT = Alanine aminotranferase.

potential phytoconstituents like flavonoids, tannins, alkaloids and triterpenoids in earlier studies (Calixto et al., 1998). Our studies also confirm the claims and observed that potent compounds like flavonoids and tannins were present in the alcoholic extract. Significant amount of total phenolic contents and fingerprinting studies has further supported the earlier claims. Qualitative and quantitative studies carried out on two potent flavonoids that is quercetin and rutin confirmed that both compounds were present in substantial quantities in PFAE. Earlier reported studies have already confirmed that flavonoids and tannins are the class of compounds which are responsible for several therapeutic activities (Iwu, 1983).

Diabetic rats are observed with increased plasma lipids, which are responsible for several cardiovascular disorders (Nikkhila and Kekki, 1973; Chaterjee and Shinde, 1994). The higher lipid profile like cholesterol and triacylglycerols are also observed in tissues of liver, pancreas, kidney and intestine of diabetic rats, which are due to increase in mobilization of free fatty acids from peripheral depots and also due to the lipolysis caused by hormones (Murray et al., 2000). The results in the present study showed that administration of PFAE have reduced the hyperlipidaemia state significantly which may be due to the control of blood glucose level and thereby control on the lipolytic hormones. The main function of the kidneys is to excrete the waste products of metabolism and to regulate the body concentration of water and salt. Significant increase of total urea and creatinine levels indicated impaired renal function of diabetic rats leading to a negative nitrogen balance, enhanced proteolysis and lowered protein synthesis (Leonard et al., 2006; Alderson et al., 2004). Treatment of alloxanized-diabetic rats with PFAE induced a fall in the level of these metabolic parameters. Similar results are observed in the earlier studies using different plants (Kedar and Chakrabarti, 1983). The improvement of renal biochemical functions with PFAE in the present investigation could be due to its antidiabetic action, resulting in alleviation of altered metabolic status in animals and by the regenerative capability of the renal tubules (Kissane, 1985). In the current study, increased activities of ALP and ALT were observed in the diabetic animals. It has already been demonstrated that tissue antioxidant status is an important factor in the development of diabetic complications (Wohaieb and Godin, 1987). The increase in the level of these enzymes in diabetes may be as a result of leakage from the tissues and migration into the bloodstream (Chaudary et al., 1993).

Administration of PFAE brought about reduction in AST, ALT and ALP. Various mechanisms of action are involved in the antidiabetic effect of oral antidiabetic agents which includes, suppressing hepatic gluconeogenesis, stimulating glycolysis and inhibition of glucose absorption from the intestine, stimulation of insulin release, inhibition of dietary disaccharides to mono-saccharides

and exerting transcription of fatty acids by activating a specific sub-class of proxisome-proliferator- activated receptor (Hardy and Nutty, 1997). It is quite possible that a combination of these applies to the whole plant PFAE in which the presence of potent phytoconsti-tuents has been demonstrated. Although, exact biological active compound responsible for the observed thera-peutic activity is yet to be discovered however, observed activity points out the role of tannins, quercetin and rutin in the pancreatic and extra pancreatic mechanism of action. Due to this, enhancement of perpheral utilisation of glucose and increase insulin release may be associated in the resultant action of PFAE.

In conclusion, standardized PFAE containing estimated amounts of phenolic contents and two potent flavonoids that is, rutin and quercetin has demonstrated potential therapeutic activity against disturbed biochemical para-meters in diabetic animals. It is therefore suggested that further work should be carried out to determine the exact mechanism of actions of the drug.

REFERENCES

Adeneye AA, Amole OO, Adeneye AK (2006). Hypoglycemic and hypocholesterolemic activities of the aqueous leaf and seed extract of *Phyllanthus amarus* in mice. Fitoterapia. 77: 511-514.

Alderson NL, Chachich ME, Frizzell N, Canning P, Metz TO, Januszewski AS (2004). Effect of antioxidants and ACE inhibition on chemical modification of proteins and progression of nephropathy in streptozotocin diabetic rat. Diabetologia. 47: 1385.

Allain CC, Poon LS, Chan CS, Richmond W, Fu PC (1974). Enzymatic determination of total serum cholesterol. J. Clin. Chem. 20: 470.

Aruna RV, Ramesh B, Kartha VNR (1999). Effect of betacarotene on protein glycosylation in alloxan induced diabetic rats. Ind. J. Exp. Biol. 37: 399-402.

Barry JAA, Hassan IAA, Al-Hakiem MHH (1997). Hypoglycemic and Antihyperglycemic effect of *Trigonella foenum-graecum* leaf in normal and alloxan induced diabetic rats. J. Ethnopharmacol. 58: 149-154.

Bartel H, Bohmer M, Heieri C (1972). Serum creatinine determination without protein precipitation. Clin. Chem. Acta. 37: 193.

Calixto JB, Santos ARS, Filho VC, Yunes RA (1998). A review of plants of the genus *Phyllanthus*: their chemistry, pharmacology and therapeutic potential. Med. Res. Rev.18: 225-234.

Chaterjee MN, Shinde R (1994). Metabolism of carbohydrates Part-II, Text book of Medical Biochemistry, Jay Pee brothers Medical Publishers Pvt Ltd: pp 421-430.

Chaudary AR, Alam M, Ahmad M (1993). Studies on medicinal herbs II. Effect of *Colchicum luteum* on biochemical parameters of rabbit serum. Fitoterapia. 64: 510-515.

Floretto P, Steffes MW, Sutherland ERD, Goetz CF, Mauer M (1998). Reversal of lesions of diabetic nephropathy after pancreas transplantation. N. Eng. J. Med. 339: 69-75.

Garg M, Dhar V J, Kalia AN. (2008). Antidiabetic and antioxidant potential of *Phyllanthus fraternus* in alloxan-induced diabetic rats. Pharma- Cog-Mag. 14 (4): 138-143.

Hardy KJ, Mc Nutty SJ (1997). Oral Hypoglycemic agents. Medicine digest. 23 (4): 247.

Higashino H, Suzuki A, Tanaka Y, Pootakham K (1992). Hypoglycemic effects of *Momordica charantia* and *Phyllanthus urinaria* extracts in streptozotocin-induced diabetic rats. Nippon Yakurigaku Zashi. 100(5): 415-421.

Iwu MM (1983). The hypoglycaemic property of *Bridelia ferruginea*. Fitoterapia. 54: 243.

Jacob NJ, Van Denmark PJ (1960). Arch. Biochem. Biophys. 88, 250-255.

Jung M, Park M, Lee HC, Kang YH, Kang KS, Kim SK (2003).

Antidiabetic agents from medicinal plants. Curr. Med. Chem. 13(10): 1203-1218.

Kedar P, Chakrabarti CH (1983). Effect of Jambolan seed treatment on blood sugar lipids and urea in streptozotocine induced diabetes in rabbits. Ind. J. Physiol. Pharmacol. 27: 135.

Khanna AK, Rizvi R, Chander R (2002). Lipid lowering activity of *Phyllanthus niruri* in hyperlipemic rats. J. Ethnopharmacol. 82: 19-22.

Kirtikar KP, Basu BD (1987). Indian medicinal Plants, vol III, 2nd edn., International book distributors, India: pp. 65-69.

Kissane JM (1985). Anderson's pathology, 8th edition. Toronto: Washington University School of Medicine: pp 754-759.

Leonard T, Thephile D, Paul DD, Acha EA, Dongmo SS, Patrice C, Jean FF, Pierre K (2006). Antihyperglycemic and renal protective activities of *Anacardium occidentale* (Anacardiaceae) leaves in Streptozotocin-induced diabetic rats. Afr. J. Trad. CAM. 3(1): 23-25.

Murray RK, Granner DK, Mayes PA, Rowell VW (2000). Harpers Biochemistry, 25th edn, Stanford CT, Appleton and Lange: pp 610-17.

Nikkhila EA, Kekki M (1973). Plasma triacylglycerols transport kinetics in diabetes mellitus. Metabolism. 22: 1-22.

Peach K, Trecy MV (1956). Modern Methods of Plant Analysis, Springer-Verlag, New York, USA. pp 121-122.

Pickup JC, Williams G (2003). Textbook of diabetes. Blackwell Science Ltd. USA, Pp. 103-114.

Renu A, Saiyada NA, Odenbach S (2004). Effect of reinstitution of good metabolic control on oxidative stress in kidney of diabetic rats. J. Diab. Compl. 5: 282-288.

Talke H, Schubert GE (1965). Enzymatic determination of urea using the coupled urease-GLDH enzyme system. Klin. Wschi. 43: 174.

Varley H (1975). Practical Clinical Biochemistry, CBS Publishers, New Delhi, p 453.

Wohaieb SA, Godin DY (1987). Alterations in free radical tissue defense mechanism in streptozotocin-induced diabetes in rat: effects treatment. Diabetes, 3: 1014-1021.

20

Utilization of dietary therapies in the alleviation of protein energy malnutrition in kwashiokor-induced rats

Obimba, Kelechukwu Clarence

Department of Biochemistry, College of Natural and Applied Sciences, Michael Okpara University of Agriculture Umudike. Abia State. Nigeria. E-mail: kechrisob@yahoo.com.

The utilization of diets prepared from available and affordable plants (soya bean, groundnut, maize, and fluted pumpkin green leaves) and animal sources (milk, catfish, and crayfish), in the alleviation of protein energy malnutrition (PEM) was studied, to determine comparative dietary efficiencies of some fortified weaning formulae. The therapeutic diets used in alleviating PEM in the kwashiorkor-induced rats were the (SGMC) (prepared at 20% dietary protein level), (SMC) (prepared at 20% dietary protein), (SGMM) (prepared at 16% dietary protein level), and Rd (reference diet, Nutrend : Nestle ®, 16% protein level) diets. The experimental design for the treatment of kwashiorkor was a Completely Randomized Design (CRD). Feeding of the kwashiokor-induced rats with the test diets caused a reversal of the earlier observed decreases in the parameters (growth performance, serum albumin, pack cell volume, NPU %) measured, in order of consecutive significant increase (p<0.05) as follows: SGMM< Rd<SGMC/SMC. Results were recorded as values of mean ± S.D unit, of the test dietary therapy groups, SGMC/SMC, as follows : growth performance : (90.4 ± 0.5/87.0 ± 2.84 grams), serum albumin : (4.11 ± 0.31/4.09 ± 0.61g/l), pack cell volume (PCV%) : (49.05 ± 0.31/50.01 ± 0.21%), aspartate aminotransferase (AST) : (11.1 ± 0.05/10.95 ± 0.05U/l) and net protein utilization (NPU%) : (93.0± 1.96/95.0 ± 0.01%), respectively. SGMC and SMC alleviation diets, elicited significant regression at 5% level, of dietary group, growth performance (grams) and corresponding Net Protein Utilization % values of the weaning formulae, and achieved the most rapid catch-up growth rates in the kwashiorkor-induced rats. SGMC and SMC test diets are effectively, preventive and curative of protein energy malnutrition.

Key words: Protein energy malnutrition, kwashiokor, therapeutic diets, weaning formulae, experimental design.

INTRODUCTION

Protein energy malnutrition (PEM) is a range of pathological conditions arising from co-incident lack, in varying proportions, of proteins and calories, occurring most frequently in infants and young children and commonly associated with infections (Roulet, 1994). Protein-energy undernutrition (PEU), previously called protein-energy malnutrition, is an energy deficit due to chronic deficiency of all macronutrients (Morley, 2007). PEU can be sudden and total (starvation) or gradual. Severity ranges from subclinical deficiencies to obvious wasting (with edema, hair loss, and skin atrophy) to starvation. Multiple organ systems are often impaired. Diagnosis usually involves laboratory testing, including serum albumin. Treatment consists of correcting fluid and electrolyte deficits with intra-venous solutions, then gradually replenishing nutrients, orally if possible (Morley, 2007).

Etukudo et al. (1999) described the broad spectrum nature of protein energy malnutrition, which ranges from marasmus through marasmic kwashiorkor to kwashiorkor, and is characterized by low weight for age, oedema, dermatitis, hair changes, mental changes, hepatomegaly and diarrhea. The Wellcome classification of protein energy malnutrition employs the weight to age percent (%) anthropometric measure, in addition to presence or absence of oedema in distinguishing kwashiorkor (65 to 80% weight for age, oedema present), from marasmic kwashiorkor (<60% weight for age, Oedema present), from marasmus (<60% weight for age, oedema absent) (Grover and Ee, 2009).

Nutritional status assessment methods used in detecting and monitoring recovery of patients suffering from PEM, include: clinical methods, morphological methods, haematological and biochemical methods, dietary survey, and anthropometry. Serum albumin and total serum proteins are markedly decreased during protein energy malnutrition (Siddiqui et al., 2007).

A major factor among the dietary causes of PEM, is deficiency in energy intake, kwashiorkor patients have slightly reduced fasting blood sugar and reduced glucose tolerance (Becker, 1983). Fatty liver and/or atrophy of the liver resulting in increased level of aspartate amino transferase enzyme in the blood, occurs in PEM subjects (Islam et al., 2007). Hormonal changes, especially in insulin and thyroid hormone secretions, occur in PEM subjects (Abrol et al., 2001). Kwashiorkor subjects have a 65-80% increase in total body water. Protein energy malnutrition results in anemia, detectable, using PCV% and Hb (haemoglobin) count haematological diagnostic tool (Siddiqui et al., 2007). Infections associated with PEM, are largely due to decrease in efficiency of patients immune system, hence the use of the WBC $_{Total}$ (white blood cell) count as a diagnostic tool of PEM (Huang and Fraker, 2003).

In dietary therapy for optimum recovery, dietary energy intake, treatment of infections, dietary protein intake, correction of electrolyte imbalance, are imperative. Disappearance of apathy, oedema, anorexia, gain in body weight and increase in nervous motor activity, are signs of recovery form kwashiorkor (Etukudo et al., 1999).

The recommended daily allowance for dietary protein is based on the requirement that dietary protein, be provided by a mixture of both animal and plant proteins. The addition of 20 to 30% of animal protein to a 7:3 combination of cereal to legume seed meal, increases, ultimately the nutritive value of the food, and is consistent with the Protein Advisory Group guidelines for weaning foods, which states that dietary protein content of weaning foods, should be at least 20% (on a dry weight basis) (FAO/WHO, 1971). The relevance of wistar albino rats in the measurement of the nutritional quality of dietary protein, as a means of scientific investigations, correlate to the human physiological condition, is founded on the fact that wistar albino rats have a dietary requirement for the same ten (10) essential amino acids as human infants. Variations in performance characteristics could occur, as a result of disease conditions such as protein energy malnutrition (Obimba, 2006). Performance characteristics of dietary protein (Tome and Cecile, 2000), measured in net protein utilization (NPU%) was used in determining the qualities of the diets. The objective of this work is to utilize diets prepared from available and affordable plants and animal sources, in the alleviation of protein energy malnutrition, and also to compare the dietary efficiencies of some dietary therapies/fortified weaning formulae.

MATERIALS AND METHODS

The experimental design of the treatment of kwashiorkor, is a single factor completely randomized design (CRD), of 20 observations per parameter, and 15 degree of freedom of error. The Linear model is:

$$Y_{ij} = \mu + T_i + e_{ij} .$$

where: Y_{ij} = Individual observations, μ = Overall mean, T_i = Effect of ith level of dietary protein treatment, e_{ij} = Random error, which is independently, identically, and normally, distributed, with zero mean, and constant variance.

About 200 g each of raw soya bean seeds, raw groundnut seeds, and raw maize seeds, were washed, and soaked, separately, in a liter of water, for 11 h, and thereafter, boiled in 800 ml of water, for 2 h. Boiled groundnut seeds and soya bean seeds were dehulled. The samples were dried in the oven for 9 h at 105°C, ground and dried for a further 4 h, at 105°C. Fresh catfish samples were dried in the oven for 24 hours at 105°C, and ground. Moisture-free crayfish, dried in industrial ovens, were ground. Fluted pumpkin vegetable leaves were washed in warm water, and dried in the oven for 1 h, and ground. Table 1 is a schematic for the diet formulation

23 weanling Wister albino rats aged 5 weeks old were weighed, and housed in stainless steel cages under 12 h light and dark cycles, under humid tropical conditions, and fed ad-libitum on a 3.47% dietary protein-kwashiokorigenic diet (Kd) for 33 days (the animals were acclimatized to the diet, within the first 3 days), during which period, kwashiorkor was induced in the rats. A control group of experimental rats were fed, during the same period, on conventional feed, prepared at 16% dietary protein level. Daily faecal deposits of the animals were collected during the last thirty days of the feeding trial, pooled, oven dried and weighed. Three experimental animals each of the kwashikor-induced diet group, and the control group were weighed and sacrificed by a sharp tap on the head with a blunt instrument. Blood samples for haematological and biochemical assays were collected in requisite blood sample bottles, and stored in a refrigerator at 4°C. The lean body mass (lungs, liver, heart, kidneys, pancreas, and spleen) were recorded. The carcasses were dried for 17 h, in an oven drier at 105°C and stored. The faecal nitrogen content and the carcass nitrogen content of the experimental animals were determined using the Kjeldahl method (AOAC, 1990).

Twenty male kwashiorkor-induced Wister albino rats were divided into five groups of 4 animals each, and housed in stainless steel cages under 12 h light and dark cycles, under humid tropical conditions, and fed ad- libitum on four different types of weaning diets (SGMC, SMC, SGMM, and Rd) for a period of 20 days. The kwashiokorigenic diet (basal diet) which served as a control was fed to another group of rat for the same period. Daily faecal deposits of the animals were collected during the 20 day period of the feeding trial, pooled, oven dried, and weighed. The experimental animals were weighed and sacrificed by a sharp tap on the head with a blunt instrument. Blood samples for haematological and bio-chemical assays were collected in requisite blood sample bottles, and stored in a refrigerator at 4°C. The lean body mass (lungs, liver, heart, kidneys, pancreas, and spleen) were recorded. The carcasses were dried for 17 h, in an oven drier at 105°C and stored. The faecal nitrogen content and the carcass nitrogen content of the experimental animals were determined using the Kjeldahl method (AOAC, 1990).

Kjeldahl Method (AOAC, 1990) was employed for the quantitative determination of nitrogen and crude protein.

The Spun microhaematocrit method of Bull and Hay (2001) was used for the determination of packed cell volume (PCV%).

Quantitative in vitro determination of albumin in serum was carried out using the method employed by Qureshi and Qureshi (2001).

Quantitative in vitro determination of serum aspartate amino transferase was carried out using the method employed by Pratt and Kaplan (2000).

Performance characteristics analysis of Net Protein Utilization (NPU%) was carried out using the method employed by Pellet and

Table 1. Diet formulation.

Dietcomponents (g/100g diet)	Kd/basal3.47% dietary protein level)	SGMC(20% dietary protein level)	SMC (20% dietary protein level)	SGMM (16% dietary protein level)
Casein	3.47	-	-	-
Soyabean seed (flour)	-	17.63	21.84	20.00
Groundnut seed (flour)	-	4.41	-	5.00
Maize seed (flour)	-	51.42	50.97	70.00
Powdered cow milk	-	-	-	5.00
Catfish	-	10.20	-	-
Crayfish	-	-	10.31	-
Vegetables (Fluted pumpkin leaves)	-	5.00	5.00	-
Palm oil	8.00	8.00	8.00	-
Vitamin-mineral premix	0.25	0.25	0.25	-
Sucrose	-	1.00	1.00	-
Garri (sourced from *manihot esculenta*)	88.28	2.09	2.68	-

Reference diet (Rd): Nutrend prepared industrially by Nestle®, of nutritional value- 16% dietary protein, 63.7% carbohydrates, 9% fat, 4% moisture, 2.3% minerals, 417.5 kcal/100g. SGMC: 19.71% dietary protein, 64.2% carbohydrates, 9.2% lipids, 3.1% moisture, 3.1% minerals, vitamins ≤ 0.69 g, 437.1 kcal/100g. SMC: 19.71% dietary protein, 64.4% carbohydrates, 9.0% lipids, 3.2% moisture, 3.1% minerals, vitamins ≤ 0.59 g, 432.1 kcal/100g. SGMM: 16% dietary protein, 71.7% carbohydrates, 6% fat, 4.5% moisture, 1.67% minerals, vitamins ≤ 0.13 g, 368.5 kcal/100 g.

Table 2. Percentage crude protein content of the experimental diets.

Diet type	Protein content (%)
SGMC	$19.71^a \pm 0.2$
SMC	$19.71^a \pm 0.1$
SGMM	$16.0^b \pm 0.05$
Bd/Kd	$4.38^c \pm 0.0$

Values are means ± S.D (n = 3).Means in the same column having the same superscripts are not significantly different at 5% level (p<0.05).

Young (1980).

Student's t-test, ANOVA (analysis of variance) and regression statistical analytical methods were used in analyzing results.

RESULTS

Table 2 shows the percentage crude protein content of the experimental diets. The percentage crude protein content of the SGMC and SMC weaning formulae, were numerically, and significantly equal (p < 0.05), and were not significantly different (p < 0.05), from the calculated value of 20% , substituted in the formulae derived by the author, for the diet formulation. The percentage protein content of each of the SGMC and SMC weaning formulae, differed significantly (p < 0.05), from those of the Rd (reference) weaning formula, and the basal diets, in consecutive order of significant decrease. The significant differences (p < 0.05), recorded of the quantitative percentage crude protein content of the weaning formulae, and the basal diet, listed in descending order, are as follows : SGMC/SMC> Rd >SGMM > Bd.

Table 3 shows the growth performance of the weaning formulae and basal diet groups of experimental animals. The growth performance recorded of the diet groups of experimental animals, in order of consecutive decrease were as follows: SMC> SGMC> Rd > SGMM. The mean value of growth performance of the SMC diet group of experimental animals was significantly higher (p<0.05) than those of the SGMM and Rd diet groups of experimental animals. The basal diet group of experimental animals suffered significant loss of weight (p<0.05).

Table 4 shows the hematological and biochemical parameters measured of the weaning formulae and basal diet groups of experimental animals. The mean values of the haematological parameter, PCV%, and the biochemical parameter, serum albumin (g/dl), measured of the weaning formulae diet groups of experimental animals, listed in sequential order of decrease were as follows: SGMC> Rd > SGMM. The mean values of the

Table 3. Growth performance of the weaning formulae and basal diet groups of experimental animals.

Diet group	Rd	SGMC	SMC	SGMM	Bd
Gain/loss of live weight (growth performance: grams)	$55.00^b \pm 13.80$	$87.00^a \pm 2.84$	$90.4^a \pm 0.5$	$54.80^b \pm 1.85$	$-30.03^c \pm 6.28$

Values are means ± S.D (n = 3).Means in the same row having the same superscripts are not significantly different at 5% level (p<0.05).

Table 4. Haematological and biochemical parameters assayed of the weaning formulae and basal diet groups of experimental animals.

Diet group	Rd	SGMC	SMC	SGMM	Bd
PCV (%)	$38.08^a \pm 0.01$	$49.05^b \pm 0.05$	$50.01^b \pm 0.01$	$36.21^c \pm 0.02$	$28.14^d \pm 0.03$
Serum albumin (g/dl)	$3.71^a \pm 0.01$	$4.11^b \pm 0.01$	$4.09^b \pm 0.02$	$3.62^c \pm 0.03$	$1.75^d \pm 0.01$
AST (U/l)	$12.05^a \pm 0.05$	$11.1^b \pm 0.05$	$10.95^b \pm 0.05$	$12.4^c \pm 0.05$	$15.4^d \pm 0.05$

Values are means ± S.D (n = 3). Means in the same row having the same superscripts are not significantly different at 5% level (p<0.05).

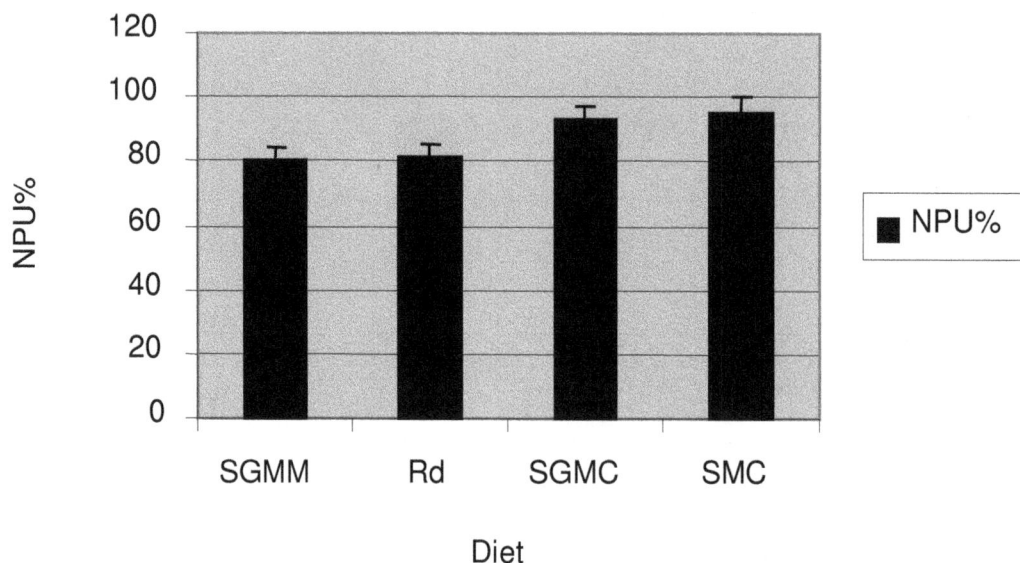

Figure 1. Net protein utilization (NPU%) of dietary therapies/weaning formulae.

PCV%, and the serum albumin of each of the SGMC and SMC diet groups of experimental animals were not significantly different (p<0.05). The mean values of the aspartate aminotransferase (AST) enzyme activity (U/l) of each of the SGMC and SMC diet groups of experimental animals were significantly reduced (p<0.05) compared with those of the Rd and SGMM. The basal diet group of experimental animals suffered the most significant decrease (p<0.05) of mean values of PCV%, and serum albumin (g/dl), and the most significant increase (p<0.05) in AST enzyme activity.

Figure 1 shows the quantitative values of the performance characteristic, net protein utilization (NPU%) of the weaning formulae and basal diet. The NPU% was used in assessing the nutritional efficiency and quality of the dietary therapies/weaning formulae. The mean values of the NPU% of the weaning formulae listed in sequence of significant decrease (p<0.05) were as follows: SGMC/SMC>Rd /SGMM.

DISCUSSION

Kwashiorkor is an acute form of childhood protein-energy malnutrition characterized by edema, irritability, anorexia, ulcerating dermatoses, and an enlarged liver with fatty infiltrates. The presence of edema caused by poor nutrition defines kwashiorkor (Ciliberto et al., 2005).

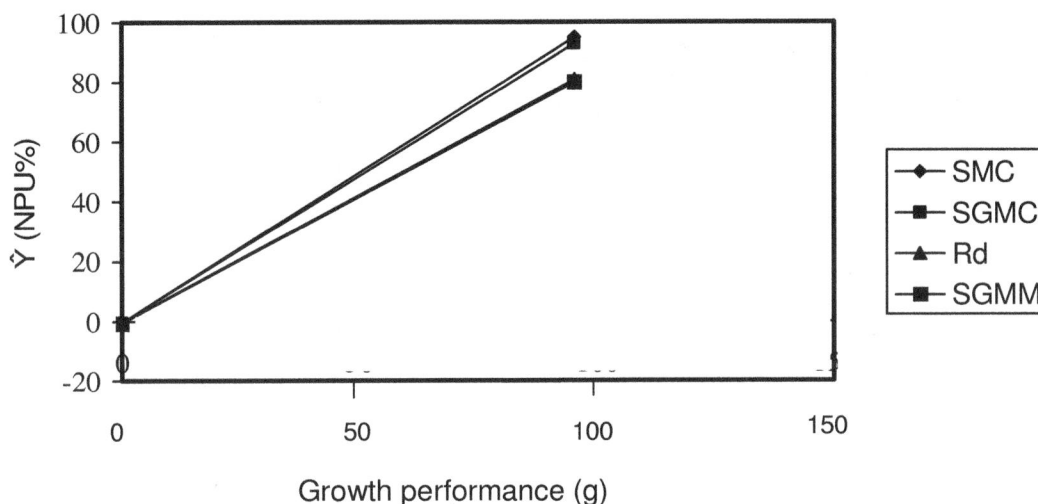

Figure 2. Regression curves of (growth performance [g]) and NPU%.

The age effect on the onset of marasmic kwashiorkor is distinct. It takes about 30 days for the induction of marasmic kwashiorkor on weanling Wister albino rats, using a dietary regimen of 3.47% of dietary protein, in keeping with the method of induction employed by Akinola et al. (2010). The significant reductions (p<0.05) observed of the growth performance (-12.5 ± 3.3 grams), lean body mass, PCV% (30.2± 0.24%), serum albumin (1.77 ± 0.97 g/l) , PER (-1.21 ± 0.04), and significant increase observed of the AST (15.53 ± 5.95 U/l) of the kwashiorkor induced rats, compared with those of the control and weaning formulae diet groups, reflect poor nutritional status at various stages of protein energy malnutrition and correspond with the findings of Collins (2003). The kwashiorkor–induced group of experimental animals was characterized by retarded growth, dermatitis, oedema, hair loss, physical inactivity, observable loss of motor co-ordination and apathy.

Dietary therapy for kwashiorkor has two primary aims which are:

(a). To reduce mortality to a minimum, achieved, partly, by controlling infections which frequently accompany PEM, with a view to affecting the most rapid recovery possible.

(b). To use therapeutic diets of PEM, which are relatively high in protein, and energy, with special emphasis placed on fluid intake, with a view to avoiding electrolyte imbalance.

The significant increases (p<0.05) in growth performance, PCV%, serum albumin, and AST activity shown in Tables 3 and 4, recorded the SGMC and SMC diet group of experimental animals, which correspond with the findings of Etukudo et al. (1999), Obatolu et al. (2003).

The efficiency of the SGMC and SMC weaning formulae in effecting rehabilitation and promoting good biological responses is consistent with the research findings of Mosha and Bennink (2004), Obatolu et al. (2003). Comparative analysis of the present study and the work of Annan and Plahar (1995), shows that protein malnutrition reduces significantly (p<0.05), the performance characteristics (NPU%) of therapeutic diets. In keeping with the observations made of the performance characteristics of the dietary therapies/weaning formulae, used in the present study, shown in Figure 1 is the evidence that cereal-legume mix and animal supplements possess a great nutritional potential to support growth and rehabilitation of protein energy malnutrition subjects (Mosha and Bennink, 2004).

Upon rehabilitation, the kwashiorkor induced group of experimental animals, were characterized by a restoration to normal dermal conditions, loss of oedema, hair growth, noticeable restoration to normal physical activities, motor co-ordination, and weight gain.

The regression between growth performance (grams) of the SGMC and SMC diet groups of experimental animals and their corresponding NPU% values of the weaning formulae were significant at 5% level (p<0.05), with Pearson's product moment correlation coefficient of 0.96 and 0.97, respectively. These figures indicate that the growth performance of the SGMC and SMC diet groups of experimental animals were a function of the respective dietary protein qualities of the SGMC and SMC weaning formulae. The regression curves are shown in Figure 2.

The ratio of the weight of dietary protein required to alleviate kwashiorkor per unit weight of human subject to the weight of dietary protein required to alleviate kwashiorkor per unit weight of rat model is 4:7 (Obimba, 2006).

Conclusion

The protein value, and efficiency of the various weaning diets, listed in sequential order of significant (p<0.05) decrease were as follows: SGMC/SMC> Rd> SGMM.

ACKNOWLEDGEMENT

The author would like to thank God for His loving kindness and mercies. He also acknowledge the technical support of Prof. E. O. Ayalogu, Prof. A. A. Uwakwe, Dr. (Mrs). I. I. Ijeh, Dr. Isma'ila A. Umar.

Abbreviations: SGMC, weaning formulae prepared at 20% dietary protein, with processed soya bean seeds, groundnut seeds, maize seeds and catfish, diet group; **SMC,** weaning formulae prepared at 20% dietary protein, with processed soya bean seeds, maize seeds and crayfish, diet group; **Rd,** Nutrend-Nestle®, weaning formulae industrially prepared at 16% dietary protein, diet group; **Bd,** basal diet (hypothetical protein-free diet), diet group; **Kd,** kwashiokorigenic diet (prepared at 3.47% dietary protein level); **PCV,** pack cell volume; **AST,** aspartate amino transferase.

REFERENCES

Abrol P, Verma A, Hooda HS (2001). Thyroid hormone status in protein energy malnutrition in Indian children. Indian J. Clin. Biochem. 16 (2): 221-223.

Annan TA, Plahar WA (1995). Development and quality evaluation of a soy-fortified Ghanaian Weaning Food. Food Nutr. Bull. 16(3): 263-269. http://www unu.edu/unu press/food 18F163e/8F163E of htm.

Akinola FF, Oguntibeju OO, Alabi OO (2010). Effects of severe malnutrition on oxidative stress in Wistar rats. Sc. Res. Essays. 5(10): 1145-1149.

Association of Official Analytical Chemists (AOAC) (1990). Protein (Crude) Determination in Animal Feed: Copper Catalyst Kjeldahl Method. (984.13) Official Methods of Analysis. 15th Edition.

Becker DJ (1983). The Endocrine Responses to Protein Calorie Malnutrition. Ann. Rev. Nutr. 3: 187-212.

Bull SB, Hay KL (2001). Is the Packed Cell Volume (PCV) Reliable? Lab. Hem. Carden Jennings Publishing Co. Ltd. 7: 191–196.

Ciliberto H, Ciliberto M, Briend A, Ashorn P, Bier D, Manary M (2005). Antioxidant supplementation for the prevention of kwashiorkor in Malawian children: randomised, double blind, placebo controlled trial. BMJ. 330 (7500): 1109.

Collins N (2003). Protein-energy malnutrition and involuntary weight loss: nutritional and pharmacological strategies to enhance wound healing. Expert Opin. Pharmacother. 4(7): 1121-1140.

Etukudo M, Agbedana O, Akang E, Osifo B (1999). Biochemical Changes and Liver Tissue Pathology in Weanling Wistar Albino Rats with Protein Energy Malnutrition. (PEM). Afr. J. Med. Sci. 28 (1-2): 43-7.

FAO/WHO (1971). Protein Advisory Group (PAG) of the United Nations. PAG Guideline No 8. Protein – Rich Mixtures for Use as Weaning Food., New York: FAO / WHO/ UNICEF.

Grover Z, Ee LC (2009). Protein energy malnutrition. Pediatr. Clin. North. Am. 56(5): 1055-68.

Huang ZL, Fraker PJ (2003). Chronic Consumption of a Moderately Low Protein Diet does not alter Hematopoetic Processes in young adult mice. J. Nutr. Sci. 133: 1403-1408.

Islam MS, Chowdhury ABM, Rahman Z, Haque M, Nahar N, Taher A (2007). Serum Aspartate Aminotransferase (AST) and Alanine Aminotransferase (ALT) Levels in Different Grades of Protein Energy Malnutrition. J. Bangladesh Soc. Physio. 2: 17-19.

Morley JE (2007). Protein-Energy Malnutrition Definition. In: The Merck Manual of Diagnosis and Therapy. Porter R (ed.). 18th Edition. Merck & Co. Inc., New Jersey.

Mosha CET, Bennink MR (2004). Protein Quality of Drum Processed Cereal–Bean–Sardine Composite Supplementary Foods for Preschool–age Children. J. Sci. Food Agric. 84 (10): 1111-1118.

Obatolu VA, Ketiku A, Adebowale EA (2003). Effect of Feeding Maize / Legume Mixtures on Biochemical Indices in Rats. Ann. Nutr. Metab. 47: 170-175.

Obimba KC (2006). Utilization of Some Dietary Therapies in the Alleviation of Protein Energy Malnutrition. M.Sc. Thesis., University of PortHarcourt. PortHarcourt. Nigeria : 145 pages.

Pellet PL, Young VR (1980). Evaluation of Protein Quality in Experimental Animals. In: Nutritional Evaluation of Protein Foods. The United Nations University, Food Nutr. Bull. Supp. 4: 41-57.

Pratt DS, Kaplan MM (2000). Evaluation of abnormal liver-enzyme results in asymptomatic patients. N. Engl. J. Med. 342(17): 1266-1271.

Qureshi MI, Qureshi Z (2001). Effect of Protein Malnutrition on the Weight and Serum Albumin of Albino Rats. J. Ayub. Med. Coll. Abottabad. 13: 8-10.

Roulet M (1994). Protein-energy malnutrition in cystic fibrosis patients. Acta Paed. 83 (395): 43-81.

Siddiqui AU, Halim A, Hussain T (2007). Nutritional Profile And Inflammatory Status Of Stable Chronic Hemodialysis Patients At Nephrology Department, Military Hospital Rawalpindi. J. Ayub Med. Coll. Abbottabad. 19(4): 29-31.

Tome D, Cecile B (2000). Dietary Protein and Nitrogen Utilization. J. Nutr. 130: 1868S-1873S.

21

Electrolyte profile and prevalent causes of sickle cell crisis in Enugu, Nigeria

E. O. Ibe[1]*, A. C. J. Ezeoke[2], I. Emeodi[3], E. I. Akubugwo[4], E. Elekwa[4], M. C. Ugonabo[5] and W. C. Ugbajah[6]

[1]Department of Chem. Path, University of Nigeria Teaching Hospital, Enugu, Nigeria.
[2]Department of Chem. Path, College of Medicine, University of Nigeria, Enugu Campus, Enugu, Nigeria.
[3]Department of Paediatrics, University of Nigeria, Enugu Campus, Enugu, Nigeria.
[4]Department of Biochemistry, Abia State University, Uturu, Nigeria.
[5]Department of Chemical Pathology, College of Medicine, UNEC, Enugu, Nigeria.
[6]Temple University Hospital, Philadelphia, PA, 19140, USA.

One hundred sickle cell patients aged between 6 - 25 years in steady state who attended Sickle Cell Clinic at University of Nigeria Teaching Hospital, Enugu, Nigeria were selected for this study. Out of this number, only thirty who were eventually admitted in crisis state within one year of this study were selected for subsequent investigations. They included 20 females and 10 males. We also selected thirty apparently healthy hemoglobin AA subjects, (17 males and 13 females) aged between 6 and 32 years to serve as secondary control. Samples were collected on the patient's initial visit to the hospital (stable state). Samples were also collected on admission and 24 h after infusion therapy. Serum electrolytes, malarial parasite count, widal agglutination, blood and urine cultures were done using standard methods. The results showed a statistically significant decrease ($p < 0.05$) in mean sodium and potassium levels in crisis when compared with those in steady state. The electrolytes were assayed 24 h after rehydration of the patients in crisis. There were significant increases ($p < 0.05$), in mean sodium and potassium levels. Considering the prevalent causes of crisis, 63% of the subjects in crisis had malarial parasitaemia. 16.7% had bacterial infection and 13.3% were infected with Hepatitis B while 7% had both malaria and bacterial infection. The significance of this study is to highlight the fact that sickle cell patients who receive hydration therapy attain electrolyte balance within 24 h of re-hydration and therefore should not be over- enthusiastically challenged especially in those localities where there are no facilities for monitoring hydration therapy. In addition, the study revealed that malaria is the major precipitating cause of sickle cell crisis in Enugu, Nigeria and governments should take a holistic approach towards the fight against malaria.

Key words: Sickle cell anaemia, electrolytes, crisis, prevalent causes, malaria.

INTRODUCTION

Epidemiology

Although the effects of sickle cell disease (SCD) on general morbidity and mortality have been studied extensively (Gladwin et al., 2004; Prasad et al., 2003), much work has not been done in determining the prevalent causes of sickle cell crisis especially in this part of the world. There is high prevalence of sickle cell disease in Africa and elsewhere (Oheme-Frempong et al., 1994). The disease is most common among people living in or originating from Sub-Saharan Africa (Akinyanju, 1989). The disorder also affects people of Mediterranean, Caribbean, Middle Eastern and Asia origin (Npat, 2002). The sickle cell gene is most common in areas where mosquito is endemic (Meremiku, 2008). Sickle cell trait affects 30% of Africa's tropical population (Oheme-

*Corresponding author. E-mail: mascot7@yahoo.com.

Frempong, 1994). It also affects an estimated 1 - 2% (120,000) of newborns in Africa annually. About 178 babies (0.28/1000 conceptions) are affected by sickle cell disease in England annually (Hickman et al, 1999). Besides, about 72,000 in USA (NIH, 2004) and 10,000 in the United Kingdom suffer from the disease (Davies et al., 1997).

Sickle cell disease refers to a group of disorders caused by inheritance of a pair of abnormal haemoglobin genes including sickle cell gene (Meremiku, 2008). It is an inherited blood disorder characterized primarily by chronic anaemia and periodic episodes of pain. The underlying problem involves haemoglobin, a component of red blood cells. Haemoglobin molecules in each red blood cell carry oxygen from the lungs to the organs and tissues and bring carbon dioxide back to the lungs. In sickle cell anaemia, the haemoglobin is defective. After haemoglobin molecules give up oxygen, some may cluster together and form rod-like structures. These structures cause red blood cells to become stiff and assume a sickle shape. Unlike normal red cells which are usually smooth and donut-shaped, sickled red cells cannot squeeze through small blood vessels. Instead, they stack up and cause blockages in arterioles thereby depriving organs and tissues of oxygenated blood. This process produces periodic episodes of pain and ultimately damage tissues and vital organs including liver, spleen, kidney, heart, bones, etc, thus leading to high morbidity and serious medical problems (NIH, 2004). Normal red cells live about 120 days in the blood stream but sickled red cells die after about 10 - 20 days. Because they cannot be replaced fast enough, the blood is chronically short of red blood cells, a condition called anaemia.

Pathophysiology

Sickle cell disease is an inherited disorder associated with abnormal haemoglobin in the homozygous state (Sergeant, 1992). The genetic abnormality is a point mutation arising from the substitution of glutamic acid by valine at the sixth position on the beta polypeptide chain (Ureme et al., 2003). It is accompanied with various clinical manifestations such as jaundice, cutaneous ulcer, skeletal changes and episodes of intravascular sickling and thrombosis resulting in painful crisis and infarcts in various organs (Kabins, 1970). Sickle cell pain crises are precipitated by infection, dehydration and hypoxia. Intercurrent infections particularly of respiratory tract, fever, abdominal, skeletal pain, haermatologic and bone pain crisis are the main causes of morbidity in sickle cell patients (Kaine, 1983). The patients may have fever, pulmonary infarction that presents as acute chest crisis and dehydration. One of the ways of managing sickle cell crisis is by re-hydration. Hydration is done using various grades of electrolytes such as 5% dextrose saline, normal saline, Darrow's solution and oral rehydration solution (ORS).

It has been demonstrated that sickling is accompanied by an intraerythrocytic loss of potassium and gain of sodium, thus creating disequilibrium in the ionic strength across the cell membrane (Statius et al., 1971). Chem et al. (1981) reported that haemolysis, intravenous potassium administration, blood transfusion among others, increase serum potassium level. This elevated potassium concentration may lead to cardiac excitability and ventricular fibrillation, weakness and ascending paralysis. There is also conformed weight loss throughout the acute phase of crisis and a negative water balance. However, the administration of fluid early in crisis not only achieved positive water balance but largely prevented weight loss from occurring. Conversely, re-hydration of patients with varying strengths of electrolytes may precipitate a shift in membrane potential and a rise to abnormal level. Infections can precipitate crisis due to underlying functional asplenia in most adults with sickle cell anemia, leading to defective immunity against encapsulated organism. An infectious disease is one caused by the invasion and multiplication of organisms (Gossel et al., 1980). The infection could be of bacterial, viral or protozoan origin. The bacterial infection may be due to *Haemophilus influenza, Streptococcus pneumonia, Staphylococcus aureus, Neisseria meningitidis* and *Escherichia coli* among others. The protozoan infection can be due to plasmodium species leading to the so-called "malaria attack" (Gossel, 1980).

The objective of this study was to follow-up 100 sickle cell patients who reported at the sickle cell clinic, University of Nigeria Teaching Hospital, Enugu, Nigeria, within one year to determine frequency of crisis and the precipitating factor in addition to assessing electrolyte balance within 24 h.

MATERIALS AND METHODS

Subjects

A total of one hundred (100) sickle cell patients who attended sickle cell clinic at University of Nigeria Teaching Hospital were selected for this study. Out of this number, only thirty (30) who were later admitted in crisis state within a period of one year was selected for further investigations and monitoring. They included 20 female and 10 male subjects aged between 6 - 25 years. Thirty apparently healthy hemoglobin AA individuals (17 males and 13 females) aged between 6 - 32 years were assessed and they served as secondary control. Ethical clearance was obtained from the Ethics Committee, University of Nigeria Teaching Hospital, Enugu while informed consent was obtained from the subjects and parents of the minors. Children below six years were excluded from this study because of the volume of blood required.

Sample collection

Blood samples were collected in EDTA bottles for Hb electrophoresis (control subjects) and malarial parasite count. Brain heart infusion (BHI) was used to collect samples for blood culture, sterile containers for urine culture while plain tubes were used for the collection of samples for Widal agglutination reaction and electrolyte

Table 1. Sodium and potassium levels of HbAA, HbSS in stable state, HbSS in crisis before, and 24 h after rehydration.

(A) Comparison of the Na$^+$ and K$^+$ values of HbAA and HbSS stable state	Sodium	Potassium
HbAA	138.37 ± 3.52	4.29 ± 0.47
HbSS stable state	136.22 ± 3.19	3.56 ± 0.42
P-value	P < 0.05	P < 0.05

(B) Comparison of Na$^+$ and K$^+$ values of HbSS stable state and crisis before re-hydration	Sodium	Potassium
HbSS stable state	136.22 ± 3.19	3.56 ± 0.42
HbSS Crisis before rehydration	135.17 ± 2.77	3.28 ± 0.35
P-value	P < 0.05	P < 0.05

(C) Comparison of Na$^+$ and K$^+$ values of HbSS crisis during rehydration and 24 h after rehydration	Sodium	Potassium
HbSS crisis before rehydration	135.17 ± 2.77	3.28 ± 0.35
HbSS 24 h after rehydration	137.57 ± 2.15	4.42 ± 0.15
P-value	P < 0.05	P < 0.05

profile. For sickle cell patients in crisis, blood and urine samples were collected on admission prior to fluid administration and another blood samples collected 24 h after re-hydration of patients. For the HbAA subjects, blood samples were collected for electrolyte estimation only. Thirty-five milliliters of blood were collected from the subjects depending on their age and state of health.

Sample analysis

The blood samples for electrolytes were allowed to clot and sera separated after centrifuging at 3000 rpm for five minutes. Sodium and potassium concentrations were analyzed using Gallenkamp flame photometer by means of flame emission. We also used quality control sera manufactured by Quimica Clinica Applicada S.A. (QCA) for the analysis. Urine culture was done using MacConkey ager and sterile wire loop calibrated to deliver 0.01 ml of urine. Blood culture was done according to WHO standard by adding 1 ml of blood per 10 ml of BHI (Vandepitte et al., 1983). Subcultures were done on Chocolate and MacConkey agar and incubated in carbon dioxide atmosphere and aerobically respectively. Hemoglobin electrophoresis was done using cellulose acetate paper in a chamber connected to the zip-zone power supply and electrophoresed for 5 - 10 min at 350 volts. Widal agglutination reaction was carried out by tube agglutination using Chromatest kit while FIBH-Tech kit was used for Australia antigen (Vaisman et al., 1960).

Statistical analysis

This was done using Z-score. Calculations were done at 5% level of significance while figures were analyzed using Microsoft Excel worksheet.

RESULTS

The mean and standard deviation (mean ± SD) of sodium and potassium of control HbAA and HbSS (stable state)

were compared as shown in Table 1A. There were significant differences in mean values of both parameters. The mean values of sodium and potassium were also compared in HbSS (stable state) and crisis before hydration (Table B) and 24 h after re-hydration as presented in Table 1(C). There was significant decrease in values in HbSS (crisis) and an increase in the values of both parameters 24 h after re-hydration (p < 0.05).

Figure 1 represents the malarial parasite count in sickle cell disease in relation to age. This analysis was done based on the fact that plasmodiasis was a major observed factor in sickle cell crisis. Children between 6 - 12 years had malarial parasite count greater than 300 parasites per microlitre of blood (57.9%). Those 13 - 19 years old had 210 - 300 parasite count per microlitre (31.6%) while those 20 - 26 years old had 101 - 200 parasite count per microlitre (10.5%).

Table 3 represents bacterial isolates in blood and urine cultures and widal agglutination test in steady state and crisis. Out of the 10 patients (stable state) who presented with bacterial infections, 60% had *Salmonella* infection, 20% (*E. coli*) 10% (*S. aureus and S. pneumonia*) respectively. Out of the 7 patients who eventually went into crisis as a result of bacterial infection, 57% had *Salmonellosis*, 29% (*E. coli*) while 14% were infected with *S. pneumonia*.

DISCUSSION

There was high incidence of sickle cell crisis as observed in the University of Nigeria Teaching Hospital, Enugu. In sickle cell anaemia, there are increased and continued obligatory losses of body fluids and electrolytes which ra-

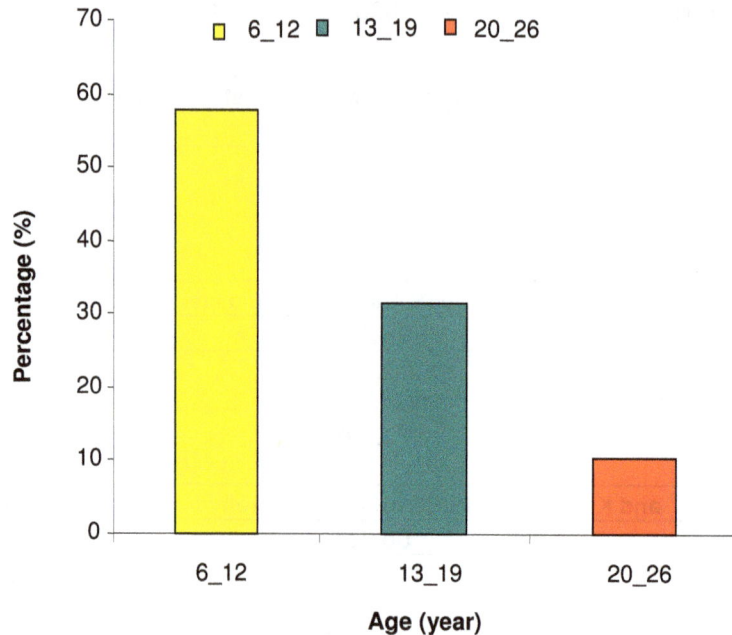

Figure 1. Represents malarial parasite count in sickle cell disease in relation to age.

Table 2. Prevailing causes of crisis in Enugu and age relationship.

Age(years)	Percentage	MP	HbsAg	Bacterial	Mixed infections
6 = 12(n = 13)	43.3	10	-	3	
13 - 19(n = 10)	33. 3	7	-	3	
20 - 36(n = 7)	23.4	2	4	1	
Prevalence in Enugu		63%	13%	17%	7%

Table 3. Bacterial isolates (steady state) and (crisis).

Source (stable state)	S. aureus	S. pneum	E. coli	S. para C	S. typhi	S. para B
Urine	1	-	2	-	-	-
B. culture	-	1	-	-	-	-
Widal test	-	-	-	1	1	4
Titre				80	80	160
Crisis state						
Urine	-	-	2	-	-	-
B. culture	-	1	-	-	-	-
Widal test	-	-	-	-	1	3
Titre					160	160

rapidly result in dehydration. Besides, there is suppression of appetite and patient may not drink. There is also high skin loss of electrolytes coupled with the obligatory urinary losses from inability to concentrate urine. For this reason, Physicians are at liberty to give intravenous electrolyte fluid continuously and it has been documented that this could cause heart failure (Serjeant, 1992). Dehydration causes salt (sodium, chloride) and other electrolyte imbalance (HCl org, 2008). It further promotes sickling, sequestration and haemolysis. Consequently, more water and electrolytes are lost from the body (NIH, 1995). Avoiding dehydration is a good way to decrease the likelihood of pain crisis. Therefore, getting plenty of fluid is extremely important, (Lewis et al., 2008).

The results showed decrease in sodium and potassium concentrations during crisis when compared with the steady state and apparently healthy HbAA subjects (p < 0.05) as represented in Table 1. The decreased levels of electrolytes are due to reduced fluid intake, accelerated influx and outflux of sodium and potassium, increased insensible loss and high incidence of hyposthenuria (NIH, 1995). Similar reports recorded in this study were made by Tosteston et al. (1955). Brugnara and cohorts observed marked decrease in potassium and chloride levels in painful crisis in the USA (Brugnara, 2000). The loss of potassium is also associated with calcium-activated potassium channel in the kidneys (Gardos pathway). There is excessive accumulation of calcium in sickle cell disease. The excessive accumulation of calcium activates the Gardos channel which then expels potassium into the renal tubules. The disturbance of electrolyte and water equilibrium has to be corrected to maintain a balance in order to ensure the normal functioning of the body system. And the major way of correcting this disturbance is by re-hydration therapy. In the course of correcting this imbalance, the patient may be overenthusiastically challenged especially in the rural and semi-urban settings where there are no facilities for monitoring hydration therapy.

Moreover, the study showed a higher malarial parasite concentration in children between 6 - 12 years old (>300 parasites per microlitre blood). This could be as a result of lowered immunity and relatively small blood volume when compared with the older age groups (Figure 1). There was also prevalence of malarial-induced crisis (63%) in relation to bacterial and viral infection, (Table 2). This could be as a result of parasite resistance to drugs as most of these patients could be indulging in self medication with monotherapy before reporting to hospital. This report differs from the findings of Buchanan who reported that bacterial infection was the most common cause of death in children with sickle cell disease in America (Buchanan et al., 1989). However, in all those who had bacterial infection, Salmonellosis was more prevalent (Table 3). The prevalence of malaria could be attributed to the endemic nature and unhealthy environment coupled with the low social standard of most people in this part of the world.

Conclusion

There was increase in electrolyte levels after re-hydration. It is therefore suggested that hydration (oral or intravenous) be monitored closely to avoid iatrogenic congestive cardiac failure and electrolyte imbalance which may result if the patient is over-challenged with different grades of electrolytes. The high incidence of sickle cell disease in Africa makes it a public health problem in many countries, but it is often not recognized as such because so many cases go undiagnosed before or even after death. Malaria and its complications further worsen the morbidity and mortality of sickle cell disease. Management of sickle cell disease includes considerations for improvement on general health conditions such as sanitation, housing, nutrition, immunization and prophylaxis against infection. Public education and programs of early diagnosis and management are needed to help prevent early mortality.

REFERENCES

Akinyanju OO (1989). A profile of sickle cell disease in Nigeria. Ann NY Acad. Sci. 565: 126-136 (Pubmed).

Buchanan GR, Mackie V, Jackson EA, Vedro DA, Hamer S, Dicerma-Holtkamp CA (1989). Splenic phagocytic function in children with sickle cell anaemia receiving hyper-transfusion therapy. J. Pediar. 115: 568-582.

Brugnara C (2000). Red cell dehydration in pathophysiology and treatment of sickle cell disease, Natl. Inst. Heart Lung Blood 23: 5-8.

Chem K, Graber MA (1981). Haematologic electrolyte and metabolic disorder in sickle cell anaemia. www.vh.org.

Davies SC, Oni L (1997). The management of patients with sickle cell disease. Br. Med. J. 315: 656-660.

HCI (2008) Dehydration and electrolyte imbalances. www.health care-informatio.org

Gladwin MT, Sachdev V, Jison ML (2004). Pulmonary hypertension as a risk factor for death in patients with sickle cell disease. N. Engl. J. Med. 350(9): 886–895. [PubMed]

Gossel TA, Stansloki DW, Krammer JA (1980). The complete medicine book. Simon and Semester, Eds. A division of Gulf Western Corporate NY pp.248-255.

Hickman M, Modell B, Greengrass P (1997). Mapping the prevalence of sickle cell and beta-thalasaemia in England: estimating and validating ethnic-specific rates. Br. J. Haematol. 104: 860-867 (Pubmed).

Kabins SA, Lerner C (1970). Sickle cell disease. JAMA 211: 467-468.

Kaine WN, (1983). Morbidity of homozygous sickle cell anaemia in Nigerian children J. Trop. Peadiatr. (2): 104-111.

Lewis HSU, William Muller (2008). Diet and Nutrition in Sickle Cell Disease http://www.drspock.com/home/0,1454,,00.html

Meremikwu MW (2008). Sickle cell disease. http//clinicalevidence.bmj. com

National Institute of Health, National Heart, Lung and Blood Institute (1995). Management and Therapy of Sickle Cell Disease. 95:2116-2117.

National Institute of Health www.nig.gov/news/pr/mar2004.

Npat GP (2002). Emergency guideline in sickle cell crisis. www.beth-pe-nhs

Ohene-Frempong K, Nkurumah FK (1994). Sickle cell disease in Africa. In Basic Principles and Clinical Practice. Raven press Ltd. New York. pp.423-435.

Prasad R, Hasan S, Castro O, Perlin E, Kim K (2003). Long-term outcomes in patients with sickle cell disease and frequent vaso-occlusive crises. Am. J. Med. Sci. 325(3): 107–109. [PubMed]

Sergeant GR (1992). Sickle Cell Disease 2nd ed. Oxford University Press pp.88-89, 429-431.

Statius LW, Schouten H, Sloof PA (1971). Sodium, potassium and calcium in erythrcytes in sickle cell anaemia. Clin. Chim. Acta 33: 475-478.

Ureme SO, Tosteson E, Darling RC, Ejezie FE, Ibegbulam GO, Ibe EO, Nwanya IJ (2003). Serum calcium inorganic phosphate and some Heamatological parameters in sickle cell disease in Enugu metropolis, Orient J. Med.

Vaisman A, Paris-Hamelin A (1969). Sero-Diagnositics per Reaction d Agglutination. Instiute Alfred Fournier pp.75-77.

Vandepitte J, Engbeak K, Piot P (1983). Basic lab procedure, Clin. Bacteriol. WHO Geneva pp.21-36.

22

The effects of aqueous extracts of the leaves of *Hibiscus rosa-sinensis* Linn. on renal function in hypertensive rats

Imafidon E. Kate[1] and Okunrobo O. Lucky[2]*

[1]Department of Biochemistry, Faculty of Life Sciences, University of Benin, Benin City, Nigeria.
[2]Department of Pharmaceutical Chemistry, Faculty of Pharmacy, University of Benin, Benin City, Nigeria.

The leaves of *Hibiscus rosa-sinensis* Linn. (Family Malvaceae) have been used in ethnomedicine for the treatment of various human diseases such as aphrodisiac, hypertension, wound healing, diabetes mellitus and cancer. In this present study, the effect of 200 mg/kg of the aqueous leaf extract on the renal function of hypertensive rats was investigated. The administration of *H. rosa-sinensis* leaves extract shows a significant ($p < 0.05$) increase in the Na^+ level of normotensive rats, thus it may interfere with the normal function of the kidney and hence produce increased salt retention. These results had shown that although *H. rosa-sinensis* leave extract reduced blood pressure; the integrity of the kidney may be compromised when this plant is used for the treatment of hypertension.

Key words: *Hibiscus rosa-sinensis,* hypertension, rats, leaves, blood pressure, kidney.

INTRODUCTION

Nature has been a source of medicinal agents for thousands of years and an impressive number of modern drugs have been isolated from natural sources, many based on their use in traditional medicine. Higher plants, as sources of medicinal compounds, have continued to play a dominant role in the maintenance of human health since ancient times (Farombi, 2003). Over 50% of all modern clinical drugs are of natural product origin (Stuffness and Douros, 1982) and natural products play an important role in drug development programs in the pharmaceutical industry (Baker et al., 1995). Some ethnomedicinal value of *Hibiscus rosa-sinensis* have been evaluated which include the followings. *H. rosa sinensis* has been used for the treatment of a variety of diseases as well as to promote wound healing. The wound-healing activity of the ethanol extract of *H. rosa-sinensis* flower was determined in rats, using excision, incision, and dead space wound models as reported by Shivananda et al. (2007). Cold aqueous extract of *H. rosa-sinensis* leaves is reported by local traditional

practioners in Western Nigeria to be aphrodisiac (Olagbende-Dada et al., 2007). The hypoglycemic activity of an ethanol extract of *H. rosa-sinensis* has been studied in glucose located rats (Sachdewa and Khemani, 1999). Antiimplantation activity of water extract of leaves of *H. rosa-sinensis* was investigated by Nivasarkar et al. (2005). It has also been investigated that *H. rosa-sinensis* extract exerts a protective effect against the tumour promotion stage of cancer development (Sharma and Sultana, 2004).

H. rosa-sinensis belongs to the family Malvaceae. The roots are cylindrical, 5 - 15 cm in length and 2 cm in diameter, off white and with light brown transverse lenticles. The roots taste sweet and are mucilaginous. The leaves are simple ovate or ovatelancolate, and are entire at the base and coarsely toothed at the apex. The flowers are pedicillate, actinomorphic, pentamerous and complete. The corolla consists of 5 petals, red coloured and about 8 cm in diameter. Traditionally this plant is used for the control of dysfunctional uterine bleeding and as an oral contraceptive. Some of the chemical constituents isolated from this plant are cyanidin, quercetin, hentriacontane, calcium oxalate, thiamine, riboflavin, niacin and ascorbic acid. Flavonoids are also present (Nair et al., 2005). The present study was

---

*Corresponding author. E-mail: bricyedo@yahoo.com.

Table 1. Effect of aqueous leave extract of *Hibiscus rosa-sinensis* (HR) on weight gain, feed intake and faecal output in rats.

Treatment	Weight gain (g)	Feed intake (g)	Faecal output (g)
Normal control (100% grower mash)	31.75 ± 7.78	64.00 ± 11.81	29.66 ± 8.02
Hypertensive rats (control)	16.70 ± 6.01*	44.32 ± 4.40*	20.31 ± 4.85*
Hyp. plus 200 mg/kg of HR	36.01 ± 4.99*	58.30 ± 8.10*	23.30 ± 1.86*
Normal plus 200 mg/kg of HR	31.01 ± 8.6	41.71 ± 1.67*	22.20 ± 2.01*

Values are mean ± S.E.M * p < 0.05, significantly different from normal control, Paired t- test (n = 5), Hyp = Hypertensive rats.

undertaken to ascertain the effect of the aqueous extract of the leaves of *H. rosa-sinensis* on the renal function of hypertensive rats knowing the fact that this plant have also been used to alleviate hypertension

MATERIALS AND METHODS

Plant collection and identification

The leaves were collected based on ethnopharmacological information. The fresh leaves of *H. rosa-sinensis* were harvested in the University of Benin main campus, at Ugbowo Benin City, Nigeria in May 2007. The botanical identification of the plant, its leaves and its authentication were done by Dr. J.F. Bamidele of the Department of Plant Biology and Biotechnology, Faculty of Life Sciences, University of Benin, Benin City, Nigeria, where a voucher specimen was deposited for future reference. The fresh leaves of *H. rosa-sinensis* were washed, rinsed, then air dried to a constant weight at room temperature, pulverised in a mixer-grinder, filtered, and the coarse powder was stored in a non-toxic polyethylene bag.

Extraction of plant

200 g of the powdered leaf was macerated in 1.5 L of distilled water and the homogenate was filtered several times through a sieve (Endecoffs London, Aperture 1.10 mm). The filtrate was concentrated to dryness with a rotary evaporator at reduced pressure. The concentrate was stored in the refrigerator until required for use in the experiment.

Treatment of animals

Twenty (20) albino rats (180 - 200 g) of both sexes were kept at the laboratory animal house of the Department of Biochemistry, University of Benin, Benin City. Nigeria, and were divided into four groups of five rats each. The animals were acclimatized for a period of two weeks. The animals were maintained under standard environmental conditions and were allowed feed (Bendel Feeds and Flour mill, Ewu, Edo State, Nigeria) and water *ad libitum*. All the procedures were conducted in accordance with the guide lines for Care and Use of Laboratory Animals published by the National Institutes of Health. After two weeks the animals were subjected to different treatments: Group 1 (normal control) received an equivalent volume of water; group 2, (hypertensive control) received 92% rat mash and 8% sodium chloride; group 3, animals were given 92% rat mash, 8% sodium chloride and 200 mg/kg body weight of the extract. Group 4, were given 100% of rat mash and 200 mg/kg body weight of the extract. They were administered these diet and extract for 4 weeks during which daily body weight,

food intake and faecal output were recorded. After the fourth week, blood pressures of the rats were measured using a two-channel recorder (Gemini, 2020). Blood was collected by cardiac puncture into sterile containers with or with anticoagulant for biochemical analysis.

Biochemical analysis

Total protein, urea, alanine and aspartate transaminase kits were products of Quimice Chnice Applicade Laboratories Spain. Total proteins were estimated using the direct Biuret method (Henry et al., 1957), urea by the modified method of Berthelot-Searcy (Searcy et al., 1967), calcium by the O-cresolphthalein method (Gitelman, 1967), sodium and potassium by the use of flame photometer (Corning 410) (Tietz, 1995). Alanine and aspartate transaminase activities were determined by the formation of corresponding hyrazones on reaction with 2, 4-dinitrophenylhyrazine (Sigma-Aldrich) (Reitman and Frankel, 1957).

Statistical analysis

Results were expressed as mean ± Standard Error of Mean (S.E.M) Statistical analysis of the data was done using one-way analysis of variance (ANOVA) followed by Dunnett's test and significance determined using P-values < 0.05.

RESULTS

There was significant (p < 0.05) reduction in food intake, weight gain and faecal output of hypertensive control rats compared with the normal control (Table 1). Hypertensive rats administered 200 mg/kg body weight of extract gained weight and feed intake significantly (p < 0.05) compared with hypertensive control rats but the increase in the faecal output was not significant.

Normal rats administered 200 mg/kg had reduced food intake and correspondingly reduced faecal output compared with the normal control (Table 2). The administration of 200 mg/kg of crude extract of *H. rosa-sinensis* significantly reduced (p < 0.05) blood pressure in normal and hypertensive rats.

Urea, ALT, AST and Na⁺ concentrations were significantly (p < 0.05) increase in normal rats administered aqueous extract of *H. rosa-sinensis* compared with normal control rats, however in the hypertensive rats administered with the extract, urea levels, AST and Na⁺

Table 2. Effect of Aqueous extract of *Hibiscus rosa-sinensis* (HR) on Blood pressure in rats.

Treatment	Systolic (mmHg)	Diastolic (mmHg)	Mean arterial pressure
Normal control (100% grower mash)	162.0 ± 9.18	119.0 ± 10.81	133.3 ± 9.12
Hypertensive rats (control)	168.0 ± 1.71*	144.0 ± 1.76*	148.1 ± 1.85*
Hyp. plus 200 mg/kg of HR	155.0 ± 4.39*	141.0 ± 2.45*	146.5 ± 3.86*
Normal plus 200 mg/kg of HR	92.0 ± 7.54*	67.0 ± 8.67*	75.7 ± 8.41*

Values are mean ± S.E.M * $p < 0.05$, significantly different from normal control, paired t- test (n = 5), Hyp =. Hypertensive rats.

Table 3. Effect of aqueous leave extract of *Hibiscus rosa-sinensis* (HR) on biochemical parameters in rats.

Treatment	Total protein	Urea (mg/dl)	ALT	AST	Na$^+$	K$^+$	Ca^{2+}
Normal control (100% grower mash)	6.51 ± 1.29	41.47 ± 2.3	44.0 ± 4.72	18.0 ± 1.12	160.0 ± 2.20	5.6 ± 1.13	9.16 ± 0.48
Hypertensive rats (control)	8.11 ± 2.12	*46.95 ± 1.56	*62.0 ± 4.54	*80.3 ± 4.31	*192.0 ± 3.81	*4.8 ± 1.98	*13.32 ± 2.41
Hyp. plus 200 mg/kg of HR	8.41 ± 2.39	*66.73 ± 5.32	*29.0 ± 1.20	*70.0 ± 5.45	*166.0 ± 9.87	*6.0 ± 1.52	*14.43 ± 0.74
Normal plus 200 mg/kg of HR	6.10 ± 1.14	*70.56 ± 6.23	*48.0 ± 7.86	*24.0 ± 4.41	*188.0 ± 1.11	*5.8 ± 1.14	10.28 ± 1.28

Values are mean ± S.E.M * $p < 0.05$, significantly different from normal control, Paired t- test (n = 5), Hyp = Hypertensive rats.

significantly increased compared with normal control.

In the hypertensive rats there were significant ($p < 0.05$) increases in the urea, ALT, AST, Na$^+$ and Ca^{++} level compared with normal control (Table 3). There was significant increase in Ca^{++} level in the hypertensive rats administered with the crude extract compared with normal control. Total protein level was not significantly affected in the test rats compared with control. Potassium ions were significantly reduced in hypertensive control rats compared with normal control.

DISCUSSION

Most people with high blood pressure are over weight, weight loss lowers blood pressure significantly in those who are both over weight and hypertensive. In fact, reducing body weight by as little as ten pounds can lead to a significant reduction in blood pressure; weight loss appears to have a stronger blood pressure-lowering effect than dietary salt restriction (Aldeman, 1994; He et al., 2000; Stevens et al., 2001).

Salt loading had earlier been shown to cause hypertension in rats (Obiefuna et al., 1992). Reduction in weight gain of hypertensive rats observed is in agreement with the report of some workers (Ebuehi et al., 1999). The administration of *H. rosa-sinensis* leave extract showed blood lowering effect in both normotensive and hypertensive rats. Blood pressure is the product of cardiac output and peripheral resistance of the blood vessels (Bowman and Rand, 1980). The administration of *H. rosa-sinensis* probably decreased the blood pressure by decreasing the heart rate, which is a major determinant of the cardiac output (Guyton and Hall, 2000).

Significant increase in the sodium level of normotensive rats administered with the crude extract in spite of significant reduction in the blood pressure of these rats compared with the control shows that *H. rosa-sinensis* may interfere with the normal function of the kidney and hence produces increased salt retention. This observation is further strengthened by the increased urea concentration, although the change in total protein concentration was insignificant in those normotensive rats administered with the crude extract. In addition, AST, and ALT concentrations were increased in these rats compared with control. These results show that the leaves of this plant may have a deleterious effect on the kidney. Hypertensive rats administered with the extract had significant increase in urea, AST and Na$^+$ concentration compared with normal control. This result therefore, shows that although, the administration of *H. rosa-sinensis* reduced blood pressure in albino rats, the use of the plant may have an unpleasant effect on the kidney.

REFERENCES

Alderman MH (1994). Nonpharmacologic approaches to the treatment of hypertension. Lancet 334: 307–311.

Baker JT, Borris RP, Carte B, Cordell GA, Soejarto DD, Cragg GM Gupta MP, Madulid DA, Tyler VEJ (1995). Natural product drug discovery and development: New perspective on international collaboration. J. Nat. Prod. 58: 1325-1357.

Bowman WC, Rand MJ (1980). Textbook of pharmacology 2nd edtion, Published 1968, Blackwell Scientific (Oxford) pp. 21-24.

Ebuehi OAT, Elekolusi O, Adegunloye BL, Mojiminiyi FBO (1999). Effect of dietary salt loading on blood pressure and erythrocytes. West Afr. J. Med. 4: 21-24.

Farombi EO (2003). African indigenous plants with chemotherapeutic potentials and biotechnological approach to the production of bioactive prophylactic agents. Afr. J. Biotech. 2: 662-671.

Gitelman HJ (1967). Estimation of Calcium. Annal. Biochem. 18: 521-531.

Guyton AC, Hall JE (2000). Textbook of medicinal physiology, 10th edition, Sanders WB and Co. Philedelphia pp. 793-844.

He J, Whelton PK, Appel LJ (2000). Long-term effects of weight loss and dietary sodium reduction on incidence of hypertension. Hypertension 35: 544–549.

Henry RJ, Sobel C, Beckman S (1957). Determination of serum proteins by the Biuret reaction. Annal. Chem. 92: 1-5.

Nair R, Kalariya T, Chanda S (2005). Antibacterial Activity of Some Selected Indian Medicinal Flora. Turk. J. Biol. 29: 41-47.

Nivasarkar M, Patel M, Padh H, Bapu C, Shrivastava N (2005). Blastocyst implantation failure in mice due to "nonreceptive endometrium": endometrial alterations by Hibiscus rosa-sinensis leaf extract. Contraception 71(3): 227-230.

Obiefuna PCM, Sofola OA, Ebeigbe AB (1992). Contractile response of normotensive rats aorta to serum from salt loaded Sprague dawley rats. Nig. J. Physiol. Sci. 8: 54-57.

Olagbende-Dada SO, Ezeobika FN, Duru FI (2007). Anabolic effect of Hibiscus rosa-sinensis Linn. Leaf extracts in immature albino male rats. Nig. Q. J. Hosp. Med. Jan- 17(1): 5-7.

Reitman S, Frankel S (1957). Determination of glutamic-oxaloacetic transaminase. Am. J. Clin. Pathol. 28: 56-59.

Sachdewa A, Khemani LD (1999). A preliminary investigation of the possible hypoglycemic activity of Hibiscus rosa-sinensis.Biomed. Environ. Sci. 12(3): 222-226.

Searcy RL, Reardon JE, Foreman JA (1967). Estimation of enzymatic urea. Am. J. Med. Tech. 33: 15-20.

Sharma S, Sultana S (2004). Effect of Hibiscus rosa-sinensis extract on hyperproliferation and oxidative damage caused by benzoyl peroxide and ultraviolet radiations in mouse skin. Basic Clin. Pharmacol. Toxicol. 95(5): 220-225.

Shivananda NB, Sivachandra RS, Orette FA, Chalapathi RAV (2007). Effects of Hibiscus rosa-sinensis L (Malvaceae) on wound healing activity: A preclinical study in a Sprague Dawley rat. Int. J. Low Extrem. Wounds. June 6(2): 76-81.

Stevens VJ, Obarzanek E, Cook NR, (2001). Long-term weight loss and changes in blood pressure: results of the Trials of Hypertension Prevention, Phase II. Ann. Intern. Med. 134: 1–11.

Stuffness M, Douros J (1982). Current status of the NCI plant and animal product program. J. Nat. Prod. 45: 1-14.

Tietz N (1995). Clinical guide to laboratory tests. 3rd Edition. W.B Saunders company, Philadelphia pp. 518-519.

23

Assessing the role of RecA protein in the radioresistant bacterium *Deinococcus geothermalis*

Haïtham Sghaier[1]*, Katsuya Satoh[2, 3], Hirofumi Ohba[2, 3] and Issay Narumi[2, 3]

[1]Research Unit UR04CNSTN01 "Medical and Agricultural Applications of Nuclear Techniques", National Center for Nuclear Sciences and Technology (CNSTN), Sidi Thabet Technopark, 2020 Sidi Thabet, Tunisia.
[2]DNA Repair Protein Group, Research Unit for Quantum Beam Life Science Initiative, Quantum Beam Science Directorate, Japan Atomic Energy Agency, 1233 Watanuki, Takasaki, Gunma 370-1292, Japan.
[3]Gene Resource Research Group, Radiation-Applied Biology Division, Quantum Beam Science Directorate, Japan Atomic Energy Agency, 1233 Watanuki, Takasaki, Gunma 370-1292, Japan.

The moderately thermophilic bacterium *Deinococcus geothermalis* exhibits extraordinary resistance to ionizing radiation. RecA protein is considered to be one of the most important participants in radioresistance. To assess the role of the RecA protein in *D. geothermalis*, the recA gene was isolated from *D. geothermalis* and over expressed in *Escherichia coli*. After the *D. geothermalis* RecA protein (GeoRecA) was purified, the recombination activity was investigated *in vitro*. GeoRecA efficiently promoted the strand exchange reaction between homologous linear double-stranded DNA and circular single-stranded DNA substrates at 50°C. Like *Deinococcus radiodurans* RecA protein (DraRecA), GeoRecA could promote DNA strand exchange reaction through normal and inverse pathways. Furthermore, GeoRecA complemented the RecA deficiency of *D. radiodurans*. These results indicate that GeoRecA is a functional homologue of DraRecA and plays an important role in radioresistance. However, unlike DraRecA, GeoRecA could not complement the RecA deficiency of *E. coli*, suggesting that GeoRecA require more strict intracellular conditions than DraRecA does to fulfill its function. This study provides new insights into the role of deinococcal RecA protein in radioresistance.

Key words: *Deinococcus geothermalis*, DNA repair, DNA strand exchange, radio resistance, RecA.

INTRODUCTION

For most organisms, DNA double-strand breaks (DSB) generated by ionizing radiation are lethal or lead to mutagenic effects on cells. However, bacteria of the genus *Deinococcus* that comprises more than 40 distinct species are known for exhibiting extraordinary resistance to ionizing radiation. Among them, only *Deinococcus geothermalis* and *Deinococcus murrayi* are thermophilic bacteria that were isolated from the hot spring at Italy and Portugal, respectively (Ferreira et al., 1997). Although the whole-genome sequence of *D. geothermalis* has been determined (Makarova et al., 2007), the radio resistance mechanisms of *D. geothermalis* remain unclear.

In *Deinococcus radiodurans* which is the best-studied species among the members of *Deinococcus*, RecA-deficient strains exhibited extreme sensitivity to ionizing radiation (Gutman et al., 1994; Narumi et al., 1999; 2001; Satoh et al., 2006).

RecA proteins are highly conserved in almost all microorganisms except for *Buchnera* spp., *Vesicomyosocius okutanii* and *Ruthia magnifica* (Shigenobu et al., 2000; Kuwahara et al., 2007; Newton et al., 2007). The *Escherichia coli RecA* protein (EcoRecA) is one of the best-studied enzymes (Stohl et al., 2002). EcoRecA plays a central role in the repair of stalled replication forks, DSB repair, general recombination, induction of the SOS response and SOS mutagenesis (Cox, 2003). *RecA* protein is one of the most important participants in the radio resistance of *D. radiodurans*. The *D. radiodurans RecA* protein (DraRecA)

---

*Corresponding author. E-mail: sghaier.haitham@gmail.com, haitham.sghaier@cnstn.rnrt.tn.

could promote the DNA strand exchange reaction and complement *E. coli* RecA deficiency (Narumi et al., 1999; Satoh et al., 2002), indicating that DraRecA and EcoRecA are functional homologues. However, EcoRecA and DraRecA promote the DNA strand exchange reaction by quite different pathway (Kim and Cox, 2002). EcoRecA is ordered such that the single-stranded DNA (ssDNA) is bound first, followed by the double-stranded DNA (dsDNA). In contrast, DraRecA binds the dsDNA first and the homologous ssDNA substrate second. It is believed that the inverse pathway of DraRecA might remarkably be related to the efficient DNA repair mechanism. However, whether the peculiar property of DraRecA is common in RecA proteins of other members of the genus *Deinococcus* is still an open question.

The aim of this study is to assess the role of the *D. geothermalis* RecA protein (GeoRecA) in radioresistance and to highlight the functional similarity and dissimilarity between GeoRecA and DraRecA. To accomplish this purpose, we purified GeoRecA and investigated recombination activity *in vitro*. Furthermore, we employed plasmid complementation assay for GeoRecA. We also employed plasmid complementation assay to assess the *in vivo* function of GeoRecA following γ irradiation.

MATERIALS AND METHODS

Strains, plasmids and growth conditions

The bacterial strains and plasmids used in this study are listed in Table 1. *D. radiodurans* and *D. geothermalis* cells were grown at 30 and 45°C, respectively, in TGY broth containing 0.5% tryptone-peptone, 0.1% glucose and 0.3% yeast extract with agitation, or on TGY agar solidified with 1.5% agar. Three μg/ml of chloramphenicol was supplemented to the medium in order to maintain plasmids in *D. radiodurans* cells. *E. coli* strains were grown in Luria-Bertani (LB) broth-Miller or on LB agar-Lennox at 37°C. For the selection of *E. coli* cells transformed with plasmids, 100 μg/ml of ampicillin was added to the medium.

Cloning and construction of expression plasmid of *D. geothermalis recA*

To amplify the *recA* gene (*GeorecA*) from the *D. geothermalis* DSM11300 genomic DNA by PCR reaction, priming was carried out using *PfuTurbo* HotStart DNA polymerase (Stratagene) with primers GeorecA-5Nde (5'-GAGAC<u>CATATG</u>AGCAAGGACAACCC-3') and GeorecA-3Bam (5'-GCCC<u>GGATCC</u>CTCACCGCTTACTCT-3') possessing *Nde*I and *Bam*HI restriction sites (underlined in the sequences), respectively. PCR product (1,077 bp) was then digested with *Nde*I and *Bam*HI to adapt the termini for in-frame insertion of *GeorecA* into the *Nde*I-*Bam*HI sites in the pET3a vector (Novagen). The resulting expression plasmid was designated pET3a-GeorecA.

The *GeorecA* expression plasmids in *D. radiodurans* were constructed as follows. The *Nde*I and *Bam*HI-digested *GeorecA* PCR product was inserted into the *Nde*I-*Bam*HI sites in the pRAD1 vector (Meima et al., 2000). The resulting plasmid was designated pRGE1. The *pprA* promoter region (208 bp) was amplified by PCR from pZA8 (Narumi et al., 2004) with primers PDpprAF2 and PDpprAR1 (Ohba et al., 2005), digested with *Nde*I, and inserted

into the *Nde*I-site in the pRGE1 to yield plasmid pGEO5.

To amplify the promoter region and *recA* gene (*EcorecA*) from the *E. coli* JM110 genomic DNA by PCR reaction, priming was carried out using *PfuTurbo* HotStart DNA polymerase (Stratagene) with primers ErecAF1 (5'-TGATGGGGAAAACTCGCA-3') and ErecAR1 (5'-CACGATCTGTGACGTCCTT-3'). PCR product (5,539 bp) was inserted into the *Sma*I site in the pUC19 vector (Takara Bio). The resulting plasmid was designated pECrecA11. The *EcorecA* expression plasmids in *E. coli* were constructed as follows. The DNA region containing *EcorecA* promoter and structural gene (2,154 bp) was amplified by PCR from pECrecA11 and primers ECrecAp-F1 (5'-GCGGCGACGGGCATATCAAC-3') plus ECrecAR2 (5'-AACAGGATCCTTAAAAATCTTCG-3'), digested with *Bam*HI, and inserted into the *Bam*HI-site in the pUC19 to yield plasmid pH-EcorecA.

The *GeorecA* expression plasmids in *E. coli* were constructed as follows. The upstream of *EcorecA* region (1,102 bp) including promoter was amplified by PCR from pECrecA11 and primers ECrecAp-F1 and ECrecAp-R1 (5'-CGATAGCCATATGTACTCCTGTCATGCC-3') and digested with *Nde*I and *Bam*HI. The *Nde*I and *Bam*HI-digested *GeorecA* and *EcorecA* promoter PCR products were inserted into the *Bam*HI-site in the pUC19 to yield plasmid pH-GeorecA.

The DNA sequence of the expression plasmids was checked to confirm the absence of errors.

Bioinformatics

Homology search was done using bioinformatics tools on the web site:http://www-archbac.u-psud.fr/genomics/GenomicsToolBox.html. FeatureMap3D helped the analysis of homologous structures in the PDB (Wernersson et al., 2006).

Protein purification

E. coli BLR(DE3) carrying pET3a-GeorecA was cultivated in LB broth-Miller containing ampicillin. At an optical density of approximately 0.4 at 600 nm, isopropyl-□-D-1-thiogalactopyranoside was added to a final concentration of 0.4 mM and growth was continued for an additional 3 h. The cells were then harvested by centrifugation. The cell pellet was resuspended in cold buffer containing 50 mM Tris-HCl (pH 8.0), 5 mM ethylenediaminetetraacetic acid (EDTA), 1 mM phenylmethylsulfonyl fluoride and 0.1% protease inhibitor cocktail (Calbiochem), and stored at −80°C. All subsequent steps were carried out at 4°C unless otherwise indicated. The suspension was lysed by sonication for 10 min on ice. The debris was removed by centrifugation at 7,000 g, 4°C for 30 min and ammonium sulfate was added slowly to the supernatant to give 20% saturation. The supernatant was stirred for 1 h and then centrifuged for 30 min. Protein was further purified to apparent homogeneity by column chromatography on Toyopearl Phenyl-650S (Tosoh), HiTrap Heparin (GE Healthcare) and Mono Q HR 5/5 (GE Healthcare). The pooled fractions were concentrated and desalted using an Amicon Ultra-15 filter 30K Centrifugal Filter Device (Millipore) with 50 mM Tris-HCl (pH 7.4).

DNA strand exchange assay

The strand exchange reaction was assayed by assembling the following reaction mixture. First, 20 μM φX174 viral DNA (ssDNA) (New England Biolabs) was preincubated in a buffer containing 20 mM Tris-HCl (pH 7.4), 1 mM dithiothreitol, 10 mM MgCl₂, 2 mM ATP, and 0.9 μM *E. coli* single-stranded DNA binding protein (SSB) (GE Healthcare) at 37°C for 5 min, after which, 34 μM GeoRecA protein

Table 1. Strains and plasmids used in this study.

Designation	Relevant description	Source or reference
D. radiodurans		
R₁	Wild-type (ATCC13939)	ATCC
MR₁	Wild-type	Moseley (1967)
rec30	Same as MR₁ but *RecA670*	Narumi et al. (1999)
D. geothermalis	Wild-type (DSM11300)	DSMZ
E. coli		
JM110	Host for plasmid subclones, *recA*⁺	Takara Bio
JM107	Host for plasmid subclones, *recA*⁺	Takara Bio
JM109	Host for plasmid subclones, *recA1*	Takara Bio
BLR(DE3)	Host for expression plasmid	Novagen
Plasmid		
pET3a	*E. coli* vector	Novagen
pET3a-GeoRecA	pET3a with *Nde*I-*Bam*HI::1,077-bp PCR product containing *GeorecA*	This study
pRAD1	*E. coli-D. radiodurans* shuttle vector	Meima et al. (2000)
pRGE1	pRAD1 with *Nde*I-*Bam*HI::1,077-bp PCR product containing *GeorecA*	This study
pZA8	pUC19 with 6,005-bp *D. radiodurans* DNA fragment containing *pprA*	Narumi et al. (2004)
pGEO5	pRGE1 containing *D. radiodurans pprA* promoter region and *GeorecA*	This study
pUC19	*E. coli* vector	Takara Bio
pECRecA11	pUC19 with 5,539-bp *E. coli* DNA fragment containing *recA*	Takara Bio
pH-ecoRecA	pUC19 containing *E. coli recA* promoter region and *EcorecA*	This study
pH-GeoRecA	pUC19 containing *E. coli recA* promoter region and *GeorecA*	This study

was added. The incubation continued for an additional 5 to 60 min at various temperatures from 37 to 70℃ and the reaction was initiated by adding 30 μM *Hinc*II-digested fragment of φX174 RF I (dsDNA) (New England Biolabs). The reaction was quenched by the addition of a buffer consisting of 20 mM Tris-HCl (pH 7.4), 0.5% SDS, 40 mM EDTA and 2 mg/ml of proteinase K (Qiagen). After being incubated at 37℃ for 30 min, the sample was subjected to 1% agarose gel electrophoresis. The substrate DNA and strand exchange products were visualized by staining with ethidium bromide (0.5 μg/ml).

Measurement of cell survival rate

D. radiodurans cells were grown in TGY broth at 30℃ with agitation to early stationary phase. Cells were harvested by centrifugation at 7,000 g, 4℃ for 5 min, washed twice with 10 mM sodium phosphate buffer (PB; pH 7.0) and resuspended in the same buffer. Aliquots (0.1 ml) of the cell suspension were dispensed into test tubes and irradiated at room temperature for 1 to 2 h with ⁶⁰Co γ rays at dose rates from 0.1 to 4 kGy per h that were regulated by changing the distance of the samples from the γ ray source. Irradiated samples were appropriately diluted with 10 mM PB,

spread onto TGY agars and incubated at 30℃ for 3 days prior to the enumeration of colonies.

E. coli cells were grown at 37℃ in LB broth-Miller, harvested, washed and resuspended as described above. Aliquots (0.1 ml) of the cell suspension were irradiated at room temperature for 1 h with ⁶⁰Co γ rays at dose rates from 0.1 to 0.2 kGy per h. Irradiated sample was diluted appropriately with the same buffer, spread on LB Agar-Lennox and incubated for 18 h at 37℃ prior to the enumeration of colonies.

RESULTS AND DISCUSSION

Sequence and structure of *D. geothermalis* RecA

To construct the expression plasmid, we cloned the *recA* gene from the thermophilic radio resistant bacterium *D. geothermalis* (GeoRecA). The predicted GeoRecA is consistent with the amino acid sequence of *D. geothermalis* DSM11300 RecA protein (GenBank monomer protein contains 358 amino acid residues and

```
EcoRecA    1   MAIDENK-----------QKALAAALGQIEKQFGKGSIMRLGEDRSMDVETISTGSLSLD    49
DraRecA    1   MSKDATKEISAPTDAKERSKAIETAMSQIEKAFGKGSIMKLGAESKLDVQVVSTGSLSLD    60
GeoRecA    1   MSKDNPKDFGTPSDSKERLKAIETAMTQIEKAFGKGSIMRLGAESKLDVQAVSTGSLSLD    60
               *  *    *          **    *   **** ******* **   **  *******

EcoRecA   50   IALGAGGLPMGRIVEIYGPESSGKTTLTLQVIAAAQREGKTCAFIDAEHALDPIYARKLG   109
DraRecA   61   LALGVGGIPRGRITEIYGPESSGKTTLALAIVAQAQKAGGTCAFIDAEHALDPVYARALG   120
GeoRecA   61   LALGVGGIPRGRITEIYGPESSGKTTLALSVIAQAQRAGGTCAFIDAEHALDPVYARSLG   120
               *** ** * *** ******* *****  *   *  ** * ************** *** **

EcoRecA  110   VDIDNLLCSQPDTGEQALEICDALARSGAVDVIVVDSVAALTPKAEIEGEIGDSHMGLAA   169
DraRecA  121   VNTDELLVSQPDNGEQALEIMELLVRSGAIDVVVVDSVAALTPRAEIEGDMGDSLPGLQA   180
GeoRecA  121   VNTDELLVSQPDNGEQALEIMELLVRSGAIDVVVVDSVAALTPRAEIEGEMGDSLPGLQA   180
               *  * ****.********  *   *  ***  ** *********** *****   *** **

EcoRecA  170   RMMSQAMRKLAGNLKQSNTLLIFINQIRMKIGVMFGNPETTTGGNALKFYASVRLDIRRI   229
DraRecA  181   RLMSQALRKLTAILSKTGTAAIFINQVREKIGVMYGNPETTTGGRALKFYASVRLDVRKI   240
GeoRecA  181   RLMSQALRKLTAILSKTGTAAIFINQVREKIGVMYGNPETTTGGRALKFYASVRLDVRKI   240
               * **** ***  *      ******  ** *****  ******** ********** * **

EcoRecA  230   G-AVKEGENVVGSETRVKVVKNKIAAPFKQAEFQILYGEGINFYGELVDLGVKEKLIEKA   288
DraRecA  241   GQPTKVGNDAVANTVKIKTVKNKVAAPFKEVELALVYGKGFDQLSDLVGLAADMDIIKKA   300
GeoRecA  241   GQPVKLGNDAVGNTVKVKTVKNKVAPPFKEVELTLLYGKGFDQLSDLVTLAADMDIIKKA   300
               *   *   *    * **** * ***  *    ** *      ** *     *   * **

EcoRecA  289   GAWYSYKGEKIGQGKANATAWLKDNPETAKEIEKKVRELLLSNPN--STPDFSVDDSEGV   346
DraRecA  301   GSFYSYGDERIGQGKEKTIAYIAERPEMEQEIRDRVMAAIRAGNAGEAPALAPAPAAPEA   360
GeoRecA  301   GSFYSYGEERIGQGKEKAIAYIAERPELEQEIRDRVLAAIKEGRD----PIAAVPETPAL   356
               *   *** * *****     *    **   **   *

EcoRecA  347   AETNEDF                                                        353
DraRecA  361   AEA----                                                        363
GeoRecA  357   AE-----                                                        358
               **
```

Figure 1. Multiple amino acid sequence alignment of EcoRecA, DraRecA and GeoRecA. Multiple alignment was determined using the CLUSTAL W program (Thompson et al., 1994), EcoRecA, *E. coli* RecA (Horii et al., 1980), DraRecA, *D. radiodurans* RecA (Narumi et al., 1999) and GeoRecA, *D. geothermalis* RecA (this study). Dashes indicate gaps in the alignment. Numbers on both sides represent the coordinates of each protein. Asterisks indicate identical residues. Boxes and over lines represent positions of polymeric domains and DNA binding domains respectively, which are proposed for EcoRecA, (Karlin and Brocchieri, 1996).

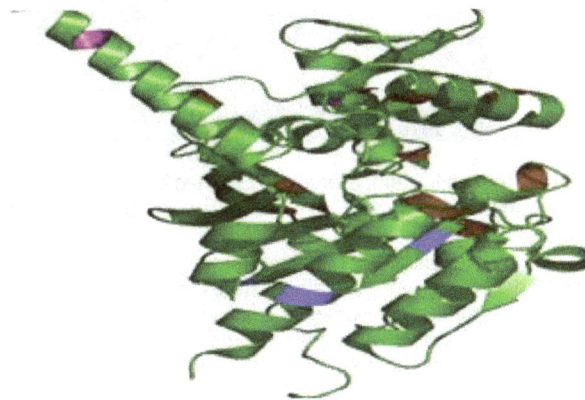

Figure 2. The PDB structure 1XP8 of the homologous protein *D. radiodurans* RecA. Green, Perfect match with GeoRecA (*D. geothermalis* RecA); Brown, Mismatch (low significance: ex, valine (hydrophobic aliphatic) → isoleucine (hydrophobic aliphatic); Violet, Mismatch (high significance: ex, alanine (hydrophobic aliphatic) → threonine (polar neutral)); Blue, Sequence gap in query sequence.

accession No. ABF46432). Figure 1 shows the alignment of amino acid sequence of GeoRecA, DraRecA and EcoRecA. GeoRecA was more similar to DraRecA (87.4% identity) than to EcoRecA (57.8% identity). However, many residues are invariants strongly attesting to the functional and structural importance of these segments containing the DNA binding and polymeric domains (Figure 1) (Karlin and Brocchieri, 1996).

The overall fold of GeoRecA, DraRecA and EcoRecA is similar but different in specific regions. We exemplify this case by the structural mismatch regions between GeoRecA and DraRecA (Figure 2). Also, for example,

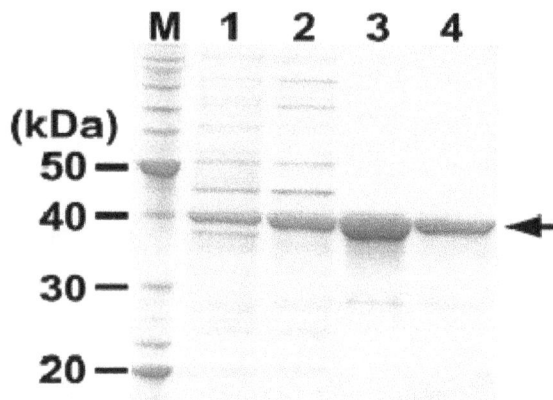

Figure 3. Purification of *D. geothermalis* RecA protein. Samples (10 µg) were subjected to 12.5% SDS-PAGE and stained with a coomassie brilliant blue. Lane M, 1 protein marker (precision plus protein standards from Bio-Rad); lane 1, total cellular proteins from *E. coli* BLR(DE3)/pET3a-GeorecA induced by IPTG; lane 2, pooled GeoRecA fractions from phenyl 650S column; lane 3, pooled GeoRecA fractions from HiTrap Heparin HP column; lane 4, pooled GeoRecA fractions from Mono S column. On the left, relative molecular masses (kDa) of the standard proteins are shown. Arrow on the right indicates the position of the 38-kDa band of GeoRecA.

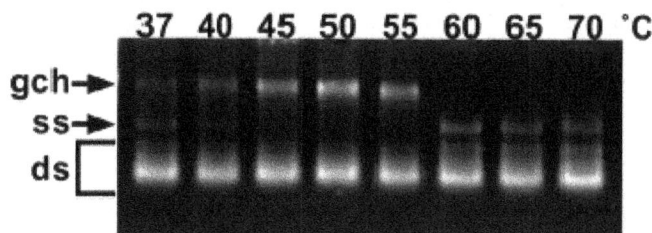

Figure 4. Optimum temperature of GeoRecA for the promotion of DNA strand exchange reaction. Reactions were carried between circular ssDNA and the linear dsDNA derived from *Hinc*II-digested φX174 RF I to yield an expected product φX174 RFII (nicked circular dsDNA) at various temperatures from 37 to 70℃ in the buffer as described in Materials and Methods. The positions of dsDNA (ds), ssDNA (ss) and the complete strand exchange product (gapped circular heteroduplex DNA, gch) are indicated.

compared to EcoRecA, the inner surface along the central axis of DraRecA protein filament has an increased positive electrostatic potential (Rajan and Bell, 2004). The theoretical pI/Mw (http://www.expasy.org/tools/pi_tool.html) for GeoRecA is 5.48/38,156.78 (358 amino acid residues including the initiating methionine), 5.45/38,144.76 (363 amino acid residues including the initiating Met) for DraRecA, and 5.09/37,973.37 (353 amino acid residues including the initiating Met) for EcoRecA.

DNA strand exchange

To investigate the recombination activity of GeoRecA *in*

vitro, we purified the protein as described in Materials and Methods. The purified protein migrated on a SDS-polyacrylamide gel with an apparent molecular mass of 38 kDa (Figure 3). This is close to the molecular mass (38,156 Da) calculated from DNA sequence data.

To check whether the purified GeoRecA possesses recombination activity *in vitro*, a DNA strand exchange assay was performed. The assays were carried out at various temperatures from 37 to 70℃. GeoRecA promoted an efficient strand exchange reaction between homologous circular ssDNA and linear dsDNA substrates to yield complete strand exchange products (Figure 4). GeoRecA most efficiently promoted the strand exchange reaction at 50℃. The optimum temperature of 50℃ for

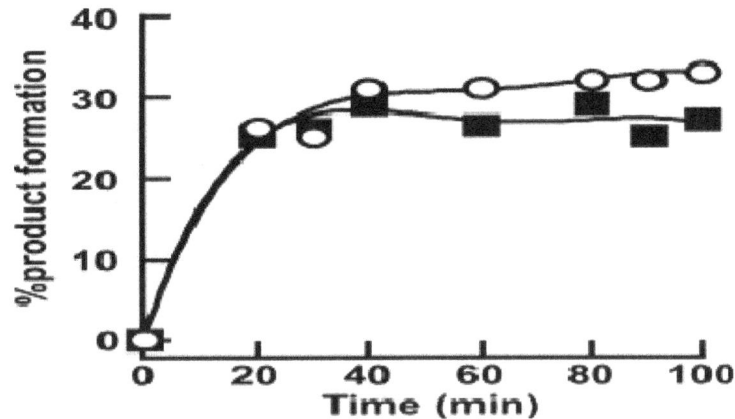

Figure 5. GeoRecA promotes DNA strand exchange reaction *via* inverse pathway. Reactions were carried out at 50°C, pH 7.4 between circular ssDNA (ss) and the linear dsDNA (ds). In the ds to ss reactions (closed squares), the GeoRecA protein was preincubated with the linear dsDNA and ATP for 5 min. The ssDNA was then added to start the reaction and the *E. coli* SSB protein was added 5 min later. In the ss to ds reactions (open circles), GeoRecA was preincubated with the ssDNA, ATP and *E. coli* SSB for 5 min. Then dsDNA was added to start the reaction.

GeoRecA protein activity as shown in Figure 4 is consistent with the fact that *D. geothermalis* can grow between 45 and 50°C (Ferreira et al., 1997).

EcoRecA initiates DNA strand exchange with a filament bound to the ssDNA, followed by uptake of the duplex substrate (normal pathway). In contrast, the pathway of DNA strand exchange promoted by DraRecA is the exact inverse (Kim and Cox, 2002). We examined which pathway of the DNA strand exchange reaction GeoRecA employed. Like DraRecA, GeoRecA promoted DNA strand exchange reaction through both pathways (Figure 5). Whereas, the inverse pathway is clearly major in DraRecA mediated DNA strand exchange pathway (Kim and Cox, 2002), GeoRecA mediated reactions showed no significant difference through both pathways under our conditions (Figure 5). However, these results suggested that the inverse DNA strand exchange pathway which is different from other bacterial RecA-mediated pathway might be related to the remarkable efficient DNA repair mechanism of the genus *Deinococcus*.

Complementation of *RecA* deficiencies of *D. radiodurans* and *E. coli*

To investigate the function of GeoRecA protein *in vivo*, we constructed the *GeorecA* expression plasmid under the control of the radiation responsive *pprA* promoter from *D. radiodurans* as described in Materials and Methods. Transcriptome analysis revealed that the *pprA* gene exhibited a *recA*-like activation pattern following γ irradiation (Liu et al., 2003). Because the *recA* promoter is not defined yet, the *pprA* promoter was used for expression plasmid instead. The *GeorecA* expression

plasmid was introduced into DNA repair deficient mutant *D. radiodurans* strain rec30 carrying *recA670* mutation (Narumi et al., 1999). Then, the sensitivity of strain rec30 carrying the *GeorecA* expression plasmid following γ irradiation was compared between strains R₁ (carrying the wild-type *recA*) and rec30. Strain rec30 exhibited much more sensitive to γ irradiation than strain R₁, consistent with previous studies (Narumi et al., 1999). In contrast, strain rec30 carrying the *GeorecA* expression plasmid was resistance to γ irradiation but slightly sensitive than that of strain R₁ carrying pRAD1 (no insert) (Figure 6), indicating that the GeoRecA protein partially complements *D. radiodurans recA670* mutation. This result suggested that GeoRecA exhibited low recombination activity under the growth condition of *D. radiodurans*, and is consistent with our observation of *in vitro* RecA-mediated DNA strand exchange reaction (Figure 4). It has also been shown that EcoRecA provides partial complementation to a *D. radiodurans recA* null mutant (Schlesinger, 2007).

The *D. radiodurans recA* gene has been shown to complement *E. coli recA1* (Narumi et al., 1999). Whether the *GeorecA* complements the deficiency was tested. For this purpose, we constructed the *GeorecA* and *EcorecA* expression plasmids under the control of the *E. coli recA* promoter as described in Materials and Methods. Then, *E. coli* strain JM109 carrying *recA1* was transformed with these expression plasmids. The *EcorecA* expression plasmid was used as control. *E. coli* strains JM107 and JM109 are isogenic except for the *recA* gene. As shown in Figure 7, the JM109 transformant carrying the *EcorecA* expression plasmid was as resistant to γ rays as strain JM107 (as for JM109 but *recA*⁺) carrying pUC19 (no insert). On the other hand, the

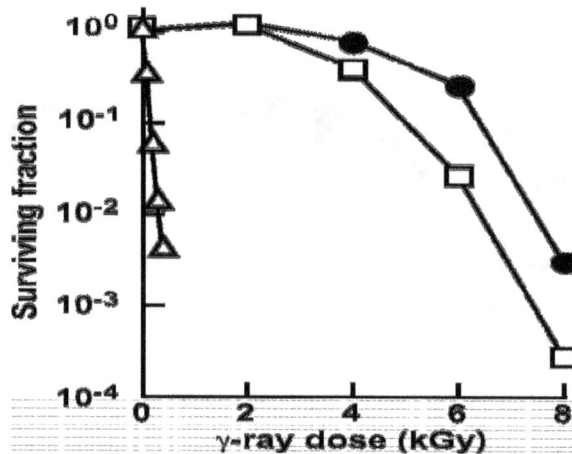

Figure 6. Sensitivity of *D. radiodurans* strains to γ-rays. Closed circles, strain R$_1$ (wild-type) carrying pRAD1; open triangles, strain rec30 (*recA670*); open squares, strain rec30 (*recA670*) carrying pGEO5.

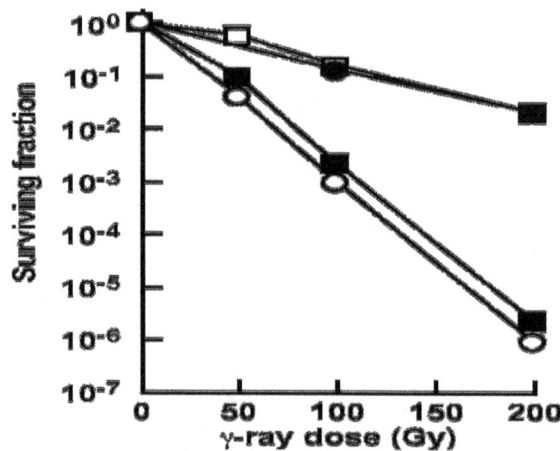

Figure 7. Sensitivity of *E. coli* strains to γ-rays. Closed circles, *E. coli* JM107 carrying pUC19; closed squares, *E. coli* JM109 (*recA1*) carrying pUC19; open squares, *E. coli* JM109 (*recA1*) carrying pH-EcorecA; open circles, *E. coli* JM109 (*recA1*) carrying pH-GeoRecA.

sensitivity of JM109 transformant carrying the *GeorecA* expression plasmid was equal to that of strain JM109 carrying pUC19, though the transformant produced abundant amounts of the plasmid-encoded GeoRecA protein (data not shown). This result indicated that GeoRecA could not complement *E. coli* RecA deficiency, whereas DraRecA could. However, it should be considered that the regulation of growth stage-specific expression level of RecA protein is critical in the DNA damage response mechanism. Therefore, the inability of complementation might be related to the expression level of GeoRecA that possibly could not be controlled enough in *E. coli*. GeoRecA may require more strict intracellular conditions than DraRecA does to fulfill its function.

Conclusion

In this study, the role of GeoRecA in radioresistance and the functional similarity and dissimilarity between GeoRecA and DraRecA were assessed through DNA strand exchange and plasmid complementation assays. The optimal temperature at which GeoRecA most efficiently promoted the strand exchange reaction between homologous linear double-stranded DNA and circular single-stranded DNA substrates was 50°C. GeoRecA could promote DNA strand exchange reaction through normal and inverse pathways, and it could complement the RecA deficiency of *D. radiodurans*, indicating that GeoRecA is a functional homologue of

DraRecA and plays an important role in radio resistance. However, unlike DraRecA, GeoRecA could not complement the RecA deficiency of *E. coli*. This result suggests that GeoRecA require more strict intracellular conditions than DraRecA does to fulfill its function. Our findings from this study provide new insights into the role of deinococcal RecA protein in radioresistance and the potential use of thermostable RecA protein as a reagent in DNA engineering such as the targeted DNA cleavage, the asymmetric linker attachment and the multiplex PCR reaction (Koob et al., 1992; Shigemori, 2005; Shigemoti et al., 2005). Further research is required to puzzle out the nature of GeoRecA proteins. Our effort is currently being directed towards generating a *GeorecA* disruptant strain to investigate gene disruption effect on radioresistance and functional complementation by EcoRecA and DraRecA.

REFERENCES

Battista JR (1997). Against all odds: the survival strategies of *Deinococcus radiodurans*. Annu. Rev. Microbiol. 51: 203-224.

Cox MM, Battista JR (2005). *Deinococcus radiodurans*-the consummate survivor. Nat. Rev. Microbiol. 3: 882-892.

Cox MM (2003). The bacterial RecA protein as a motor protein. Annu. Rev. Microbiol. 57: 551-577.

De Groot A, Dulermo R, Ortet O, Blanchard L, Guérin P, Fernandez B, Vacherie B, Dossat C, Jolivet E, Siguier P, Chandler M, Barakat M, Dedieu A, Barbe V, Heulin T, Sommer S, Achouak W, Armengaud J (2009). Alliance of proteomics and genomics to unravel the specificities of Sahara bacterium *Deinococcus deserti*. PLoS One 5: e1000434.

Ferreira AC, Nobre MF, Rainey FA, Silva MT, Wait R, Burghardt J, Chung AP, da Costa MS (1997). *Deinococcus geothermalis* sp. nov. and *Deinococcus murrayi* sp. nov., two extremely radiation-resistant and slightly thermophilic species from hot springs. Int. J. Syst. Bacteriol. 47: 939-947.

Gutman PD, Carroll JD, Masters CI, Minton KW (1994). Sequencing, targeted mutagenesis and expression of a *recA* gene required for the extreme radioresistance of *Deinococcus radiodurans*. Gene 141: 31-37.

Horii T, Ogawa T, Ogawa H (1980). Organization of the recA gene of *Escherichia coli* Proc. Natl. Acad. Sci. USA 77: 313-317.

Karlin S, Brocchieri L (1996). Evolutionary conservation of RecA genes in relation to protein structure and function. J. Bacteriol. 178: 1881-1894.

Kim JI, Cox MM (2002). The RecA proteins of *Deinococcus radiodurans* and *Escherichia coli* promote DNA strand exchange via inverse pathways Proc. Natl. Acad. Sci. USA 99: 7917-7921.

Koob M, Burkiewicz A, Kur J, Szybalski W (1992). RecA-AC: single-site cleavage of plasmids and chromosomes at any predetermined restriction site. Nucleic Acids Res. 20: 5831-5836.

Kuwahara H, Yoshida T, Takaki Y, Shimamura S, Nishi S, Harada M, Matsuyama K, Takishita K, Kawato M, Uematsu K, Fujiwara Y, Sato T, Kato C, Kitagawa M, Kato I, Maruyama T (2007). Reduced genome of the thioautotrophic intracellular symbiont in a deep-sea clam, *Calyptogena okutanii*. Curr. Biol. 17: 881-886.

Liu Y, Zhou J, Omelchenko MV, Beliaev AS, Venkateswaran A, Stair J, Wu L, Thompson DK, Xu D, Rogozin IB, Gaidamakova EK, Zhai M, Makarova KS, Koonin EV, Daly MJ (2003). Transcriptome dynamics of *Deinococcus radiodurans* recovering from ionizing radiation Proc. Natl. Acad. Sci. USA 100: 4191-4196.

Makarova KS, Omelchenko MV, Gaidamakova EK, Matrosova VY, Vasilenko A, Zhai M, Lapidus A, Copeland A, Kim E, Land M, Mavrommatis K, Pitluck S, Richardson PM, Detter C, Brettin T, Saunders E, Lai B, Ravel B, Kemner KM, Wolf YI, Sorokin A,

Gerasimova AV, Gelfand MS, Fredrickson JK, Koonin EV, Daly MJ (2007). *Deinococcus geothermalis*: the pool of extreme radiation resistance genes shrinks. PLoS One 2: e955.

Meima R, Lidstrom ME (2000). Characterization of the minimal replicon of a cryptic *Deinococcus radiodurans* SARK plasmid and development of versatile *Escherichia coli–D. radiodurans* shuttle vectors. Appl. Environ. Microbiol. 66: 3856-3867.

Moseley BEB (1967). The isolation and some properties of radiation sensitive mutants of *Micrococcus radiodurans*. J. Gen. Microbiol. 49: 293-300.

Narumi I (2003). Unlocking radiation resistance mechanisms: still a long way to go. TrendsMicrobiol. 11: 422-425.

Narumi I, Satoh K, Cui S, Funayama T, Kitayama S, Watanabe H (2004). PprA: a novel protein from *Deinococcus radiodurans* that stimulates DNA ligation. Mol. Microbiol. 54: 278-285.

Narumi I, Satoh K, Kikuchi M, Funayama T, Kitayama S, Yanagisawa T, Watanabe H, Yamamoto K (1999). Molecular analysis of the *Deinococcus radiodurans* recA locus and identification of mutation site in a DNA repair-deficient mutant, rec30. Mutat. Res. 435: 233-243.

Newton ILG, Woyke T, Auchtung TA, Dilly GF, Dutton RJ, Fisher MC, Fontanez KM, Lau E, Stewart FJ, Richardson PM, Barry KW, Saunders E, Detter JC, Wu D, Eisen JA, Cavanaugh CM (2007). The *Calyptogena magnifica* chemoautotrophic symbiont genome. Science 315: 998-1000.

Ohba H, Satoh K, Yanagisawa T, Narumi I (2005). The radiation responsive promoter of the *Deinococcus radiodurans* pprA Gene. 363: 133-141.

Rajan R, Bell CE (2004). Crystal structure of RecA from Deinococcus radiodurans: insights into the structural basis of extreme radioresistance. J. Mol. Biol. 344: 951-963.

Satoh K, Narumi I, Kikuchi M, Kitayama S, Yanagisawa T, Yamamoto K, Watanabe H (2002). Characterization of RecA424 and RecA670 proteins from *Deinococcus radiodurans*. J. Biochem. 131: 121-129.

Satoh K, Ohba H, Sghaier H, Narumi I (2006). Down-regulation of radioresistance by LexA2 in *Deinococcus radiodurans*. Microbiology 152: 3217-3226.

Schlesinger DJ (2007). Role of RecA in DNA damage repair in *Deinococcus radiodurans*. FEMS Microbiol. Lett. 274: 342-347.

Shigemori Y (2005). A novel means of asymmetric linker attachment to DNA molecule by RecA-mediated multistranded DNA formation. Anal. Biochem. 338: 143-150.

Shigemori Y, Mikawa T, Shibata T, Oishi M (2005). Multiplex PCR: use of heat-stable *Thermus thermophilus* RecA protein to minimize non-specific PCR product. Nucleic Acids Res. pp. 33-126.

Shigenobu S, Watanabe H, Hattori M, Sakaki Y, Ishikawa H (2000). Genome sequence of the endocellular bacterial symbiont of aphids *Buchnera* sp. APS. Nature 407: 81-86.

Stohl EA, Blount L, Seifert HS (2002). Differential cross-complementation patterns of *Escherichia coli* and *Neisseria gonorrhoeae* RecA proteins. Microbiology 148: 1821-1831.

Thompson JD, Higgins DG, Gibson TJ (1994). CLUSTAL W: improving the sensitivity of progressive multiple sequence alignment through sequence weighting, position-specific gap penalties and weight matrix choice. Nucleic Acids Res. 22: 4673-4680.

Wernersson R, Rapacki K, Staerfeldt HH, Sackett PW, Molgaard A (2006). FeatureMap3D–a tool to map protein features and sequence conservation onto homologous structures in the PDB. Nucleic Acids Res. 34: 84-88.

White O, Eisen JA, Heidelberg JF, Hickey EK, Peterson JD, Dodson RJ, Haft DH, Gwinn ML, Nelson WC, Richardson DL, Moffat KS, Qin H, Jiang L, Pamphile W, Crosby M, Shen M, Vamathevan JJ, Lam P, McDonald L, Utterback T, Zalewski C, Makarova KS, Aravind L, Daly MJ, Minton KW, Fleischmann RD, Ketchum KA, Nelson KE, Salzberg S, Smith HO, Venter JC, Fraser CM (1999). Genome sequence of the radioresistant bacterium *Deinococcus radiodurans* R1. Science 286: 1571-1577.

24

Biochemical effects of drinking *Terminalia catappa* Linn. decoction in Wistar rats

Ibegbulem, C. O.[1]*, Eyong, E. U.[2] and Essien, E. U.[2]

[1]Department of Biochemistry, Federal University of Technology, Owerri, Nigeria.
[2]Department of Biochemistry, University of Calabar, Calabar, Nigeria.

The biochemical effects of drinking *Terminalia catappa* Linn. decoction in place of water using weanling wistar rat (*Rattus norvegicus*) models of both sexes was studied. In folklore, the decoction is taken as a medicine by sicklers. Some of the phytochemical and chemical constituents detected in the decoction included tannins, flavonoids, saponins, β-carotene, thiocyanates, cardiac glycosides, alkaloids, cyanogenic glycosides, vitamin C, ρ – hydroxybenzoic acid (ρ – HBA), alkaloids, catechins, free amino acids and monosaccharides. The decoction was acidic, pH 5.94 ± 0.01. It was administered to the rats in place of water, *ad libitum*. Test for its biochemical effects lasted for 35 days and was ingested at 18.3 mg/ml. Results of its effects on liver function parameters showed that most of the parameters were not significantly ($p > 0.05$) affected. However, it increased ($p < 0.05$) the alanine aminotransferase activity and serum total protein contents of the female rats and significantly reduced ($p < 0.05$) the serum total bilirubin levels of both male and female rats. It also significantly ($p < 0.05$) reduced the serum total cholesterol levels of the female rats and the serum LDL – cholesterol levels of both the male and female rats. The haematological indices were not significantly ($p > 0.05$) altered. Ingestion of the decoction in place of water significantly ($p < 0.05$) increased the body weight gained, fluid and feed intakes and did not interfere with the nutrient adequacy of the feed by reducing the feed conversion ratios (FCR's) of the test rats. In conclusion, the study established the safety of the decoction when drunk in place of water.

Key words: Biochemical effects, decoction, drink, rats, *Terminalia catappa*.

INTRODUCTION

The universal role of plants in the treatment of disease is exemplified by their employment in all major systems of medicine irrespective of the underlying philosophical premise. There is a great wealth of knowledge concerning the medicinal, narcotic and other properties of plants that is still transmitted orally from generation to generation by tribal societies, particularly those of Tropical Africa, North and South America and the pacific countries (Evans, 2002).

Information on the use of medicinal plants has been obtained from herbalists, herb sellers and indigenous people of Africa over many years (Sofowora, 2002). The United Nations Development Programme estimates the value of pharmaceutical products derived from developing world plants, animals and microbes to be more than \$30 billion per year. Consider the success story of vinblastine and vincristine. These anticancer alkaloids are derived from the Madagascar periwinkle (*Catharanthus roseus*) (Cunningham et al., 2005). The use of single pure compounds, including synthetic drugs, is not without its limitations and in recent years there has been an immense revival in interest in the herbal and homoeopathic systems of medicine, both of which rely heavily on plant sources (Evans, 2002).

Terminalia species tree has a characteristic pagoda shape because it sends out a single stem from the top centre. Oliver-Bever (1986) reported that it is made up of 250 species. Oboh et al. (2008) also reported the phenotypic diversity of *Terminalia catappa* from southwestern Nigeria where variability in leaf shape and ripe fruit colour were observed. The fruit is a sessile, laterally compressed ovoid to ovate, smooth-skinned

*Corresponding author. E-mail: ibemog@yahoo.com.

drupe. Ahmed et al. (2005) reported that the leaves contained several flavonoids, tannins, saponins, triterpinoid and phytosterols. Due to the above chemical richness, the leaves and bark are used in different traditional medicines for various purposes worldwide. They also reported the biochemical effects of administrating *T. catappa* Linn. aqueous and cold leaf extracts, intraperitoneally and showed that it caused the regeneration of the β-cells of the islets of Langerhans, decreased blood sugar, serum cholesterol, triglycerides, low density lipoprotein (LDL), creatinine, urea and alkaline phosphatase levels, while increasing the high density lipoprotein (HDL) level in diabetes mellitus (DM). Ram et al. (1997) who had earlier worked on a sister species (*Terminalia arguna*) reported that the oral administration of its aqueous tree bark extract did not have any effects on the liver, kidneys, lipid profile and haematological parameters of rats. The aqueous extracts of *T. catappa* leaves have been reported to have strong free radical scavenging activities (Kinoshita et al., 2007).

Moody et al. (2003) and Ibegbulem et al. (2010), respectively, reported the *in vitro* antisickling property of the extracts and decoction of *T. catappa*. The decoction is administered in folklore medicinal practice as a home-made prophylaxis against sickle-cell crises; without any reported side effect. Most of our traditional medicines are in forms of decoctions. This paper presents the biochemical effects of drinking *T. catappa* decoction in place of water in albino wistar rats.

MATERIALS AND METHODS

Preparation of decoction

Dried leaves of T. catappa Linn. (pink mesocarp fruited) were picked from under its tree situated at the Nekede Village, Owerri, Imo State. The leaves were authenticated by Professor S.E. Okere, a plant taxonomist, of the Department of Plant Science and Biotechnology, Imo State University, Owerri, Imo State. They were cleaned of debris, washed in distilled water, mopped dry of water then chopped into bits. One 100 g of the chopped leaves was put in an aluminum pot and 4.6 L of distilled water was poured into the pot. This was brought to boil and allowed to simmer for 20 min. The decoction produced was filtered off the leaves using a muslin cloth and stored in a refrigerator at 4°C until used.

Analyses of decoction for constituents and property

Tests for the presence of catechins, β-carotene, cardiac glycosides, flavonoids and alkaloids were carried out using to the methods of Evans (2002). Test for the presence of cyanogenic glycosides was done using the method of AOAC (1990). Test for the presence of saponins was carried out using the method of Sofowora (2006). An aspect of the method used for the estimation of tannins by AOAC (1984) was adopted for the detection of tannins: 1.0 ml of the decoction was mixed with 0.5 ml of Folin - Denis reagent and 1.0 ml of 17% Na_2CO_3. The mixture was stood at room temperature (30°C) for 3 min for colour development. The decoction tested positive for the presence of tannins when it developed a blue colour (intensity varying with the concentration of tannins in the test sample). Test for the presence of thiocyanates was carried out by modifying the

alkaline picrate paper method of Haque and Bradbury (1999). The modification made was that the orange/ brick-red paper strip was not washed in distilled water and the colour intensity measured spectrophotometrically.

Test for the presence of ρ -hydroxybenzoic acid (ρ – HBA) was run using the method of ASEAN (2005). Test for the presence of vitamin C was evaluated using the method of Lambert and Muir (1968). The presence of free amino acids and monosaccharides were detected using the methods of Plummer (1971). The pH and concentration of the decoction were determined using the methods of AOAC (1990).

Testing for the biochemical effects

A total of 32 weanling white albino rats (*Rattus norvegicus*) of the wistar strain (of both sexes) were used for the animal feeding experiment. They were purchased from the animal colony of the Department of Biochemistry, University of Port Harcourt, River State, Nigeria and were aged between seven (7) and eight (8) weeks. They weighed between 42 and 75 g. The rats were allotted to 4 groups of 8 rats each (2 groups for males and 2 groups for females). The groups, on sex bases, were equalized as nearly as possible on weight basis. Each rat was housed in a wire-screened cage with provisions for feed and fluid. Acclimatization of the rats to their new environment (at the laboratory of the Department of Biochemistry, Federal University of Technology, Owerri, Imo State, Nigeria) lasted for 4 days and the test period lasted for 35 days. All the rats were maintained under the same conditions of light and dark cycles (circadian rhythm) and ambient room temperature.

Administration of the decoction was done according to the method of Pepato et al. (2001) who administered *Eugenia jambolana* leaf decoction in place of water on streptozotocin induced diabetic rats. The duration of administration here was however for 35 days. Two groups of the test rats (of the respective sex) were respectively placed on the decoction from the fallen dried leaves of *T. catappa*. The decoction was served to the rats in place of water, *ad libitum*, for their respective rat groupings while distilled water served as the only source of fluid for the control group. All the rats had growers mash (guinea feed) (produced by Bendel Feed and Flour Mill Limited, Sapele, Delta State, Nigeria) as the only source of solid feed, *ad libitum*.

On day 35 of the experiment, each rat was re-weighed (and weight gained calculated) before being anaesthetized with chloroform ($CHCl_3$) vapour. Incisions were then made into their thoracic cavities. Blood samples were collected by heart aorta puncture using 10 ml hypodermic syringes. One millilitre of each blood sample was quickly transferred into sequestering bottle (containing EDTA as anticoagulant) and the rest put in a test tube and allowed to clot before the serum was collected. The whole blood samples and sera were used for the clinical assays. Diagnostic test kits for total protein, albumin, total bilirubin and the lipid profile parameters (with the exception of very low density lipoproteins, VLDL) were purchased from BioSystems® (S.A. Costa Brava of Barcelona, Spain) and diagnostic test kits for the estimation of the activities of alanine and aspartate aminotransferases (ALT and AST) were purchased from Randox® (Randox Laboratories Ltd., Antrim, United Kingdom). The assays were performed according to their manufacturers' instructions. All other chemicals were of good analytical grades. VLDL concentration was estimated using the methods of Burnstein and Sammaille (1960). Haemoglobin (Hb) concentration was estimated by the cyanmethaemoglobin method described by Bain and Bates (2002). Packed cell volume (PCV), mean cell haemoglobin concentration (MCHC) and visual white cell count (WBC) were estimated by the techniques of Baker et al. (2001).

Total fluid and feed intakes were also calculated and daily intakes evaluated. Feed conversion ratio (FCR) was calculated as a

Table 1. Phytochemical, chemical and property of the decoction*.

Parameter	Result
Tannins	+
Flavonoids	+
Saponins	+
Vitamin C	+
β-carotene	+
Catechins	+
Thiocyanates	+
ρ – HBA	+
Cyanogenic glycosides	+
Cardiac glycosides	+
Alkaloids	+
Amino acids	+
Monosacchrides	+
pH	5.94 ± 0.01
Concentration (mg/ ml)	18.3 ± 0.26

*Values are means of triplicate determinations. Key: + = present.

ratio of daily feed intake to daily weight gained as described by Church and Pond (1988). Results are presented as means ± SD of triplicate values for eight rats.

Statistical analysis

Data generated were evaluated by the use of the students' t – test of significance at 95% confidence limit. The p – values for the statistical analyses between the test and their control rats were reported while the statistical differences between the male and female rats for a parameter were shown using superscript letters. Ratios were not tagged.

RESULTS AND DISCUSSION

All the phytochemical and chemical constituents detected in the decoction (Table 1) are known to be beneficial. Their health benefits have been espoused by Ram et al. (1997), Wardlaw and Kessel (2002) and Adeneye et al. (2008). Balagopalan et al. (1988) reported the benefits of eating foods that contain thiocyanates and cyanates to sickle cell anaemia (SCA) patients while Akojie and Fung (1992) reported the antisickling activity of ρ–HBA in *Cajanus cajan*. Ibegbulem et al. (2010) attributed the antisickling property of the decoction to the presence of ρ–hydroxybenzoic acid, flavonoids, thiocyanates and some antisickling amino acids. The acidity of the decoction showed that most of our traditional medicines may be acidic.

The oral route of administration is the most popular route for administering decoctions. A decoction can also be referred to as a drug since it is a substance that is taken as a medicine. Zakrzewski (1991) reported that this

route had the advantage that drugs could mix with food, acid, gastric enzymes and bacteria which could alter their toxicity either by influencing absorption or by modifying the compound. He added that sex, age and body weight were also contributing factors to how individuals reacted to drugs. Results of the liver function indices (Table 2) showed that the female rats were more sensitive (p< 0.05) to ingesting the decoction in place of water for the duration of the study. They were shown to have had higher serum AST activities and serum total protein levels and lower serum total bilirubin levels than their controls. Their serum AST activity and total bilirubin level did not however increase (p > 0.05) more than those of their male counterparts. The AST: ALT ratios of the rats showed that the pathologic condition, if any, was more of the liver. This may have been indicative of hypertrophy. The liver is normally the first port of call of nutrients and toxins alike that enter the body. In the course of trying to detoxify these foreign materials, there may be a hypertrophy of this organ as an adaptive measure.

Schoen (1999) said that hypertrophy was a compensatory mechanism for increased stress; even in increased physical exertion. It is the neutralization of noxious stimulus by the cells or one of its organelles (Cotran et al., 1999). However, the AST: ALT ratios of the test rats compared favourably with those presented by the control rats. The liver function indices also showed that the test rats were not jaundiced as shown by their total bilirubin levels. This meant that the decoction was not haemolytic. The test rats did not also develop oedema because they did not exhibit hypoalbuminemia (and hyperalbuminemia) as well as hypoproteinemia (and hyperproteinemia) as their albumin and total protein levels generally compared favourably with values presented by the control rats. The total protein and albumin levels of the test rats were by extension indicative of the maintenance of kidney integrity. The total serum protein levels of the female test rats were though higher (p< 0.05) than those of their male counterparts and control, indicating that they may have mobilized enzymes for detoxification, as the decoction was a xenobiotic. The decoction was found to have contained a wide spectrum of phytochemicals, chemicals and nutrient moieties that may have generally endangered the well-being of the liver. Antioxidants like tannins, flavonoids, vitamin C, β-carotene, saponins may have also contributed to the maintenance of the health of the liver. Ram et al. (1997) showed that liver and renal function parameters were not adversely affected when the tree bark extract of *T. arguna* was administered to rats.

Ingestion of the decoction generally reduced (p< 0.05) the susceptibility of the test rats to atherosclerosis (Table 3). The differences in the levels of some parameters noticed between the male and female rats may have been results of the interplay of sex hormones as they grew. Though lower values were noticed for some of the parameters on ingestion of the decoction, the reductions

Table 2. Effect of decoction on liver function indices of test rats.

Group	Parameter*											
	AST (U/l)		ALT (U/l)		AST: ALT		Total Bilirubin (mg/dl)		Total Protein (g/l)		Albumin (g/l)	
	M	F	M	F	M	F	M	F	M	F	M	F
T. catappa	21.00^a±2.83	21.08^a±0.30	27.48^b±8.88	29.45^b±5.16	0.76	0.72	0.97^c±0.15	1.07^c±0.33	59.40^d±5.39	66.83^e±5.64	35.35^f±4.53	36.80^f±1.55
Control	20.98^b±0.45	20.75^b±0.25	24.73^c±2.36	24.48^c±5.00	0.85	0.85	1.13^d±1.06	1.16^d±0.06	58.58^f±4.15	61.05^f±1.91	36.30^g±1.81	36.30^g±1.81
t$_{cal}$	0.02	2.39	0.85	1.96	2.81	6.40	2.81	6.40	0.34	2.74	0.55	0.59

*Values are means ± S.D of eight determinations. M = male rats, F = female rats. Values on the same row bearing the same superscript letter for a parameter are not significantly different (p>0.05).

Table 3. Effects of decoction on lipid profile of test rat serum.

Group	Parameter (mg/dl)*													
	Triacylglycerol		VLDL- Cholesterol		Total Cholesterol(TC)		HDL- Cholesterol		LDL- Cholesterol		TC: HDL-C		LDL-C:HDL-C	
	M	F	M	F	M	F	M	F	M	F	M	F	M	F
T. catappa	58.81^a±6.57	72.12^b±7.28	10.80^c±1.39	14.93^d±1.99	98.60^e±9.49	51.10^f±34.78	57.87^g±6.86	64.79^g±17.96	47.22^h±5.66	38.89^h±6.41	1.70	0.79	0.82	0.60
Control	63.47^c±14.56	75.00^f±18.45	12.70^f±2.92	15.00^f±3.69	100.00^g±9.06	100.00^g±15.70	59.42^h±1.58	57.87^h±1.20	63.89^k±3.21	65.28^k±2.78	1.64	1.73	1.08	1.13
t$_{cal}$	0.15	0.41	1.66	0.03	0.30	3.62	0.62	1.09	7.25	10.68				

*Values are means ± S.D of eight determinations. Values on the same row bearing the same superscript letter for a parameter are not significantly different (p>0.05).

were not significant (p> 0.05). The total cholesterol levels of the female rats and the LDL – cholesterol levels of both the male and female rats were significantly (p< 0.05) affected by the decoction when it was drunk in place of water. Gender differences also affected (p< 0.05) the levels of the triacylglycerol, VLDL – cholesterol, total serum cholesterol and LDL – cholesterol of the rats on ingestion of the decoction. The female rats seemed to have had their serum triacylglycerol, VLDL – cholesterol and HDL - cholesterol levels elevated (p< 0.05) after drinking the decoction, when compared with those of their male counterparts. This partly corroborated the findings of Kayali et al. (2009) who reported that aged female rats had reduced serum total

cholesterol and elevated serum LDL – cholesterol levels than their aged male counterparts. This may have meant that the aged female rats were prone to atherosclerosis. The Kayali et al. (2009) type of elevation of serum LDL – cholesterol was noticed in the female control rats when compared with their male counterparts (though not significantly different, p> 0.05). Seidell et al. (1991) had earlier reported that men had higher serum triacylglycerol and total cholesterol and lower HDL – cholesterol compared to women. Our study showed that ingestion of the decoction made the male rats seem more (p< 0.05) prone to atherosclerosis, since their serum total cholesterol and serum LDL – cholesterol levels were raised when compared with their female counterparts.

However, most of the values presented by the test rats compared favourably with those presented by their controls. The LDL–cholesterol: HDL-cholesterol ratios of the test rats showed that their sera seemed to have contained more phospholipids than cholesterol. The decoction seemed to have engendered the production of HDL and phospholipids. This meant that it may have led to the reduction of the cholesterol contents of the sera; especially for the female rats. Berg et al. (2002) postulated that a serum esterase that degraded oxidized lipids had been found to be associated with HDL. They went further to say that the HDL-associated protein possibly destroyed the oxidized LDL, accounting for HDL's ability to protect against coronary

Table 4. Effect of decoctions on haematological indices of test rats*.

Group	Hb (g/dl)		PCV (%)		MCHC (g/dl)		WBC (number/L)	
	M	F	M	F	M	F	M	F
T. catappa	9.83[a]±1.77	8.50[a]±2.05	32.25[b]±6.13	27.25[b]±4.79	0.31	0.31	4637.50[c]±1978.37	3675[c]±830.16
Control	8.98[b]±0.46	8.05[b]±1.00	34.50[c]±1.73	31.75[c]±6.24	0.26	0.25	4175.00[d]±1074.3	5000[d]±1802.31
t_{cal}	1.31	0.56	1.00	1.62			0.58	1.89

*Values are means ± S.D eight determinations.

disease. Glew (2006) said that the detergent properties of phospholipids, especially phosphotidylcholine, were important in bile to aid in solubilizing cholesterol. Quercetin, found in plants and related substances found in tea, as well as phenolics in wines act as antioxidants and reduce LDL oxidation (Wardlaw and Kessel, 2002). The lipid profile also showed that ingesting the decoction would discourage the development of atherosclerosis as shown by their LDL- cholesterol lowering effects and encourage the formation of the good cholesterol (HDL-cholesterol).

The risk of developing atherosclerosis is directly related to plasma LDL-cholesterol and inversely related to HDL-cholesterol levels (Glew, 2006). Pigments like tannins and flavonoids may have been responsible for this cholesterol lowering action. Phytochemicals like tannins, saponins and flavonoids are constituents of plant extracts that have lipid lowering effects (Ram et al., 1997; Adeneye et al., 2008). These phytochemicals also prevent the oxidation of LDL, preventing it from being atherogenic (Wardlaw and Kessel, 2002). Ram et al. (1997) showed that T. arguna tree bark extract reduced the total cholesterol, LDL-cholesterol, HDL-cholesterol, triacylglycerols and cholesterol: HDL ratio as well as the LDL-cholesterol: HDL-cholesterol ratio of the rats.

Ugonabo et al. (2007) and Essien (2008), respectively reported that SCA patients physiologically had lower total cholesterol, free cholesterol, cholesteryl esters, HDL-cholesterol, LDL-cholesterol, total phospholipids and triglycerides (except free fatty acids) than normal patients. Ugonabo et al. (2007) attributed this to increased synthesis of red cell membranes; hence cholesterol mobilization. Phospholipids are also major components of RBC membranes. The reduction in serum cholesterol by the decoctions may also be seen as a double edged sword. While it may discourage atherosclerosis, it may also deplete the cholesterol base needed for the synthesis of red cell membranes in SCA patients.

The haematological parameters were found not to have been affected (p>0.05) when the decoction was ingested in place of water (Table 4). Though the Hb levels were observed to have been elevated, this would seem beneficial to sicklers, as their Hb levels are normally low. PCV levels were also not found to have been altered (p>0.05). PCV levels are also low in SCA (Baker et al., 2001). The MCHC values were found to have been increased (though not significantly (p>0.05)). Increase in MCHC had been reported to be one of the conditions that precipitated sickling (Schechter et al., 1987; Cotran et al., 1999). Schechter et al. (1987) also reported that the MCHC, MCH and MCV values were not strickingly different between sickle and normal cells but for the existence of a significant number of very dense (corpuscular haemoglobin concentration >37 g/dl), small (corpuscular volume <80 μm^3) cells. In this study, the increase in MCHC may not necessarily result in the sickling of HbSS red blood cells because of the presence of the antisickling agents (like thiocyanates and ρ – HBA) that may always be present in the blood stream and/ or red cells at steady states. Intracellular polymerization leads to changes in the internal milieu of the cells. But when polymerization is supposedly inhibited by these agents, the homeostatic balance will be maintained. So, increase in MCHC may not be synonymous with sickling but may result in increased oxygen carrying capacity of the red blood cells. However, this is suggested for further studies. The WBC's were also found not to have been elevated (p>0.05). This may mean that the decoction did not present any immunologic challenge. Leucocytes are known to help man the body's immune system alongside macrophages and lymphocytes (Aster and Kumar, 1999; Nelson and Cox, 2000). These findings corroborated those of Ram et al. (1997).

Ingestion of the decoction (Table 5) generally increased (p< 0.05) the daily body weight gained, feed and fluid intakes leading to non interference with the feed's nutrient adequacy and improved their efficiency in feed utilization (for instance, its FCR) of the test rats. Church and Pond (1988) reported that lowering of FCR meant that less of total feed consumption was used for maintenance and more was available for gain. The test rats may also have gained more weights because of the added nutrients from the decoction like amino acids and monosaccharides that were detected in it (Table 1).

Conclusion

Our study showed that the decoction was safe when drunk in place of water.

Table 5. Effects of the decoctions on daily weight gained, fluid and feed intakes and feed conversion ratios of test rat*.

Group	Weight gained (g/day)		Fluid intake (ml/day)		Feed intake (g/day)		Feed conversion ratio (FCR)	
	M	F	M	F	M	F	M	F
T. catappa	$1.64^a \pm 0.19$	$1.76^a \pm 0.19$	$12.89^b \pm 0.40$	$13.11^b \pm 0.43$	$10.75^c \pm 0.34$	$10.11^d \pm 0.44$	6.55	5.74
Control	$1.20^b \pm 0.26$	$1.09^b \pm 0.38$	$11.57^c \pm 0.17$	$11.46^c \pm 0.22$	$9.59^f \pm 0.56$	$9.66^f \pm 0.48$	7.99	8.86
t_{cal}	3.86	4.46	8.52	9.59	5.04	1.95		

*Values are means ± S.D of eight determinations. Values on the same row bearing the same superscript letter for a parameter are not significantly different (p>0.05).

REFERENCES

Adeneye AA, Adeneke TI, Adeneye AK (2008). Hypoglycemic and hypolipidemic effects of the aqueous leaves extracts of *Clerodendrum capitatum* in wistar rats. J. Ethanopharmacol., 116(1): 7-10.

Ahmed SM, Vrushabendra SBM, Dhanapal PG, Chandrashekara VM (2005). Anti-diabetic activity of *Terminalia catappa* Linn. leaf extracts in alloxan-induced diabetic rats. Iranian J. Pharmacol. Ther., 4: 36-39.

Akojie FO, Fung LW (1992). Antisickling activity of hydroxybenzoic acid in *Cajanus cajan*. Planta Med., 58 (4): 317-320.

AOAC (1984). Official method of analysis, 13th edn. Association of Official Analytical Chemists, Virginia.

AOAC (1990). Official methods of analysis, 15th edn. Association of Official Analytical Chemists, Virginia.

ASEAN (2005). Identification and determination of 2-phenoxy-ethanol, methyl, ethyl, propyl, and butyl 4-hydroxybenzoate in cosmetic products by TLC and HPLC. Association of Southeast Asian Nations (ASEAN). Retrieved December 2, 2007 from http://www. aseansec.org/MRA-Cosmetic/Doc-4.pdf.

Aster JC, Kumar V (1999). White cells, lymph nodes, spleen, and thymus. In: Cotran et al. (eds) Robbins pathologic basis of disease, 6th edn. W.B. Saunders, Philadelphia. Pp. 644-695.

Bain B, Bates I (2002). Basic haematological techniques. In: Lewis et al. (eds) Dacie and Lewis practical haematology, 9th edn. Churchill Livingstone, London. Pp. 19-46.

Baker FJ, Silverton R E, Pallister CJ (2001). Baker and Silverton's introduction to medical laboratory technology, 7th edn. Sam-Adex Printers, Ibadan.

Balagopalan C, Padmaja G, Nara SK, Moorthy SK (1988). Cassava: In food, feed and industry. CRC Press, Florida.

Berg JM, Tymoczko JL, Stryer L (2002). Biochemistry, 5th edn. W.H. Freeman, New York.

Burnstein MA, Sammille J (1960). A rapid determination of cholesterol bound to A and B-lipoprotein. Clin. Chem. Acta., 5: 601-609.

Church DC, Pond WG (1988). Measurement of feed and nutrient utilization and requirement by animals. In: Basic animal nutrition and feeding, 13th edn. John Wiley and Sons, New York, pp 49-62.

Cotran RS, Kumar V, Collins T (1999). Robbins pathologic basis of disease, 6th edn. W.B. Saunders, Philadelphia.

Cunningham WP, Cunningham MA, Saigo BW (2005). Environmental science: A global concern, 8th edn. McGraw-Hill, Boston.

Essien EU (2008). The good, the bad and the ugly: A mimicry of the role of lipids in humans and implications for human society. 40th Inaugural Lecture of the University of Calabar, Nigeria, February 20.

Evans WC (2002). Trease and Evans pharmacognosy, 15th edn. W.B. Saunders, Edinburgh.

Glew RH (2006). Lipid metabolism II: pathways of metabolism of special lipids. In: Devlin TM (ed) Textbook of biochemistry with clinical correlations, 6th edn. Wiley Liss, New Jersey, pp. 695-741.

Haque R, Bradbury JH (1999). Simple kit method for determination of thiocyanate in urine. Clin. Chem., 45: 1459-1464.

Ibegbulem CO, Eyong EU, Essien EU (2010). Antipolymerization effect of *Terminalia catappa* decoction on sickle cell triat haemoglobin. Nig. J. Biochem. Mol. Biol., 25(2): 40-45.

Kayali R, Aydin S, Çakatay U (2009). Effect of gender on main clinical chemistry parameters in aged rats. Curr. Aging Sci., 2: 67-71.

Kinoshita S, Inoue Y, Nakama S, Ichiba T, Aniya Y (2007). Antioxidant and hepatoprotective actions of medicinal herb, *Terminalia catappa* L. from Okinawa Island and its tannin corilagin. Phytomedicine, 14(11): 755-762.

Lambert J, Muir TA (1968). Practical chemistry, 2nd edn.

Heinemann, London.

Moody JO, Segun, FI, Aderounmu O, Omatade OO (2003). Antisickling activity of *Terminalia catappa* leaves harvested at different stages of growth. Nig. J. Nat. Prod. Med., 7: 30-32.

Nelson DL, Cox MM (2000). Lehninger principles of biochemistry, 3rd edn. Worth Publishers, New York.

Oboh B, Ogunkanmi B, Olasan L (2008). Phenotypic diversity in *Terminalia catappa* from southwestern Nigeria. Pak. J. Biol. Sci., 11(1): 135-138.

Oliver-Bever B (1986). Medicinal plants in tropical West Africa. Cambridge University Press, Cambridge.

Pepato MTI, Folgado VBB, Kettelhut IC, Brunetti IL (2001). Lack of antidiabetic effect of a *Eugenia jambolana* leaf decoction on rat streptozotocin diabetes. Braz. J. Med. Biol. Res., 34: 389-395.

Plummer DT (1971). An introduction to practical biochemistry. McGraw-Hill, London.

Ram A, Laura P, Gupta R, Kumar P, Sharma VN (1997). Hypocholesterolaemic effects of *Terminalia arguna* tree bark. J. Ethanopharmacol, 55(3): 165-169.

Schechter AN, Noguchi CT, Rodgers GP (1987). Sickle cell disease. In: Stamatoyannopoulos et al. (eds) The molecular basis of blood disease, 1st edn. W.B. Saunders, Philadelphia, pp. 179-218.

Schoen FJ (1999). The heart. In: Cotran et al. (eds.) Robbins pathologic basis of disease, 6th edn. W.B. Saunders, Philadelphia, pp. 543-599.

Seidell JC, Ciqolini M, Charzewska J, Elisinger BM, Björntorp P, Hautvast JG, Szostak W (1991). Fat distribution and gender differences in serum lipid in men and women from four European communities. Atherosclerosis, 87(2-3): 203-210.

Sofowora A (2002). Plants in African traditional medicine-an overview. In: Evans WC (ed) Trease and Evans pharmacognosy, 15th edn. W.B. Saunders, Edinburgh, pp. 488-496.

Sofowora A (2006). Medicinal plants and traditional medicine in Africa.Spectrum Books Limited, Ibadan.

Ugonabo MC, Onwuamaeze IC, Okafor EN, Ezeoke ACJ (2007). Plasma cholesterol of sickle cell anemia patients in Enugu, Nigeria. Bio-res., 5(2): 241-243.

Wardlaw GM, Kessel MW (2002). Perspective in nutrition, 5th edn. McGraw-Hill, Boston.

Zakrzewski SF (1991). Principles of environmental toxicology: ACS professional reference book. American Chemical Society, Washington D.C.

25

Phytochemical screening and histopathological effects of single acute dose administration of *Artemisia annua* L. on testes and ovaries of Wistar rats

P. O. Ajah[1]* and M. U. Eteng[2]

[1]Institute of Oceanography, University of Calabar, Calabar, Cross Rivers, Nigeria.
[2]Department of Biochemistry, University of Calabar, Calabar, Cross Rivers, Nigeria.

Artemisia annua L. is a medicinal plant whose derivatives are used in the treatment of malaria. Phytochemical screening and histopathological effects of single acute intraperitoneal dose of the alcoholic extract of *A. annua,* were investigated in Wistar rats in this study. Thirty five adult albino Wistar rats (109 - 307 g) of both sexes used for the study were randomly assigned on the basis of weight and litter origin into seven study groups of five rats per group. The control (group A) was administered placebo (2 ml of pure olive oil) while test groups B to G received single graded doses of 0.18, 0.35, 0.52, 0.70, 0.86 and 1.05 g/kg body weight of extract in olive oil vehicle, respectively. Twenty four hours after the single acute dose administration, animals were sacrificed and testes and ovary tissues obtained for histological evaluation using standard methods. Results of phytochemistry of the alcoholic extract assessed by standard methods identified the presence of alkaloids, glycosides (presence of these in *A. annua* is questionable), flavonoids, reducing compounds and polyphenols present in different concentrations. Histopathological changes showed adverse lesions on the ovary ranging from atretic-degenerating corpus luteum, with loss of connective substance, arrest of ovarian follicle maturation to complete absence of ovarian follicle. No adverse histopathological changes were observed in the testes. The result suggests arrest of ovulation and a predisposition to infertility (add in the female animals). The possible roles of phytoestrogenic constituents in the favonoid fraction of the phytochemicals are discussed.

Key words: *Artemisia annua*, phytochemistry, histopathology, testes and ovaries.

INTRODUCTION

Artemisia annua L. (add,) also known as sweet wormwood, annual wormwood or sweet Annie (add,) is an aromatic annual herb which is the source of artemisinin and essential oils (Simon et al., 1990). The secondary plant product artemisinin known in Chinese folk medicine as qinghasu is an antimalarial with reduced side effects compared to quinine, chloroquine or other antimalarials (Klayman, 1985). The quinolines and sulphadozine-pyrimethamine have reduced clinical effectiveness in Africa owing to biocidal resistance to them by malarial parasites. Artemisinin is effective in cases where there is biocidal resistance.

However, monotherapy with artemisinin results in high recrudescence rates and fears of possible biocidal resistance developing have been expressed. The World Health Organisation therefore recommended the use of artemisinin drugs in combination with other effective antimalarials. Such artemisinin based combination therapies are increasingly in use in Africa (WHO, 1981).

Jellin et al. (2000) has reported that the herb is unsafe for use during pregnancy due to its uterine and menstrual stimulating effects. And although there is no sufficient or reliable information in this regards, it should not also be used during lactation. In the light of the above, the present study undertook the phytochemical screening and evaluation of the possible histopathological effects of the herb on the testes and ovary, the primary reproductive organs of albino Wistar rats and hence the

---

*Corresponding author. E-mail: ajapaulo@yahoo.com.

fertility status.

MATERIALS AND METHODS

Plant material

Fresh leafy biomass of *A. annua* L. was obtained courtesy of Prof. E. A. Brisibe of the Department of Genetics and Biotechnology, University of Calabar, Calabar, Nigeria, who has worked extensively on the plant and help to authenticate it for this study. The sample was air dried and the leaves were harvested and ground into coarse powder to give a pulverised sample of 50 g. This was extracted using 200 ml of 95% ethanol in a soxhlet equipment. The extract obtained was concentrated in vacuo at 4°C to give a 10.3 g mass representing 29.43% yield. The extract was divided into two portions, one of which was immediately subjected to phytochemical screening based on Sofowora (1982) to test for the presence of different phytochemical constituents such as alkaloids, glycosides, tannins, saponins, (Sofowora, 1982), flavonoids, polyphenols, reducing compounds (Harbone, 1973), phlobatannins, anthraquinines, and hydroxymethyl (Trease and Evans, 1978). The second portion of the extract was suspended in pure olive oil as vehicle and from these stock appropriate graded doses was obtained for administration.

The single intraperitoneal administration was done using a sterile syringe and needle between 08.00 and 09.00 a.m. daily.

Animals and treatments

Permission for use of animal was obtained from the College of Medical Sciences animal ethics committee. Handling and treatment of animal's protocol adhered strictly to the laid down rules in the ethical guide. Thirty two adult albino Wistar rats consisting of 19 females and 13 males weighing between 109 and 307 g were obtained from the animal facility of Department of Zoology and Environmental Sciences, Faculty of Science, University of Calabar, Calabar, Nigeria and used for the study. The animals were transferred to the Animal House facility of the Animal Science Department, Faculty of Agriculture, University of Calabar, Calabar, Nigeria and following acclimatisation for one week, they were randomly assigned into seven study groups of five rats per group with each group consisting of three females and two males. They were housed in plastic cages with wire screen top and kept under adequate ventilation and the environmental temperature (28 ± 2°C) and relative humidity (50 ± 5%) with a 12 h light/dark cycle. The animals were maintained on a commercial rat chow with tap water and food provided *ad libitum* throughout the experimental period.

Animals in group A (control) were treated with a placebo of pure olive oil as a vehicle while those in groups B to G were administered graded doses of the *A. annua* ethanolic extracts of 0.18, 0.35, 0.52, 0.70, 0.86 and 1.05 g/kg body weight, respectively. The single acute dose intraperitoneal administration was done using syringe and needle.

Collection of tissues and histological analysis

Twenty four hours after the single acute dose administration, male and female animals were collected from each group, anaesthetised in chloroform vapour and dissected. The ovaries and testes were collected and immediately fixed in Bouins fluid for 6 h and transferred to 70% alcohol for histological processing according to Drunny and Wallington (1990). Briefly, following fixation of the right side testes and ovaries from both control and test animals, tissue sections were processed by dehydration in 95% and absolute alcohol, cleared in xylene and embedded in pure clean molten

Table 1. Phytochemical profile of *Artemisia annua* L. extract.

Chemical constituent	*Artemisia annua*
Alkaloids	+
Glycosides (free anthraquinones)	(This is questionable)
Saponins	-
Tannins	-
Flavonoids	++
Reducing compounds	+
Polyphenols	+++
Phlobatannins	(This is questionable)
Anthraquinones	-

Key: +++ = Much excess, ++ = excess, + = slight presence, - = not present.

paraffin wax from which blocks of tissues were made for sectioning. Ribbon slices of about 5.0μm in thickness were made with the aid of a microtome (delete machine) and the sections picked with slides which were dried in oven. The slices were then stained with Haemotoxylin and Eosin, and then mounted using DPX onto a light microscope (delete magnification x 40 for testes and x 10 for ovary) for histopathological and morphological changes. The changes observed were recorded and photomicrographs of the most prominent pathological lesions taken.

RESULTS

Phytochemical screening

The phytochemical profile of *A. annua* is summarized in Table 1. The results identified (delete with exception of saponins, tannins, phlobatannins or anthraquinones) the presence of alkaloids, glycosides (presence of these compounds is questionable), flavonoids, reducing compounds and polyphenols with flavonoids and polyphenols present in excess and much excess (add levels), respectively.

Histological studies

Graded concentrations (add of *A. annua* ethanolic extract namely:) 0.18, 0.35, 0.52, 0.70, 0.86 and 1.05g per kg body weight were administered to the rats in a single acute dose and histological changes evaluated in ovarian and testicular tissues obtained from treated and control rats. The histological changes are summarised in Figures 1 to 3.

Plate 1a. The control Group A shows well layered cells of the seminiferous tubules which vary greatly in size and shape of cells. The nuclei lack defined laminal margin and inconspicuous cell boundary. The photo-micrograph shows the effect of pure olive oil on the histology of gonad section of Group A (control).

In Plate 1a, the control testes, the interstitial cells are prominent, well distinct and vascularized with interstitial

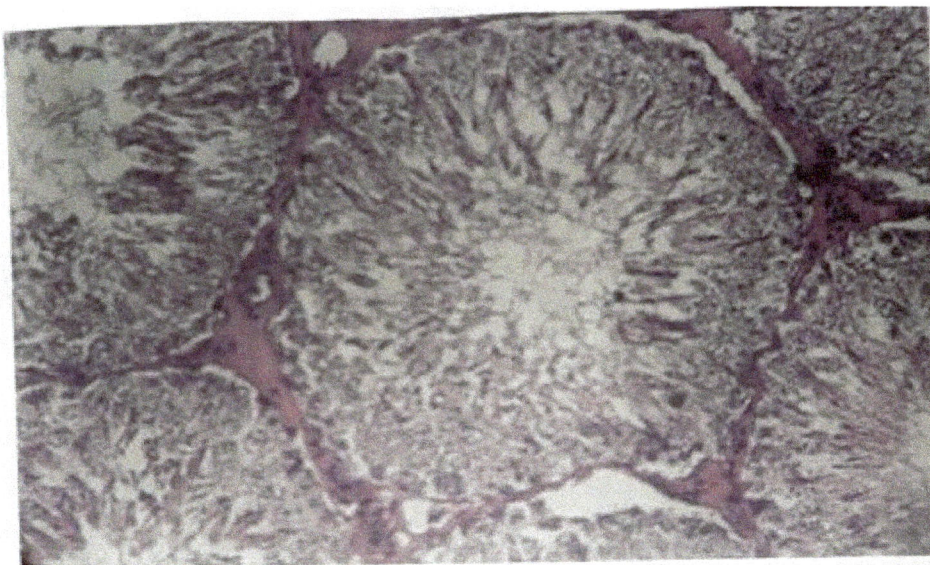

Plate 1a. Control male albino rat (X 10).

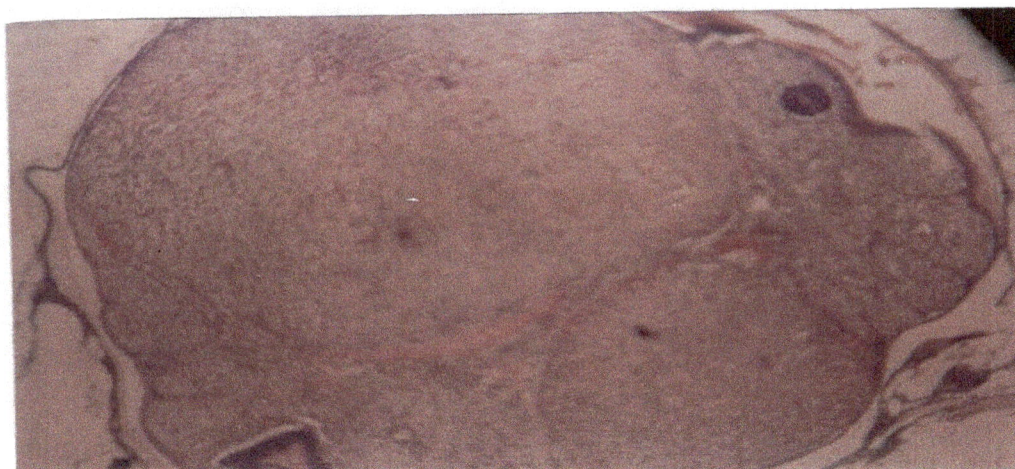

Plate 1b. Control female albino rat (X 10).

spaces adequately filled up with connective tissue. The seminiferous tubules vary greatly in size and shape of the cells and nuclei (Plate 1a).

Histological effect of ethanolic extracts of *A. annua* on gonadal tissues (testis of albino Wistar rats)

Plate 2a treated with *A. annua* at dose 0.52 g/kg body weight showed spermatogenic series of cells seen at the various stages of transformation and differentiation with central tubule showing marked spermatogenic transformation of spermatids into spermatozoa. The supporting cells (Sertoli cells) are intact.

Plate 3a (delete photomicrograph) shows that the basal lamina placed spermatogonia are intact and spermatids at various stages of differentiation filled up the central lumen of the tubules (add when the extract is administered at a) dose of 1.05 g/kg body weight. The features all point to an increased activity of spermatogenesis.

Histological effect of ethanolic extracts of *A. annua* on gonadal tissues (ovum of albino Wistar rats)

The control ovary Plate 1b also shows normal architecture with no lesion. In Plate 2c, regressing corpus

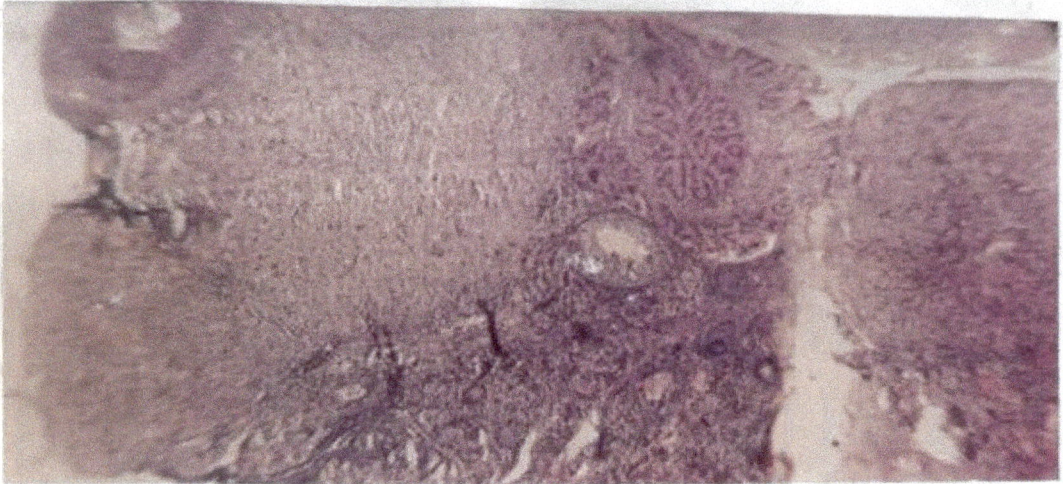

Plate 2a. Treated with A. annua at dose 0.52 g/kg body weight showed spermatogenic cells with central tubule showing marked spermiogenic transformation.

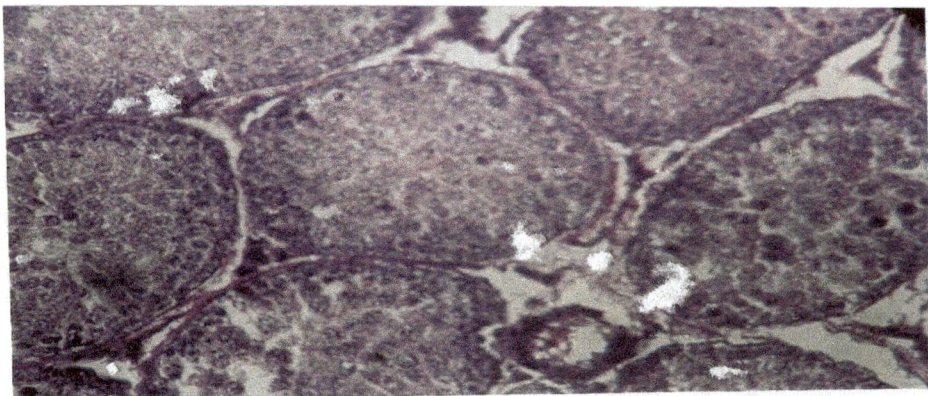

2b

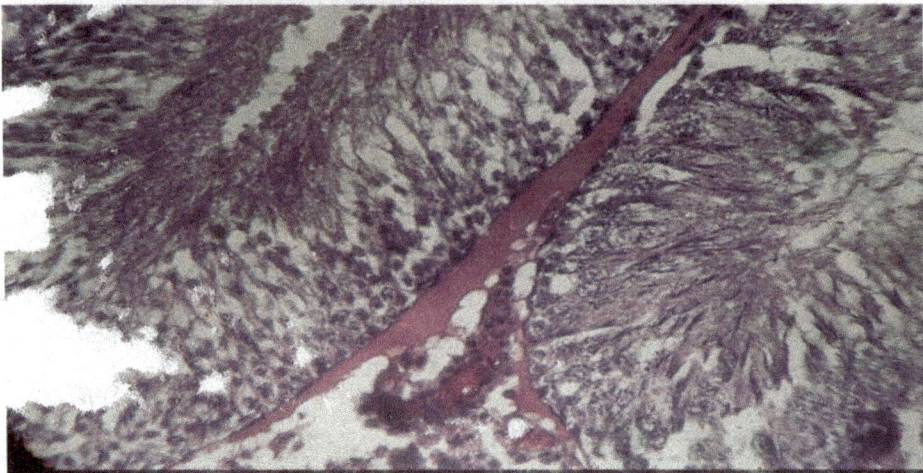

2c

Plate 2b. 0.18 g/Kg body weight of extract. In plate 2c, regressing corpus luteum is observed; cells are gradually shrinking with atretic and degenerating follicles even at 0.18 mg/kg body weight of the extract administered.

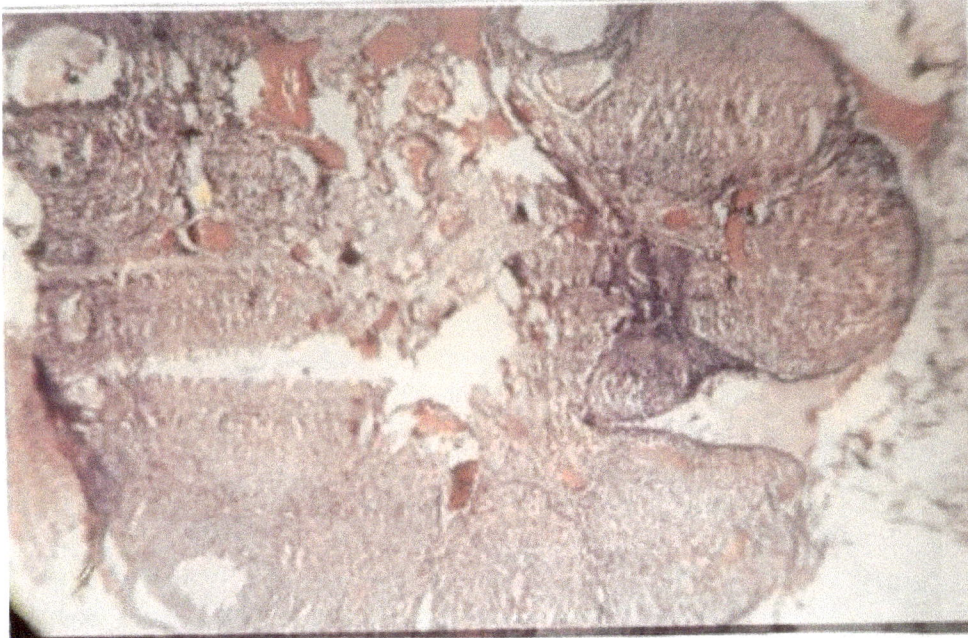

Plate 3a. Photomicrograph shows that the basal lamina placed spermatogonia are intact and spermatids at various stages of differentiation filled up the central lumen of the tubules at dose of 1.05 g/kg body weight. The features all point to an increased activity of spermatogenesis.

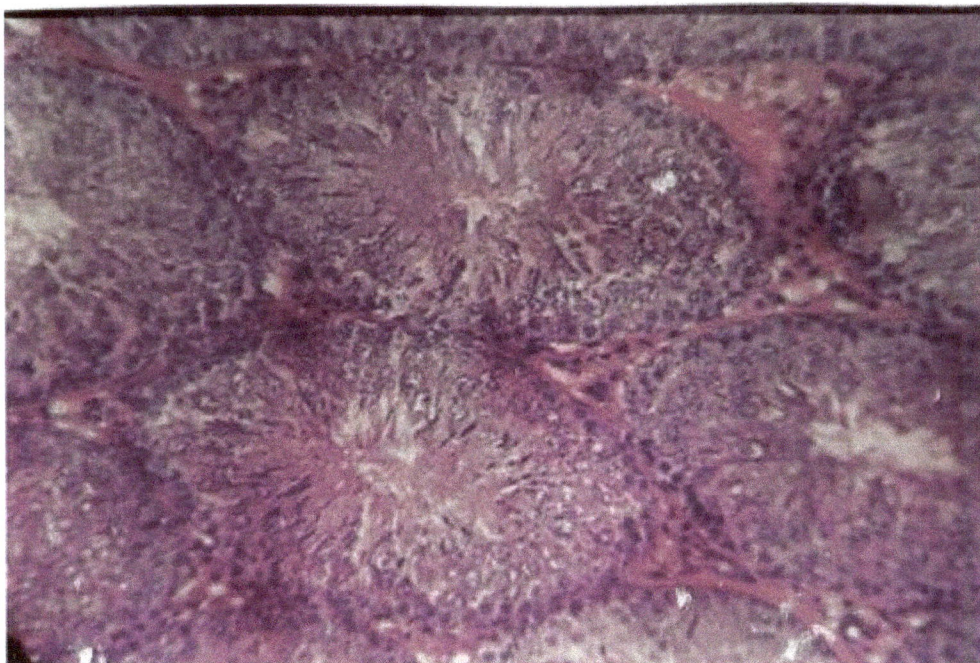

Plate 3b. Dosage of 0.35 g/kg body weight.

luteum is observed; cells are gradually shrinking with atretic and degenerating follicles even at 0.18 mg/kg body weight of the extract administered.

Plate 3b shows degeneration and avulsion of the ovum from most follicles at dose of 0.35 g/kg body weight.

Plate 4 shows further degeneration and avulsion of the ovum from most follicles at dose of 1.05 g/kg body weight.

Plate 5 shows almost total degeneration and avulsion of the ovum from most follicles at dose of 2.5 g/kg

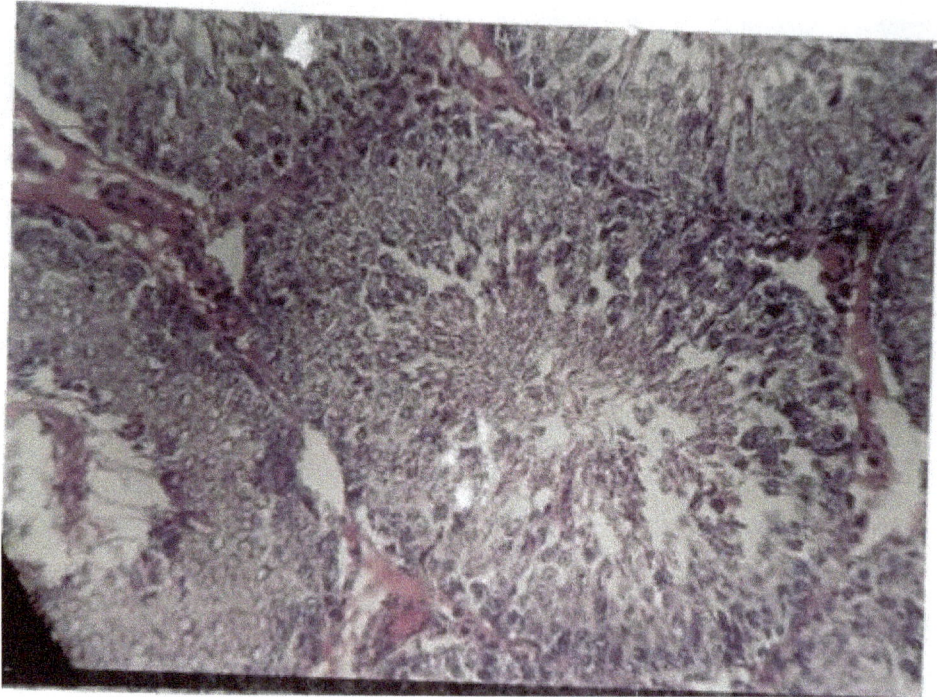

Plate 4. Dosage of 1.05 g/kg body weight.

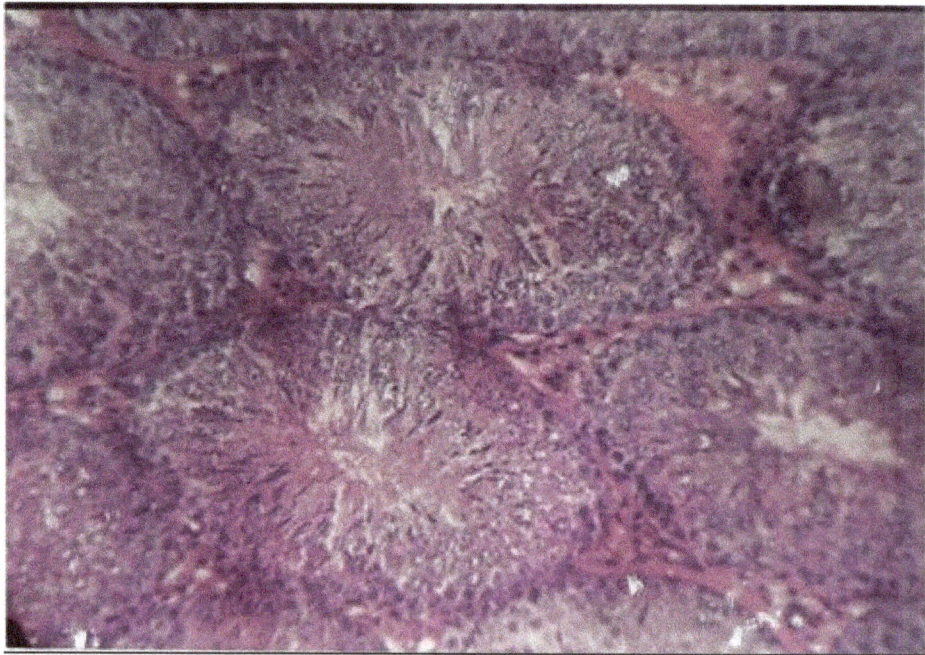

Plate 5. Dosage of 2.5 g/Kg body weight.

g/kg body weight.

Histological effect of ethanolic extracts of *A. annua* on gonadal tissues (testis and ovum) of albino Wistar rats are given in Table 1.

DISCUSSION

Neurological damage following administration of artemisinin in high doses has been reported in animal studies (Zhu and Woerdenbag, 1995). Although toxicity is a function of dose, the neurological effects have not yet been reported in humans using artemisinin for managing ailments at the therapeutic dose. However, the use of the herb is not safe during pregnancy because of its uterine and menstrual stimulating effects (Jellin et al., 2000). From the results of our study, in the male testis of all treated groups, there was no adverse histopathological change; rather there was marked proliferation of spermatogenic cell lines, transformation from primodial cells to spermatids and further to spermatozoa at a high rate. The seminiferous tubules were moderately enlarged, the central lumen inclusive. There was increased vascularity of interstitial and leydig cells which secrete the male hormone-testosterone. The hyperplasia of leydig cells correlate with increased spermatogenic activity.

On the other hand, the medulla of the ovum in the ovary was highly vascularized but with no ovarian follicle. The corpus luteum appeared degenerative, signifying lack of evidence of proliferation of the ovarian follicle. The follicles in most instances were alterative and gradually merging in the connective substance of the ovary. The stomal cells and the primodal follicles were indistinct. The picture connotes a lack of increased reproductive activity, tending towards degeneration of follicles (corpus luteum) arrest of maturation of ovarian follicle and hence ovulation. A comparison of the histological changes of testis and ovary following a single acute dose of administration shows that there are adverse lesions on the ovary but not on the testes. This may predispose (the female animals) to infertility.

Histology provides conclusive evidence from internal cell characteristics that other techniques fail to highlight. The failure of the gonads to mature as shown by degeneration of follicles may be due to insufficient stimuli for normal gonadal development particularly for the ovary (Msiska, 2002).

The probable reason for the observed histological effects (add in the ovaries) may be due to phytoestrogenic constituents in *A. annua* which may interact with steroids sex hormone metabolism and also the hypothalamic-pituitary-gonadal axis (HPA) of the female reproductive tract. Phytoestrogenic activity is due to such bioactive secondary metabolites as isoflavonoid, sterols, lignans and essential oils (Miksicak, 1993). This may explain a tendency towards atrophy and degeneration. This may not be observed in the testes because of higher estrogenic activity in females compared to males. It is also probable that the ovary may have target receptors for binding with the phytochemical components in the extract which may be lacking in the testes.

In conclusion, intraperitoneal administration of graded single acute doses of *A. annua* L. induced adverse histological changes on the ovaries which may predispose (add the animals) to infertility, but the testes were not affected and retained their normal architecture and normal spermatogenic activity. The phytochemistry revealed the presence amongst others of flavonoids in the extract whose bioactivity may explain the observed effects. The present result suggests that caution should be exercised in the use of *A. annua* and its derivatives in pregnancy.

REFERENCES

Drunny RAB, Wallington EA (1990). Histological Techniques 5th edition London, Oxford University Press, pp 199-220.

Jellin J, Batz F, Hitchens K (2000). Natural Medicines Comprehensive Database. Third Edition. Stock, California: Therapeutic Research Faculty.

Klayman DL (1985). Quinghasosu (artemisinin), an antimalarial drug from China. Science, 228: 1049-1055.

Miksicak R (1993). Commonly occurring plant flavonoids have estrogenic activity. Mol. Pharmacol, 44(1): 37-43.

Msiska DV (2002). The histology of *Oreochromis* (Nyasalapia) Karongal (Trewaves). Afr. J. Ecol. 40(2): 164.

Simon KV, Charles DJ, Wood KV, Heinstein P (1990). Germplasm variation in artemisinin content of *Artemisia annua* L. using an alternative from crude plant extract. J. Nat. Prod., 3: 157-160.

Sofowora EA (1982). Medicinal Plants and Traditional Medicine in Africa. New York. John Wiley and Sons.

Trease GE, Evans EC (1978). Pharmacognosy Cassel and Coller. London. Macmillan Publishers.

WHO (1981). Report of the Fourth Meeting of the Scientific Working Group on the Chemotherapy of Malaria. Beijing, People's Republic of China, October 6-10.

Zhu YP, Woerdenbag HJ (1995). Traditional Chinese Herbal Medicine. Pharm. World Sci., 17: 103-112.

26

Studies on separation techniques of pomegranate seeds and their effect on quality of *Anardana*

Amit Parashar[1,2]*, S. K. Gupta[3] and Ashok Kumar[4]

[1]Department of Chemistry, Eshan College of Engineering, Agra-Mathura Highway Agra-282001, India.
[2]Department of Chemistry, Institute of Engineering and Management, Mathura, India, 281406 India.
[3]Department of Chemistry, Bipin Bihari College, Jhansi, India.
[4]Department of Chemistry, St. John's College, Agra, India.

The dehydrated seeds of wild pomegranate fruits (*Anardana*) are used as acidulent in culinary preparations and in making various medicines. For easy separation of seeds, fruits were subjected to sand roasting, hot water dipping and soaking in cold water. Hot water dip for 2 min saved time required for separation of seeds from fruits (80% excluding treatment time and 74% including treatment time). Due to the high temperature during sand roasting, quality of seeds and *Anardana* was inferior when compared to other treatments. The desirable color and chemical composition of the product *Anardana* was retained in the treatment hot water dipping for 2 min.

Key words: Wild pomegranate, seeds, treatments, separation, quality.

INTRODUCTION

Pomegranate (*Punica granatum L.*) is an important fruit plant of tropical and subtropical regions. It is extensively cultivated in Iran, Spain, Egypt, Russia, France, Argentina, China, Japan, USA, and in India (Patil and Karade, 1996). The versatile adaptability, stable and therapeutic values and better keeping quality are the features responsible for its cultivation on a wide scale (Dhandar and Singh, 2002). In addition to dessert (sweet type) purposes, pomegranate is also found in its wild form in sub mountainous and outer Himalayas of Himachal Pradesh, Jammu and Kashmir, and Uttaranchal in India, up to an elevation of 1800 m above mean sea level (Parmar, 1981; Saxena et al., 1987).

The conventional utilization of wild pomegranate fruit lies in the drying seeds along with pulp (seeds), which constitute the product "Anardana" (Pruthi and Saxena, 2005).The dehydrated seeds are acidic (7.8 - 15.4%) and help in improving mouth-feel and digestion. *Anardana* is widely used as acidulent in culinary preparations. Mahajan, Chopra, and Sharma (2004) reported its richness in vitamin C and minerals (Ca, Zn, Mn), and usefulness for making various digestive and other ayurve-

dic medicines. The removal of seeds for preparation of anardana is a difficult process, since the seeds of wild pomegranate are tightly adhered to each other. Removal of seeds manually results in staining of hands and dress of the workers. Moreover, it is labor consuming and affects the yield and quality of the anardana. Manual extraction of seed is a tedious process and cause drudgery to the worker.

Mahajan et al. (2004) studied the effect of sand roasting on separation of seeds. Due to the high temperature, they observed that the quality of product was inferior. Also, the gelatinization of whole mass due to heat induced water diffusion caused difficulties in separation of seeds. Due to disadvantages in high temperature thermal treatments, various treatments were tried in this present study to loosen the peel of the wild pomegranate fruit for easy separation of seeds.

The objectives of the study were (i) to study the effect of different pretreatments on separation of seeds from wild pomegranate (ii) to study the effect of method of separation on physico-chemical composition and quality of anardana.

MATERIALS AND METHODS

Sampling and treatments

Wild pomegranate fruits were harvested at fully ripened stage from

*Corresponding author. E-mail; parashar.amit1@gmail.com.

Table 1. Effect of treatments on separation of seeds from wild pomegranate and their chemical characteristics.

| Treatments | Time taken for separation (min) | Labour saving (%) | | Titratable reducing ascorbic acid | | | |
		Excluding treatment time	Including treatment time	pH	Acidity (%)	Sugar (%)	(mg/100 g)
CT	13.6 (±5.4)	–	–	6.63 (±0.39)	2.8 (±0.20)	7.2 (±0.2)	19.20 (±0.65)
SR1	8.3 (±3.7)	39.0	3.40	6.16 (±0.16)	3.0 (±0.20)	7.8 (±0.4)	17.70 (±0.71)
SR2	5.8 (±2.6)	57.0	6.02	6.13 (±0.11)	3.1 (±0.23)	7.9 (±0.4)	17.20 (±0.38)
HW1	3.4 (±2.0)	75.0	60.0	6.03 (±0.10)	3.1 (±0.21)	7.8 (±0.2)	19.60 (±0.60)
HW2	2.5 (±1.7)	80.0	74.0	6.10 (±0.13)	3.0 (±0.10)	7.8 (±0.1)	19.47 (±0.55)
WS1	10.0 (±3.0)	–	–	6.70 (±0.16)	2.9 (±0.15)	7.2 (±0.3)	19.30 (±0.75)
WS2	12.1 (±4.1)	–	–	6.80 (±0.20)	2.8 (±0.20)	7.1 (±0.1)	19.56 (±0.45)

field-grown trees at Solan (Himachal Pradesh), India. The fruits were washed with fresh water and subjected to the following treatments:

(i) Sand roasting for 5 min (sand temperature: 165 ± 2°C) (SR1)
(ii) Sand roasting for 7 min (sand temperature: 165 ± 2°C) (SR2)
(iii) Hot water dip for 1 min (water temperature: 80 ± 2°C) (HW1)
(iv) Hot water dip for 2 min (water temperature: 80 ± 2°C) (HW2)
(v) Soaking in water for 5 min (water temperature: 30 ± 2°C) (SW1)
(vi) Soaking in water for 10 min (water temperature: 30 ± 2°C) (SW2)
(vii) Control (without any treatment) (CT).

Ratio of fruit weight to that of sand and water in sand roasting and water dip was kept as 1:8 (Mahajan et al., 2004). The temperature during sand roasting, hot water dip and soaking in water was measured using a digital thermometer (0 - 250°C). In each treatment, 1 kg of the fruit was taken for experimentation. The number of fruits per kg ranged from 25 - 28. The experiments were conducted in triplicate to avoid experimental error and the average values were taken for further analysis.

Separation of seeds

After the treatments, the seeds were separated manually using a stainless steel knife, giving vertical cut to the fruits and collecting the seeds. The time taken for separation of seeds was noted down using a stopwatch.

Chemical characteristics of seeds

The chemical characteristics of seeds after separation was determined as described below. Titratable acidity was determined by the procedure given by Ranganna (2000).

Reducing sugars, pH, and ascorbic acid were determined by AOAC (2002).

Preparation of anardana

The seeds separated by various treatments were dried to prepare anardana in an air circulatory tray dryer (0 - 300°C, Narang Scientific Works, New Delhi) at a temperature of 60 ± 2°C (Singh and Sethi, 2003) with an air flow rate of 1.2 m/s. The samples were evenly placed in the aluminum trays with a loading rate of 1.50 kg/m². The final moisture content of anardana was kept at 9 ± 1%. The dehydrated samples were packed in lots of 100 g each in Low-density polyethylene (LDPE) bags of 200 gauge thickness and kept at ambient temperature (30 ± 2°C) for physico-chemical analysis.

Physico-chemical characters of anardana

The size (length and breadth) of the randomly selected 20 anardana samples were measured using a micrometer of 0.01 least count. Ascorbic acid, reducing sugars and titratable acidity were determined as mentioned above in Section 2.3.

RESULTS AND DISCUSSION

All the treatments saved time and labour over the manual practices to a significant level (Table 1). Since the seeds were tightly adhered to the rind and with the peel, the manual separation in control treatment was observed to be difficult. The minimum time of separation as 2.5 min/kg of fruit was observed in case of hot water dip of fruits for 2 min. Highest percentage of labor saving (80%) over conventional method excluding treatment time and including treatment time (74%) was also recorded in same treatment.

Sand roasting reduced the seed separation time than the control treatment, but the percentage time taken was more than that of hot water dipping. Due to the high temperature during sand roasting the inner temperature of fruits has increased and lead to gelatinization. This affected the separation of seeds from the fruits, which were treated by sand roasting. A similar effect was reported by Mahajan et al. (2004). Soaking of fruits in water at ambient temperature was not found economical in reducing the separation time.

The chemical constituents of fresh seeds produced by various treatments are reported in Table 1. The titratable acidity of seeds separated by hot water dip and sand roasting treatments was less (6.03 - 6.16%) than in the seeds separated by cold water dip method and control treatment (6.63 - 6.80%). Hot water dip and sand roasting increased the pH level and was found in the range of 3.0 - 3.1, whereas, it was comparatively on lower side in seeds separated by soaking in cold water.

Different treatments affected the reducing sugar content of the seeds separated from the fruits. The reducing sugar content was higher in case of hot water treatment and sand roasting (7.8 - 7.9%) while it was minimum in seeds separated by dipping in cold water for 10 min (7.1%). Ascorbic acid content of the seeds varied signify-

Table 2. Chemical composition of *Anardana* produced from seeds separated by various treatments.

Treatments	Titrable acidity (%)	Ascorbic acid (mg/100 g)	Reducing sugars (%)	Size of seed	
				Length (cm)	Breadth (cm)
CT	6.9 (±0.10)	16.8 (±0.60)	36.5 (±0.95)	0.81 (±0.24)	0.69 (±0.12)
SR1	7.0 (±0.20)	16.0 (±0.47)	36.7 (±0.60)	0.72 (±0.21)	0.62 (±0.11)
SR2	7.1 (±0.17)	16.1 (±0.42)	36.9 (±1.10)	0.68 (±0.17)	0.61 (±0.11)
HW1	6.9 (±0.21)	16.7 (±0.39)	37.2 (±0.76)	0.79 (±0.16)	0.63 (±0.09)
HW2	6.9 (±0.11)	16.8 (±0.50)	37.2 (±0.80)	0.76 (±0.30)	0.63 (±0.20)
WS1	6.4 (±0.18)	16.8 (±0.51)	36.4 (±1.00)	0.81 (±0.11)	0.70 (±0.13)
WS2	6.5 (±0.13)	16.7 (±0.53)	36.4 (±0.75)	0.80 (±0.18)	0.70 (±0.17)

cantly according to treatments. In case of control treatment, cold water dip and hot water dip, ascorbic acid was in the range of 19.20 - 19.60 mg/100 g. But, ascorbic acid content of seeds separated by sand roasting was less. The high temperature during sand roasting has resulted in reduction of ascorbic acid content, as it is heat sensitive. Similar result is also reported by Singh et al (2007), where steaming for 5 min has resulted in loss of ascorbic acid.

The physico-chemical character of *Anardana* prepared from the seeds separated by various treatments is presented in Table 2. The results reveal that length (0.80 - 0.81 cm) and breadth (0.69 - 0.70 cm) of *anardana* prepared from the seeds separated by control and cold water dip treatment was higher. Hot water dip and sand roasting resulted in reduction in size of *Anardana,* which may be due to deformation of seeds and excessive loss of moisture during treating the fruits at high temperature. Gelatinization of seeds was also observed in the fruits treated by sand roasting, which has affected the size of the product. Various treatments have significantly affected the titratable acidity content of anardana produced from seeds separated from fruits. The acidity was found maximum in case of sand roasting of fruits for separation whereas; it was minimum in coldwater dip.

Ascorbic acid content of the product *Anardana* showed similar trend as that of the fresh seeds separated by various treatments. There was no significant difference in the ascorbic acid content of anardana produced from seeds separated manually, hot and cold water dip and the value range between 16.7 and 16.8 mg/100 g, whereas it was minimum in sand roasting treatment (16.0 and 16.1 mg/100 g for roasting for 5 and 7 min respecttively). Reducing sugar content was higher in the product *Anardana* than the fresh seeds. A similar observation has been reported by Pruthi and Saxena (2005) and Singh and Sethi (2003).

The reducing sugar content was higher in the *anardana* prepared from the seeds separated by hot water dipping. Color is an important parameter in deciding the quality of anardana. The chromaticity values L*, a*, and b* of the anardana significantly varied depending on the treatments (Table 3). *Anardana* prepared from the seeds separated by hot water dip and control treatments were

Table 3. Color of *Anardana* produced from seeds separated by various treatments.

Treatments	L	a	b
CT	33.07 (±3.8)	11.69 (±1.7)	13.19 (±2.2)
SR1	29.11 (±5.2)	10.18 (±0.4)	12.42 (±2.9)
SR2	29.90 (±2.6)	10.96 (±2.1)	12.88 (±3.2)
HW1	33.92 (±2.9)	11.37 (±1.8)	13.87 (±1.7)
HW2	33.32 (±4.3)	11.70 (±1.3)	14.36 (±4.1)
WS1	29.81 (±3.8)	11.39 (±1.7)	15.13 (±2.3)
WS2	29.82 (±3.2)	11.42 (±0.6)	15.15 (±1.4)

able to retain the desirable luminance comparatively. Other than sand roasting, rest of the treatments was able to produce the desirable reddish color in *anardana*. Treatments of water soaking and hot water dip produced *anardana* having comparatively more yellowness having 'b' value in range of 13.87 - 15.15, whereas in sand roasting 'b' value was low showing reduction in yellowness.

Conclusion

The effect of treatments on time taken for separation of seeds from wild pomegranate fruits was studied. Fruits dipped in hot water for 2 min required less time for separation of seeds. The acidity and ascorbic acid content of the seeds separated by this treatment was also high. High temperature treatment of sand roasting affected the chemical constituents of fresh seeds and also the product *Anardana*. Size of the anardana prepared from fresh seeds also reduced due to gelatinization during hot water dipping and sand roasting. But, it was found that hot water dipping of fruits for 2 min was economical in terms of labour saving and retained the desirable chemical composition.

REFERENCES

Dhandar DG, Singh DB (2002). Current status and future needs for the development of pomegranate. In: Programme and discussion papers, National Horticulture Conference, New Delhi p. 12.

Mahajan BVC, Chopra SK, Sharma RC (2004). Processing of wild pomegranate (Punica granatum L) for anardana: Effect of thermal treatments and drying modes on quality. J. Food Sci. Technol. 29(5): 327–328.

Parmar C (1981). Wild fruits of Sub Himalayan region. Ludhiana, India: Kalyani Publications p. 32.

Patil AV, Karade AR (1996). In Bose TK, Mitra SK (Eds.), Fruits: Tropical and subtropical. Calcutta, India: Naya Prakash.

Pruthi JS, Saxena AK (2005). Studies on anardana (dried pomegranate seeds).J. Food Sci. Technol. 21(5): 296.

AOAC (2002). In official methods of analysis of the association of official analytical Chemists 19th ed. Maryland, USA : AOAC International.

Ranganna S (2000). In Handbook of analysis and quality control for fruit and vegetable products (2nd ed.). New Delhi, India: Tata McGraw-Hill.

Saxena AK, Manan JK, Berry SK (1987). Pomegranate post harvest technology, chemistry and processing. Indian Food Packer pp41- 43.

Singh RP, Gupta AK, Berry SK (2007). Utilization of wild pomegranate in North West Himalayan-status and problems. In: Proceedings of the national seminar on production and marketing of indigenous fruits, New Delhi pp. 100–107.

Singh D, Sethi V (2003). Screening of pomegranate genotypes for the preparation of quality grade anardana. J.Food Sci. Technol. 40(2): 236–238.

27

Epidemiological and biochemical studies of human lymphatic filariasis and associated parasitoses in Oguta, South-Eastern, Nigeria

Okey A. Ojiako* and G. O. C. Onyeze

Department of Biochemistry, Federal University of Technology, Owerri, Nigeria.

Possible organ infections associated with human filariasis, helminthiasis and malaria in Oguta Local Government Area of Imo State, South-Eastern Nigeria were investigated. Blood, urine and stool samples were collected in appropriate containers from 200 male and female respondents aged 31 – 85 years. Parasitological studies were carried out on blood samples for malaria and/or microfilariae parasites while stool samples were tested for the presence of some intestinal parasites. The study showed a prevalence of intestinal protozoa (*Entamoeba histolytica*), *Wuchereria bancrofti*, the intestinal helminthes *Ascaris lumbricoides* and Hookworms. Biochemical parameters of liver integrity were also studied across the various infection cohorts among the respondents. Results obtained show that these parasitic infections depressed the hematological parameters relative to 'normal' respondents. Comparative biochemical analyses showed significant ($p < 0.05$) differences in some liver function parameters obtained for infected respondents relative to those not infected. There was also a positive correlation between age brackets with highest filarial infection (with no malarial co infection) and age groups with elevated markers of liver dysfunction. This study can be of immense diagnostic value in the clinical management of the filariases especially in malaria-endemic and resource-poor areas.

Key words: Filariasis, co-parasitoses, Nigeria, prevalence patterns, liver dysfunction.

INTRODUCTION

Till date the dream of eliminating some parasitic diseases like malaria and filariasis has not been fully realised. Instead, there appears to be a recrudescence of these old endemic debilitating parasitic diseases in some parts of developing countries. Malaria is still known to be the major cause of mortality and morbidity in the tropical and subtropical regions of the world (WHO, 2004) and is caused by *Plasmodium* species that have mosquitoes as their intermediate hosts and also serve as vectors of infective parasite stages to man. An added danger to malarial infections is that its effects and even recrudescence after intervention are usually worse with children (Borrmann et al., 2008).

Filariasis on its own is a major public health problem in many parts of Asia, Africa, the Western Pacific and the Americas (Anosike et al., 2005). Filariases are a group of vector – borne parasitic diseases of humans and other animals, caused by long, threadlike worms (hence the name "filaria" from Latin) that in their mature adult stages reside in the lymphatics or in connective tissue. Of the eight filarial parasites that commonly infect man three species account for most of the pathology associated with these infections: the lymphatic dwelling filariae *Wuchereria bancrofti* and *Brugia malayi* and the skin dwelling *Onchocerca volvolus* (Ottesen, 1984). Infection by *W. bancrofti* is the most common (Anosike et al., 2005) and accounts globally for approximately 90% of all infections. Worldwide, over 120 million people are infected with lymphatic filariasis and in Africa; the prevalence is especially striking, affecting over 40 million people in the sub-Saharan region alone. Overall, Africa is thought to account for 40% of all global cases (Lenhart et al., 2007).

The health, social and economic burdens of endemic tropical parasitic diseases have been assessed to include direct disease-related costs to individuals and households, costs to government-funded healthcare systems, lost productivity of infected individuals, and reduced pro-

*Corresponding author. E-mail: okeyojiako@yahoo.com.

ductivity from structural changes in the economies of endemic villages (Evans et al., 1993; Gyapong et al, 1996; Haddix and Kestler, 2000).

In Nigeria and elsewhere where diseases like helminthic infections, malaria and filariasis are endemic, the conventional diagnostic techniques are invasive and often times repetitive as different blood samples will be required for parasitological examination of filarial and malaria patients and confirmation is based on positive parasitaemia. They are also highly technical, time consuming, expensive and in most cases fraught with poor cooperation from patients who are already anaemic. Yet due to several factors including poverty, development projects without environmental impact assessment and other activities that favour the breeding of the mosquito vectors (Haddix and Kestler, 2000) the burden of these parasitic diseases has virtually bent the back of the inhabitants of the tropics and does not seem to be abating as evidenced by recent studies (Anosike et al., 2005; Okoye and Onwuliri, 2007). Incidentally, both malaria and filariasis have feverish symptoms and in rural areas it is often difficult to distinguish drug-resistant falciparum malaria and periodic fever due to filaria infection.

These facts justify the need for more and possibly faster tools of diagnosis, management and control of these parasitic diseases. This work is therefore aimed at studying the patterns of some of these parasitic infections especially lymphatic filariasis as well as assessing the possibility of using biochemical parameters in its diagnosis in the presence or absence of malaria and intestinal helminthes whose febrile and nauseous symptoms oftentimes confuse with those of filariasis.

MATERIALS AND METHODS

Study area

The study was carried out in Oguta Local Government Area, one of the 27 Local Government Areas of Imo State in the South-eastern part of Nigeria. The administrative headquarters, Oguta, is a suburban town. Most of the residents of the local government area (LGA) are predominantly farmers, fishermen and civil servants. Oguta town has some recreational facilities like an international golf course, the Oguta Lake and the Oguta Lake Motel among others in addition to a natural scenic beauty accentuated by the presence of rivers which nearly encircle the town and the adjoining towns in the LGA. The presence of these rivers which serve as mosquito breeding sites coupled with the continual interaction with the water bodies in normal daily activities and chores makes the area potentially endemic for mosquito- and water-associated parasitic infections.

Study group

A total of 200 potential patients were recruited from over 1000 volunteers. Informed consent of the volunteers was obtained as demanded by WHO (TDR, 2001). With the assistance of individuals recognised by WHO (2004a) as key informants like school teachers, private medical practitioners, patent medicine dealers and traditional healers, certain villages were mapped out for study after which possible respondents to a preliminary questionnaire were

identified. Male and female respondents aged between 30 - 85 years with clinical symptoms of filariasis like presence of eye worm, skin depigmentation, itching, diarrhoea, fever, hydrocoele, lympho-edema and elephantiasis (TDR, 2002) etc were selected for the study. There were no language and cultural limitations. Blood, urine and faecal samples were collected from volunteers among the selected respondents who were fully aware of the objectives of the study and the possible benefit of free drugs should eventually be confirmed positive.

Sample collection

Urine, faecal and blood samples from each respondent were collected by qualified health personnel using appropriate containers, which were then covered in black polythene bags and stored in small plastic food coolers loaded with ice before analyses.

Mid-stream urine samples were collected and examined for the presence/absence of cysts, ova etc as recommended by Zeibig (1997). A 5 ml portion of blood was collected from each respondent using a disposable syringe and a needle. A 2 ml portion of the blood sample was put into an EDTA bottle to prevent clotting while the remaining 3 ml was allowed to clot in the syringe. Sera from the clotted samples were used for liver function tests after centrifuging for 10 min at 3000 rpm with a Wisperfuge Centrifuge (Ojiako and Nwanjo, 2006) while the unclotted samples were used for parasitological examination for malaria and filaria parasites as described by Zeibig (1997).

Urinalyses using Medi-Test Combi 9[TM] (Macheray-Nagel, Duren) test strips designed to detect and roughly estimate the following parameters: pH, glucose, ascorbic acid, protein, ketone, nitrite, bilirubin, urobilinogen and blood were carried out in all urine samples prior to other analysis. Urine samples were further analysed microscopically for the presence of epithelial cells, white blood cells (WBC), red blood cells (RBC), casts, crystals and organisms such as yeast cells, Trichomonas vaginalis, ova of Schistosoma haematobium, according to the methods of Baker et al. (2000) and Zeibig (1997). All samples that tested positive or slightly positive for infection or another disease condition different from those under study were discarded to avoid confusion with the possible role of filariasis and the associated diseases under study.

Parasitological tests

The direct wet smear technique of stool analysis was used to determine the presence of ova or larvae of helminthes. Thin and thick blood films stained by Giemsa staining technique were used to detect and confirm malaria and filaria parasites (Zeibig, 1997; Udonsi, 1999; Wanji, 2001).

Biochemical and haematological tests

Total and conjugated bilirubin levels in serum samples were measured according to the modified (Henry, 1974) method of Malloy and Evelyn (1937). The activity of alkaline phosphatase E.C. 3.1.3.1 (orthophosphoric monoester phosphohydrolase) was measured according to the method of King and Armstrong (1934) as modified in Tietz (1983) while aspartate aminotransferase (ASAT) E.C.2.6.1.1 and alanine aminotransferase (ALAT) E.C.2.6.1.2 activities were measured spectrophotometrically (Camspec M210 spectrophotometer) using Randox[TM] kits based on the method of Reitman and Frankel (1957). Haemoglobin levels were determined spectrophotometrically using Drabkins's solution (potassium ferricyanide, BDH) while packed cells volume (PCV) was determined using the microhaematocrit method (Tietz, 1976; Baker et al., 2000).

Table 1. Age profile of respondents and patterns of infection of studied parasitic diseases.

Age (Years)	Sex	Number examined	Filariasis	Plasmodiasis	Intestinal helminthes	Intestinal protozoa
31-35	M	6	-	2	-	2
	F	7	-	-	1	-
36-40	M	-	-	-	-	-
	F	12	-	2	-	2
41-45	M	17	7	-	1	-
	F	13	9	1	-	-
46-50	M	7	-	-	1	3
	F	11	3	-	1	1
51-55	M	19	5	2	-	-
	F	13	6	-	-	1
56-60	M	23	3	-	-	2
	F	11	-	-	-	1
61-65	M	22	-	2	-	-
	F	7	-	2	-	-
66-70	M	18	2	1	-	-
	F	5	4	1	-	-
71-75	M	-	-	-	-	-
	F	6	-	2	-	-
76-80	M	-	-	-	-	-
	F	-	-	-	-	-
81-85	M	-	-	-	-	-
	F	3	1	-	-	-
Total		200	40	15	4	13

- = Not detected.

All the biochemical parameters were also obtained for a set of persons (matched for age and gender) in the same area that tested negative to all the parasitic infections investigated (n = 10).

Statistical analysis

Chi-square test of independence was used to test the association of measured parameters with respect to patterns of parasitic infection or age group of respondents. Chi-square test of homogeneity was used to ascertain any difference within cohorts. Differences at $P < 0.05$ were considered significant (Bailey, 1994). Sign test was used for comparing liver function parameters between infected and non-infected volunteers.

RESULTS

Table 1 shows the patterns of infection of the studied parasitoses according to age while table 2 shows the relative prevalence of the parasitic infections in the studied population. Table 3 shows the patterns of infections and coinfections of the diseases while Tables 4 and 5 show the results of the liver diagnostic and haematological parameters respectively. Results of the parasitological examination of blood and stool samples show that of the 200 chosen respondents, 40 (20%) had filariasis while 15% had malaria. Of the filaria and malaria-positive cases there were mutual coinfections as well as coinfections by helminthiasis (*Ascaris lumbricoides* and hookworms) and protozoasis (*Entamoeba histolytica*) as shown in tables 1 and 3.

The respondents fall within the age bracket of 30 and 85 even though some of the respondents did not know their exact age and we had to use some well established historical events like the second World War, Nigerian Independence in 1960 and the Nigerian Civil War (1967 - 1970) among others to place their age. Infection of lymphatic filariasis was highest among the age brackets 41 - 45 and 51 - 55 and no infection was observed among the age brackets 31 - 35, 36 - 40, 61 - 65, 71 - 75 and 76 - 80. There were however positive parasitoses in all the age groups even in the absence of filariasis except for age group 76 - 80 which had no sample representative in the study.

Analyses of gender-related prevalence (Table 2) show that though there were more males (112) than females (88) among the respondents, there were more positive cases of plasmodiasis, filariasis and helminthiasis among the female respondents in the study area. The difference in prevalence of filarial infection between the female (26.1% of total sample) and the male respondents

Table 2. Prevalence of studied parasitic diseases.

Disease	Sex	Number examined	Number infected	Prevalence (%)
	Male	112	17	15.18
Filariasis	Female	88	23	26.14
	Total	200	40	20.00
	Male	112	7	6.25
Malaria	Female	88	8	9.10
	Total	200	15	7.5
	Male	112	1	0.89
Intestinal helminthiasis	Female	88	3	3.41
	Total	200	4	2
	Male	112	9	8.04
Intestinal protozoa	female	88	4	4.55
	Total	200	13	6.5

Table 3. Patterns of infection and co-Infection of the studied parasitic diseases.

Parasitic infections	Number affected	Percentage of sampled population
Filariasis	40	20
Malaria	8	4
Filariasis + Malaria	7	3.5
Filariasis + Intestinal helminthiasis	4	2
Filariasis + Intestinal protozoasis	13	6.5

(15.2% of total sample) was at the limits of statistical significance (P = 0.054). What about the gender differences in other parasitc infections?

Results of the liver function tests (Table 4) show that of the 40 filaria-positive cases 23 (57.5%) had elevated (above adult normal levels) levels of bilirubin and also elevated levels of alanine and aspartate aminotransferase. Statistical analyses of these results using sign test show that there were significant differences (p < 0.05) in the values of liver function parameters of respondents infected with filariasis and malaria and the corresponding values for those not infected. Bilirubin levels, for instance, were highest in the 41 - 45, the age range with the highest prevalence of filariasis, followed by the age range 61 - 66, the age range with the highest prevalence of malaria. The values for the haematological parameters, packed cells volume (PCV) and haemoglobin level however showed no significant differences among the various groups.

DISCUSSION

The findings of this study showed that 20% of all the respondents examined were infected with filariasis. This is ordinarily a confirmation of endemicity. The sampling procedure however was not a blind and random one and the aim of the study was not to determine endemicity.

There was already a statewide distribution of ivermectin (Mectizan[R]) and prevalence of disease has already been confirmed in different parts of the State (Nwoke, et al., 1994; Dozie, 2003, Unpublished PhD thesis). The respondents were recruited from a group of volunteers with clinical symptoms of filariasis as well as other parasitic diseases, who were already feeling sick and responded to invitation of health personnel with the hope of receiving free medication.

Results of the liver function tests show that most (57.14%) of the individuals infected with filariasis had elevated levels (above adult normal ranges) of liver diagnostic parameters. Infection prevalence pattern also correlated with the levels of liver parameters of respondents aged 41 to 60 years. Highest prevalence coincided with the age group having the highest levels of bilirubin and highest activities of alanine and aspartate aminotransferases as well as serum alkaline phosphatase which are all markers of liver integrity. Elevated levels of these parameters are indicative of liver disease (Peters, 1989). This would indicate therefore, that filariasis has a direct influence on the functions of the liver.

The association of malaria with liver damage is not in doubt and our results corroborate that the association of filariasis with renal damage is also known. Ottesen (1984) had earlier associated filariasis with impairment of kidney function. The possible association of filariasis with

Table 4. Results of liver diagnostic parameters.

Age (Years)	ALP (IU/L)	ALAT (IU/L)	ASAT (IU/L)	Total Bilirubin (µmol/L)	Conjugated Bilirubin (µmol/L)
31-35	65.2 ± 18.38	8.00 ± 0.01	15.01 ± 1.44	12.83 ± 3.59	5.99 ± 1.20
36-40	66.01 ± 5.65	8.00 ± 0.02	15.5 ± 0.71	13.68 ±2.39	7.70 ±1.20
41-45	89.3 ± 32.33	14.01± 6.92	26.0 ± 12.48	22.23 ± 7.70	7.18 ± 0.68
46-50	78.4 ± 12.75	6.60 ± 0.89	16.6 ± 2.40	14.36 ± 0.86	11.29 ± 5.47
51-55	78.6 ± 15.51	10.40 ± 4.39	18.0 ± 6.04	15.05 ± 4.45	8.55 ± 2.91
56-60	69.3 ± 25.78	9.30 ± 4.43	17.3 ± 5.46	14.54 ± 8.04	7.35 ± 3.93
61-65	85.0 ± 22	12.2 ± 4.26	22.4 ± 7.16	18.13 ± 6.16	9.92 ± 6.33
66-70	68.5 ± 10.24	8.80 ± 1.25	16.3 ± 1.71	11.12 ± 2.22	6.50 ± 0.86
71-75	72.01 ± 0.01	8.01 ± 0.02	15.0 ± 0.06	13.68 ± 0.17	6.84 ± 0.17
76-80	ND	ND	ND	ND	ND
81-85	94.01 ± 0.01	14.01 ± 0.02	20.0 ± 0.08	15.39 ± 0.17	8.55 ± 0.17
Control	65.60 ± 12.83	8.40 ± 1.74	15.26 ± 1.08	13.25±3.45	6.86 ± 1.21

Values are means ± S.D.
ALP = Alkaline phosphatase, ALAT = Alanine aminotransferase, ASAT = Aspartate aminotransferase and ND = Not Determined.

Table 5. Results of measured haematological parameters.

Age (Years)	PCV (%)	Haemoglobin (g/dL)
31 - 35	35.01 ± 7.07	10.30 ± 2.12
36 - 40	33.52 ± 6.36	10.30 ± 2.40
41 - 45	32.31 ± 5.50	9.80 ± 1.44
46 - 50	33.20 ± 3.11	10.06 ± 1.23
51 - 55	32.40 ± 2.88	9.80 ± 1.04
56 - 60	34.72 ± 5.13	9.60 ± 1.93
61 - 65	34.62 ± 2.79	10.20 ± 1.33
66 - 70	34.84 ± 4.57	10.40 ± 1.08
71 - 75	30.01 ± 0.01	8.80 ± 1.01
76 - 80	ND	ND
81 - 85	27.00 ± 0.02	8.40 ± 0.07
Control		
Male	40.54 ± 7.81	13.61± 1.95
Female	36.44 ± 9.05	11.54 ± 5.32

Values are means ± S.D
ND = Not Determined, PCV = Packed cell volume.

liver function is novel and deserves further investigation in different locations to eliminate other confusing factors like nutrition, environmental factors and drug abuse among others.

This study has shown that filariasis has influence on liver and kidney functions at least in the studied area. Therefore the estimation of liver function parameters and the subsequent interpretation will be of diagnostic value if it also accommodates the possibility of filarial infection even in proven cases of malaria parasites from microscopy. This will improve the clinical diagnosis and management of filariasis especially as most fever pa-tients that report for treatment assume that they had malaria (as we found out through our questionnaire) and it is a common medical practice to treat malaria in almost all patients that present symptoms of fever.

REFERENCES

Anosike JC (1994). Studies on filariasis in Bauchi State, Nigeria V: The distribution and prevalence of mansonellosis with special reference to clinical signs. Appl. Parasitol. 35: 189-192.

Anosike JC, Nwoke BE, Ajayi EG, Onwuliri CO, Okoro OU, Oku EE, Asor JE, Amajuoyi OU, Ikpeama CA, Ogbusu FI, Meribe CO (2005). Lymphatic filariasis among the Ezza People of Ebonyi State, Eastern Nigeria, Ann. Agric. Environ. Med. 12:181-186.

Bailey NTJ (1994). Statistical methods in biology, 3rd edn., University Press, Cambridge.pp. 38-87.

Baker FJ, Silverton RE, Pallister CJ (2000). Introduction to Medical Laboratory Technology. Bounty Press Limited Nigeria. Pp 353-362.

Borrmann S, Matsiegui PB, Missinou MA, Kremsner PG (2008). The effects of Plasmodium falciparum parasite population size and patient age on early and late parasitological outcomes of antimalarial treatment in children. Antimicrobial Agents Chemotherapy.

Dozie INS (2003). Onchocerciasis and its socioeconomic effects in some parts of Imo State, Nigeria. Unpublished PhD Thesis, University of Jos, Nigeria, 240-243.

Evans DB, Gelband H, Vlassoff C (1993). Social and economic factors and ther control of lymphatic filariasis. Acta Tropica 53:1-2.

Gyapong M, Gyapong JO, Adjei S, Vlassoff C, Weiss M (1996). Filariasis in northen Ghana: some cultural beliefs and practices and their implications for disease control. Soc. Sci. Med. 43:235-242.

Haddix AC, kestler A (2000). Elimination of lymphatic filariasis: economic aspects of the disease and programmes for its elimination. Transactions of the Royal Society of Tropical medicne and Hygiene 94:592-593.

Henry JB (1974). Clinical diagnosis and management by laboratory method, W.B. Saunders Co., Philadelphia, 332-335.

Lenhart A, Eigege A, Kal A, Pam D, Miri ES, Gerlong G, Oneyka,J, Sambo Y, Danboyi J, Ibrahim B, Dahl E, Kumbak D, Dakul A, Jinadu MY, Umaru J, Richards FO, Lehmann T (2007). Contributions of different mosquito species to the transmission of lymphatic filariasis in central Nigeria: Implications for monitoring infection by PCR in mosquito pools. Filaria .J. 6:14 doi:10.1186/1475-2883-6-14.

King EJ, Armstrong AR (1934). A convenient method for determining serum and bile phosphatase activity. Can. Med. Ass .J. 31:376-381.

Malloy HT, Evelyn KA (1937). The determination of bilirubin with the photoelectric colorimeter. J. Bio. Chem. 119:481.

Mathieu E, Amann J, Lammie P (2008). Collecting baseline information for national morbidity alleviation programs: Different methods to estimate lymphatic filariasis morbidity prevalence. Am. J. Tropical Med, Hyg. 78(1):153-158.

Nwoke BEB, Edungbola LD, Mencias BS, Njoku AJ, Abanobi OC, Nkwogu F, Nduka, FO, Oguariri RM (1994). Human onchocerciasis in the rainforest zone of southeastern Nigeria. The. Nig. J. Parasitol. 15:7-18.

OjiakoOA, Nwanjo HU (2006). Is *Vernonia amygdalina* hepatotoxic or hepatoprotective? Response from toxicity and biochemical studies. Afr. J. Biotechnol. 5(18): 1648-1651.

Okoye IC, Onwuliri CO (2007). Epidemiology and psycho-social aspects Of onchocercal skin diseases in northeastern Nigeria Filaria .J. 6:15 doi:10.1186/1475-2883-6-15

Ottesen E A (1984). Filariasis and tropical eosinophilia In: Tropical and Geographical Medicine. Warrenks, Mahmoud A A F (eds). McGraw-Hill Inc. New York. 390 – 422

Ottesen E A (1998). Towards elimination of lymphatic filariasis. In: Infectious diseases and public health. Angelico M, Rocchi G (eds), Balaban Publishers, Philadelphia, 58-64.

Peters W (1989). Liver enzymes and diagnosis, (6th edn), Golden Press, Philadelphia, 8-9.

Reitman S, Frankel S (1957). A colorimetric method for the determination of SGPT and SGOT, Am. J. Clin. Pathol. 25: 56-62.

TDR (2001). Handbook of good laboratory practice (GLP): quality practices for regulated non-clinical research and development, TDR/PRG/GLP/01.2

TDR (2002). Guidelines for rapid assessment of *Loa loa*. UNDP/World Bank/WHO Special programme for research and training in tropical diseases, TDR/IDE/RAPLOA/02.1.

Tietz N W (1976). Fundamentals of Clinical Chemistry, W.B. Saunders Company, Philadelphia, 7.

Tietz NW (ed.) (1983). Study Group on Alkaline phosphatase: a reference method for measurement of alkaline phosphatase activity in human serum. Clin. Chem. 29:751.

Udonsi JK (1999). Parasites and Parasitic Diseases, A Textbook of General and Medical Parasitology in Tropical Africa, Josany Press, Port Harcourt, 127.

Wanji S (ed) (2001).Rapid assessment procedures for loaiasis: report of a multi-centre study, UNDP/World Bank/WHO Special Programme for Research and Training in Tropical Diseases (TDR), (TDR/IDE/RP/RAPL/.01.1).

World Health Organisation (2004) a. Community participation and tropical disease control in resource-poor settings, TDR/STR/SEB/ST/04.1

World Health Organisation (2004) b. Report of the Scientific Working Group on Malaria, TDR/SWG/03, 7-8.

Zeibig EA (1997). Clinical Parasitology: A Practical approach. W.B. Saunders Company, Philadelphia, 9ff.

28

Effect of falciparum malaria on some plasma proteins in males: With special reference to the levels of testosterone and cortisol

Muawia A. Abdagalil[1]* and Nabiela M. ElBagir[2]

[1]Department of Biochemistry, Faculty of science, University of West Kordufan, Sudan.
[2]Department of Biochemistry, Faculty of Veterinary Medicine, University of Khartoum, Sudan.

Sex-associated hormones were evidenced to modulate immune responses and consequently directly influence the outcome of infection. Testosterone is known to influence both protein metabolism and the level of cortisol which is the hormone of stress. This work was conducted to explore the influence of the degree of parasitemia in *Plasmodium falciparum* malaria on the male sex hormone testosterone, plasma proteins and the stress hormone cortisol in male patients. The study targeted male subjects whose ages ranged between 20 and 40 years old. The subjects were divided into three groups: lightly-infected patients (Infected with *P. falciparum* density 1 - 10 asexual form of the parasite per 100 fields), heavily-infected patients (Infected with *P. falciparum* density 11 - 99 asexual form of the parasite per 100 fields) and a group of malaria-free individuals who were used as a control group. Blood samples were taken from the median cephalic vein to investigate for malaria parasite, plasma proteins and hormones. The effect of the degree of parasitemia was considered for all parameters studied. The study revealed that low parasitemia in malaria-infected patients resulted in significantly ($p < 0.05$) higher level of plasma total proteins and it was found to be due to a significant ($P < 0.05$) increase in the total globulins fraction which reached 5.85 ± 1.03 g/dl compared to 3.44 ± 0.4 g/dl in the control group. The opposite was true for the heavily-infected group as it reported significantly ($P < 0.05$) lower total plasma proteins value which was found to be due to a significant ($p < 0.05$) reduction of the total globulins fraction reported only 2.96 ± 0.20 g/dl. The albumin fraction maintained levels similar to that of the control group in both infected groups. The levels of the hormones tested also showed significant changes, manifested as significantly ($p < 0.05$) lower values in both groups of patients compared to the control group for the testosterone hormone, with significant ($p < 0.05$) difference between the two groups. Thus, high parasitemia resulted in the least testosterone level in the heavily-infected group of 2.64 ± 0.28 ng/ml, compared to 6.03 ± 0.86 ng/ml in the control group. In contrast, the stress hormone, the cortisol, showed the highest level in the heavily-infected patients of 191.03 ± 18.17 ng/ml, with significant ($p < 0.05$) difference compared to 166.28 ± 10.63 ng/ml in the control group.

Key words: Malaria, testosterone, cortisol, parasitemia.

INTRODUCTION

Malaria is one of the most widespread transmissible diseases distributed throughout the world. Although the cause of malaria was identified more than hundred years ago, it still remains one of the leading causes of morbidity and mortality in the tropics. The worldwide incidence of malaria is estimated to be 300 - 500 million clinical cases and 1.5 – 2.7 million deaths (mostly children) annually, representing 2 - 3% of the overall global disease burden. Malaria is a serious problem e.g. in Sudan. It accounts for 32% of all cases and an estimated 7 – 7.8 million cases of the disease occur annually with a 20% mortality rate (Robert et al., 2004). Several studies have shown the relation between malaria-induced stress and the level of proteins. Abdelgadir (2002) reported that no significant

---

*Corresponding author. E-mail: abdalgalil@yahoo.com.

difference was observed between males and females in total proteins and total globulins, but the albumin showed significantly higher values in male patients compared to females with the same pattern being observed in normal individuals. Adebisi et al. (1998) reported significant decrease in plasma total proteins in malaria patients compared to the normal individuals.

Numerous epidemiological and clinical studies have noted differences in the incidence and severity of parasitic diseases between males and females. Although in some instances this may be due to gender-associated differences in behavior, there is overwhelming evidence that sex hormones can also modulate immune responses and as a consequence, directly influence the outcome of parasitic infection (Benten et al., 1997). However, Bello et al. (2005) evaluated the pattern of infection in *P. falciparum* malaria cases at nested sentinel points in northern Nigeria and he reported that there were significantly more female patients than male.

In the present study, the male sex hormone testosterone and cortisol were assayed during *P. falciparum* infection in males and the levels of total proteins, albumin and total globulins were also measured. This is to investigate the effect of malaria as a stressing disease on the level of testosterone which is known to influence both the protein metabolism (Fahey, 1998) and cortisol which is the stress hormone (Resumo, 1998). Determining how all this is influenced by the degree of parasitemia was done by studying malaria patients as lightly-infected (one cross) and heavily-infected (two crosses), their results compared to values from malaria-free subjects of the same age.

MATERIALS AND METHODS

Forty-five men (age range 20 - 40 years) were employed in this study. Thirty of them were malaria patients, 15 were lightly-infected and other 15 were heavily-infected with *P. falciparum*. The immuno-chromatographic (ICT) malaria Pf test described by Garcia and Marlborough (1996) was used for rapid diagnosis of *P. falciparum* malaria. The other fifteen were free from infection and included as a control group. All individuals were divided into three groups: Group A as the control group (malaria-free individuals); Group B (one-cross patient) infected with *P. falciparum* density 1 - 10 asexual form of the parasite per 100 fields and Group C (two-cross patients) infected with *P. falciparum* density 11 - 99 asexual form of the parasite per 100 fields.

Blood samples for the laboratory test were collected in the morning after subjects signed a consent form and 5 ml of blood from the medium cephalic vein was collected into heparin bottles. Plasma was obtained by centrifugation at 4°C (x 800 g, 15 min) and kept frozen at – 20°C until used for the quantification of hormones, total proteins and albumin and the difference between total protein and albumin was calculated as total immunoglobulin.

Microscopic examination

Thick and thin films were prepared and examined microscopically for malaria parasite identification. Universal precautions were followed during preparation of the smears. The thick smears blood

films were left to dry and the samples here were not fixed with methanol. This allowed the red blood cells to be hemolysed and leukocytes and any malarial parasites presented were the only detectable elements. The presence and relative parasite count of *P. falciparum* in each blood sample was determined from Giemsa stained thin and thick films after staining for 30 min. The identification of the species of human parasites in the blood films was carried out according to (WHO, 1980). A slide was scored as negative when 100 high power fields (at 1000x magnification) had been examined for about 30 min without seeing any parasites. Then blood films were stained by Giemsa stain and left to dry and examined microscopically at 10×100 magnification by oil immerse objective. Parasites count was estimated according to the plus system described by Dayachi et al. (1991). The number of asexual forms of parasite (rings, trophozoites and schizonts) was counted against 100 fields as follows, one cross as 1 - 10 and two cross as 11 - 99 per 100 thick fields, respectively.

Biochemical measurements

Kits used for biochemical measures of total protein, albumin, testosterone and cortisol were obtained from Linear Chemicals Laboratory in Spain. Total protein was determined by the method described by Henary et al. (1974) using Biuret reagent kit. Albumin was determined by the method described by Doumas and Waston (1971) using Bromocresol green kit. Testosterone was determined by using radioimmunoassay (RIA) according to the method described by Soini and Kojola (1983). Cortisol was measured by RIA according to the method described by Prasad (1979).

Statistical analyses

Data were entered into an access database that was then imported into SAS for statistical analysis with SAS/STAT software. Differences in parasitemia categories between male cases were assessed using the Fisher's exact X^2. Differences in variable levels between cases and the control group were assessed using Wilcoxon rank sum exact test, the non-parametric counterpart of the independent t test. For the repeated measures analyses, mixed analyses of variance (ANOVA As) was used. Power analyses were calculated for all tests that revealed significant results ($p < 0.05$).

RESULTS

P. falciparum malaria infection and the degree of parasitemia in male individuals, showed clear changes in all parameters measured compared to malaria free subjects. The mean value of total protein, in all infected individuals as a group and the group of one-cross patients, was significantly ($p < 0.05$) higher than in the patients with two crosses and the control group as shown in Table 1 and Figure 1, while albumin levels, presented in Table 1 showed similar levels in the two groups of patients and the control group, with slightly lower levels in the groups of patients specially those of one-cross parasitemia.

Table 1 and Figure 1 present the effect of malaria infection on total globulins. The results showed significantly ($p < 0.05$) higher level in the group of all infected patients and in the one-cross parasitemia patients than the patients of two-cross parasitemia and control group. In two-cross patients the mean values showed significantly ($p < 0.05$) lower level than in the control group.

Table 1. The effect of Plasmodium falciparum malaria on plasma total proteins, albumin, total globulins, testosterone and cortisol in male individuals (N = 15).

Subjects	Total proteins g/dL	Albumin g/dl	Total globulins g/dL	Testosterone ng/mL	Cortisol ng/ml
The control group	7.97 ± 0.11^a	4.53 ± 0.07^a	3.44 ± 0.4^a	6.03 ± 0.86^a	166.28 ± 10.63^a
One-cross patients	10.28 ± 0.24^b	4.43 ± 0.09^a	5.85 ± 1.03^b	3.93 ± 0.57^b	184.77 ± 14.70^{ab}
Two-cross patients	7.42 ± 0.23^c	4.45 ± 0.12^a	2.96 ± 0.20^c	2.64 ± 0.28^c	191.03 ± 18.17^b

Means within columns followed by different letters are significantly different at (p < 0.05). N: Number of individual.
One cross = 1 - 10 asexual form of parasites per 100 leukocytes. Per 100 fields
Two crosses = 11 - 100 asexual form of parasites per 100 leukocytes. 11 - 99 asexual form of parasite per 100 fields
Means above columns followed by different letters are significantly different at (p < 0.05)

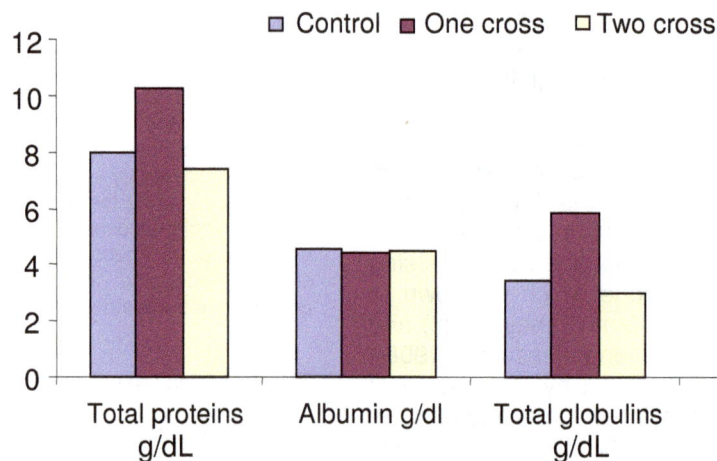

Figure 1. Effect of degree of parasitemia on plasma total proteins, albumin and total globulins in male (Means ± S.E).

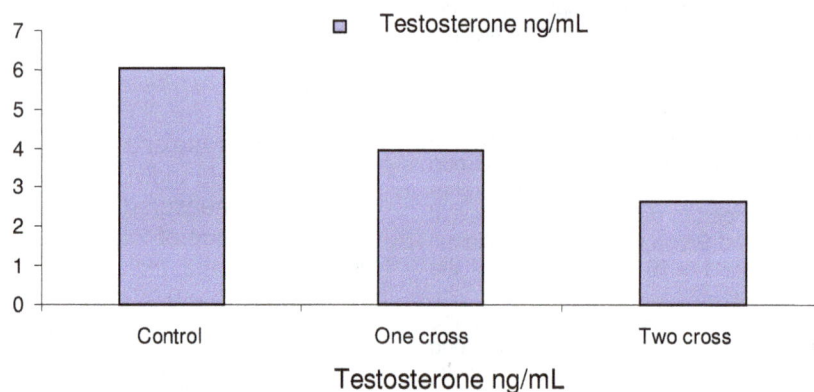

Figure 2. Effect of degree of parasitemia on testosterone in males (Means ± S.E).

The effect of presence and density of infection on the hormones is also presented in Table 1 and on the testosterone and cortisol in Figures 2 and 3 respectively. Testosterone mean value showed significantly (p < 0.05) lower level in the mean value of all infected individuals and each of the two groups of patients compared to the control group. The mean value in the patients with two crosses was significantly (p < 0.05) lower compared to

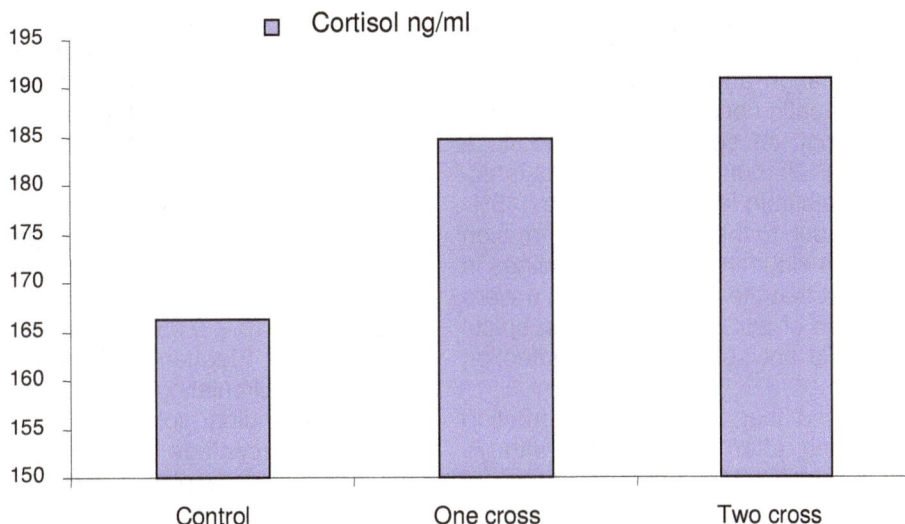

Figure 3. Effect of degree of parasitemia on cortisol in males (Means ± S.E).

the patients with one cross. In contrast, the cortisol showed higher values in the two groups of patients compared to the control group. Also significantly ($p < 0.05$) higher level of cortisol was observed in the two-cross patients compared to the control group, while the level of this hormone was just higher in the one-cross patients than that of the control group.

DISCUSSION

Effect of *P. falciparum* malaria infection on plasma proteins

Several studies showed that the level of total proteins in plasma decrease after the infection with *P. falciparum* malaria. Adebisi et al. (1998), Abdelgadir (2002) and Adeosun et al. (2007) reported significant decrease of plasma total proteins in malaria patients compared to non-infected individuals. They explained their results by the fact that the concentration of plasma proteins determines the colloid osmotic pressure of plasma and this is influenced by the nutritional status, hepatic and renal function. Chang and Herzog (1976) mentioned that malaria has an effect on all these functions and that it results in decreased plasma total proteins. It is also well known that hepatic protein biosynthesis shifts during inflammation from albumin synthesis to the synthesis of proteins involved in the acute inflammatory response such as C-reactive protein, coagulation factors, fibrinogen and complement components (Mac and Whaley, 2001). In the present study the plasma total proteins in all infected patients as one group and the two-cross patients group agree with these finding and were found to have significantly ($p < 0.05$) low levels compared to the control group subjects and the one-cross patients. But the level

of total proteins was found to be significantly ($p < 0.05$) high in one-cross patients compared to the control group. These findings showed that the values of total proteins can be influenced by the degree of parasitemia. However, It is well known that the first attack of the infection results in severe clinical symptoms e.g. diarrhea and vomiting (Dayachi, 1991). This might result sometimes in hemo concentration and elevated plasma proteins. In the present study the estimation of total globulin fraction showed significantly ($p < 0.05$) higher level in one-cross patients compared to the normal individuals, while the albumin was of similar levels in the patients and the control group. The very low levels of total proteins in the two-cross patients agree with the results of Chang and Herzog (1976) who mentioned that patients of high parasitemia develop liver disease, chronic kidney failure, malnutrition, or decrease in the immunoglobulins fractions. Abdelgadir (2002) found that the total protein values in children patients with one-cross parasitemia were within the same levels of the patients with two-cross parasitemia, so it was suggested that since all patients studied were children and supposed not to be suffered high frequency malaria infection (as in adult individuals) and the synthesis of the protein in the liver is not yet affected. The level of plasma albumin is very important. Matsuda et al. (1986) reported that a patient with lower levels of albumin has an unfavorable outcome, due to the fact that the decreased protein level may be a reflection of the severity of systemic illness rather than nutritional deficiency. In the present study, no changes were found in albumin levels of all groups. Lower albumin levels were observed in the two groups of patients compared to the control group individuals (Table 1) though the difference was not significant. This may be because the pronounced decline in plasma albumin (hypoalbuminemia) is usually known to follow prolonged malnutrition due to inadequate dietary intake of protein, impaired digestion of protein, chronic loss of protein or inability to synthesize albumin in chronic liver disease (Change and Herzog, 1976). This may also indicate that hepatic condition in all patients in the present study was not yet seriously affected. In a previous work, Golden (1982) found that in patients infected with malaria, serum albumin levels dropped by 15%. This was explained to be due to the fact that inflammation leads to rapid decrease in albumin levels. The findings in the present study suggest that the levels of albumin were not affected by the degree of parasitemia, or all subjects employed in this work did not suffer repeated infection with malaria.

Previous studies showed that total globulins fraction levels increase in plasma after the infection with *P. falciparum* malaria. Lunn et al. (1966) reported a rise in the γ-globulins level at the initial period of the disease which was correlated with the increasing of malaria antibodies. He also reported that stable levels of hyper gamma globulinaemia are observed in populations of endemic malarious areas. Many authors reported that cytokines and other proinflammatory mediators initiate both cellular and hormonal changes that induce skeletal muscle proteolysis, all of which are components of the metabolic responses observed following injury and infection (Cannon et al., 1990; Hill et al., 1996; Ling et al., 1997; Mitch and Goldberg, 1996; Michie et al., 1988). In the present study the plasma total globulin in all infected patients as one group and in the one-cross patients agree with these findings and were found to be significantly (p < 0.05) higher compared to the control group individuals. But the level was found to be significantly (p < 0.05) lower in the two-cross patients compared to the control group. These findings suggested that the level of total globulins is influenced by the degree of parasitemia. Very high levels of total globulins in patients are known to be due to liver disease, acute or chronic infection (Bams and Miranda, 1985). It is also known that in the first attack of the infection the immune system stimulates the lymphocytes to produce specific antibodies (Roitt, 1993). Infection with malaria in the present study increased the level of total globulins in the group of low infection. Whereas, the total globulins level in the patients of heavy parasitemia in the present work showed values significantly (p < 0.05) lower than in the non-infected group. This finding is similar to the results obtained by Cohen et al. (1974) and Benten et al. (1992) who reported that total globulin levels decreased in heavy malaria cases which can be explained as immunity-suppression that is caused by acute *P. falciparum* malaria parasite infection.

Effect of *P. falciparum* malaria infection on plasma testosterone

Several studies showed a characteristic hormonal response pattern when the body is subjected to chronic or acute stress (Bessey et al., 1984; Bessey and Lowe,

1993; Watters et al., 1986). Hilary (2002) reported that the hypothalamic-pituitary-adrenal (HPA) axis which is an important hormonal system is acted upon by the testosterone and this inhibits the release of the stress hormone cortisol. He suggested that this is the reason why males react differently to stress than females. In the present study testosterone levels showed significantly (p < 0.05) lower values in the two-cross patients compared to the patients with one-cross parasitemia (Table 1). These findings suggested that testosterone levels decrease significantly (p < 0.05) with the increase in the degree of parasitemia. Stephen (1999) reported that approximately 50% of the circulating testosterone is tightly bound to sex hormone binding globulin (SHBG) produced by the liver and high percentage bound to albumin, so that increased or decreased levels of SHBG or albumin influenced testosterone level. He also reported that the most accurate indicator of hypogonadism is the concentration of unbound testosterone. However, the very low levels of testosterone in malaria patients in the present study may be due to impaired biosynthesis of SHBG in the liver and this might increased the unbound testosterone (Matsuda et al., 1986; Stephen, 1999). Zhi et al. (2000) explained the possible modulation by male sex hormone of Th1/Th2 function in protection against *Plasmodium chabaudi* infection in mice. He suggested a possible counterbalancing between the immune and the endocrine systems in the response of a host to malarial infection. Benten et al. (1991) reported that testosterone has been shown to inhibit the ability of a host to overcome *P. chabaudi* infection, they also reported that although the mechanisms of testosterone-mediated inhibition are not clear, malaria-specific T and non-T cell suppression were observed with an increase in the number of CD8+ cells and are all thought to be involved in this immune suppression. It has also been shown that certain cytokines modulate male sex hormone production. Orava (1989) demonstrated that productions of interferons (IFNs) inhibit testosterone production *in vitro* in porcine leydig cells. Meikle et al. (1992) showed that IFN-γ, interleukin-2 and tumor necrosis factor-α when used as therapy to treat chronic viral hepatitis, serum testosterone levels decreased. All these results clearly suggest that not only does male hormone inhibit immune responses, but also that immune responses can in turn modulate hormone production, resulting in the formation of an endocrine immune circuit. However, it is well known that malaria stimulate the immune system (Roitt, 1993) which may explain the low levels of testosterone in the present study which is in agreement with all these findings. On the other hand, the effect of testosterone on protein biosynthesis may be mediated by stimulation of intramuscular insulin like growth factor-1 (IGF-1) system (Fahey, 1998) and it is well known that cortisol inhibits (IGF-1) expression by stimulating the release of somatostatin (Resumo, 1998). In the present study, increased levels of cortisol and decreased levels of testosterone may demonstrate the low levels of total proteins in heavy infection patients.

Benten et al. (1992) stated that the influence of gender itself on the severity of malaria in human remains a matter of contention, the impact is clearly evident in certain rodent models of disease. Benten et al. (1997) found that following infection with *P. chabaudi*, female C57BL/10 mice self-cure, whereas male mice do not. The susceptibility of male mice is dependent on testosterone, as castration prior to infection makes male mice resistant, whereas testosterone administration to females impairs their ability to self-cure. Cernetich et al. (2006) examined the hormonal and immunological mechanisms that mediate sex differences in susceptibility to malaria infection, intact and gonadectomized (gdx) C57BL/6 mice were inoculated with *P. chabaudi* AS-infected erythrocytes and the responses to infection were monitored. In addition to reduced mortality, intact females recovered from infection-induced weight loss and anemia faster than intact males.

Effect of *P. falciparum* malaria on plasma cortisol

Resumo (1998) reported that when persons experience stress, our bodies release cortisol. He also reported that levels of cortisol in serum sample collected from malaria patients were significantly higher than those of normal subjects. He concluded that corticosteroid may interfere with initial response of *P. falciparum*-infected patients to treatment. In the present study also cortisol showed higher levels in the whole infected patients as one group and the two groups of patients compared to the control group. The highest cortisol level in his study was observed in the group of two-cross parasitemia and was significantly ($p < 0.05$) higher compared to the control group. These findings suggested that cortisol levels were increased with the increase of the degree of parasitemia and reach significant ($p < 0.05$) levels in heavy infection. Hilary (2002) reported that infection with *P. falciparum* malaria increases the secretion of pro-inflammatory hormones and mediators which induce resistance to cortisol, such as tumor necrosis factors (TNFs) and antimicrobial agents, also the infection reduce the synthesis of cortisol receptors that increase plasma cortisol level. It is also well known that the degree of parasitemia has effects on the immune system (Deans and Scohen, 1983). Very high levels of cortisol found in patients of two-cross parasitemia in the present study are in agreement with all these findings (Table 1).

Conclusions

This study showed the clear effect of falciparum malaria infection and degree of parasitemia on the levels of testosterone and cortisol. Though the number of the studied subjects was very little, still the differences between groups were very significant. The results showed that the higher the parasitemia the lower the testosterone levels

and the higher the cortisol levels. This indicates that the degree of infection by the disease and the produced cytokines burden greatly on testosterone production. This can influence its effect on the HPA axis hormonal system and result in higher cortisol levels in malaria patients, increasing with the increase of parasitemia. Though findings concerned plasma proteins cannot easily explained, both hormones studied are known to influence protein metabolism and immunity. The influence of gender itself on the severity of malaria in humans remains a matter of contention and is an area of research that is likely to grow. Therefore, careful consideration should be given to these factors when designing and operating programs to study falciparum malaria infection.

Authors' contributions

Nabiela M. ElBagir conceived, designed and coordinated the study. M. A. Abdagalil obtained and prepared data and contributed to policy and manuscript preparation. All authors read and approved the final manuscript.

ACKNOWLEDGEMENTS

Authors acknowledge Dr. Abd ElWahab, Faculty of Agriculture, University of Khartoum, for assistance with the statistical analysis.

REFERENCES

Abdelgadir NE (2002). Effect of plasmodium falciparum malaria on some plasma proteins and immuoglobulins. M.Sc. research. Department. of Biochemtry. Veterinary Medicine. University of Khartoum Khartoum North Sudan.

Adebisi SA, Soladoye AQ, Adekoya D, Odunkanmi OA (1998). Serum protein fractions of Nigerian with plasmedium infection: ILRON Experience pp. 82-84.

Adeosun OG, Oduola T, Akanji BO, Sunday AM1, Udoh SJ, Bello IS (2007). Biochemical alteration in Nigerian children with acute falciparum malaria Afr. J. Biotechnol. 6 (7): 881-885.

Bams JL, Miranda DR (1985). Outcome and costs of intensive Care. Intensive. Med. 11: 234-241.

Bello SO, Muhammad BY, Bello AY, Ukatu AI, Ahmad BM, Adeneye AA, Cherima JY (2005). The pattern of infection and in vivo response to Chloroquine by uncomplicated Plasmodium falciparum malaria in northwestern Nigeria. Afr. J. Biotechnol. 4 (1): 79-82.

Benten WP, Bettenhaeuser U, Wunderlich F, Massmann H (1991). Testosterone induced abrogation of self-healing of P. Chabaudi malaria. Infection and Immunity 59: 4486-4490.

Benten W, Wunderlich F, Mossman H (1992). Testosterone-induced suppression of self-healing *Plasmodium chabaudi*: and effect not mediated by androgen receptors. J. Endocrinol. 135: 407-413

Benten WPM, Ulrich P, KuhnVelten WN, Vohr HW, Wunderlich F (1997). Testosterone-induced susceptibility to *Plasmodium chabaudi* malaria: persistence after withdrawal of testosterone. J. Endocrinol. 153: 275-281

Bessey PQ, Lowe KA (1993). Early hormonal changes effect the Catabolic response to trauma. Ann. Surg. 218: 476-489.

Bessey PQ, Watters, JM, Aoki TT, Wilmore DW (1984). Combined hormonal infusion stimulate the metabolic response to injury. Ann. Surg. 200: 264-281.

Cannon JG, Tomokins RG, Gelfond JA, Michic HR, StanfordGG, Vander Meer JW, Lonnemann GJ, Corsetti B, Chemow,

Wilmore DW, Wolff SM, Burke JF, Dinarello CA (1990). Circulating interleukin-1 and tumor necrosis factor In septic shock and experimental endotoxin fever, J. Infect. Dis. 161: 79-84.

Cernetich A, Garver LS, Jedlicka AE, Klein PW, Kumar N, Scott A L, Klein SL (2006). Involvement of Gonadal Steroids and Gamma Interferon in Sex Differences in Response to Blood-Stage Malaria Infection. Infect. Immunity 6(74): 3190-3203.

Dayachi F, Kabongo L, Ngoie K (1991). Decreased mortality from Malaria in children with symptomatic HIV infection. Int. Cont. AIDS 2: 164.

Chang FC, Herzog B (1976). Burn morbidity: A follow-up Study of physical and psychological disability. Ann. Surg. 183: 34-37.

Deans JA, Scohen (1983). Immunology of malaria. Ann. Rev. Microbiol. 37: 25-49.

Doumas B, Waston W (1971). Clin. Chim. Acta 13: 87. Publisher Spinreact, S.A. Spain.

Fahey TD (1998). Anabolic-androgenic steroids: mechanism of action and effects on performance. Encyclopedia of Sports Medicine and Science. Internet Society: http: 11 sportsci.org.

Garcia M, Marlborough D (1996). Arapid immunochromatoraphic test (ICT) for the diagnosis of Plasmodium Falciparum malaria. J Parasitic Dis.20 (1):64.

Golden M (1982). Transport proteins as indices of protein status. Am. J. Clin. Nutr. 35: 1159-1165.

Henary RJ, Cannon DC, Winkelam (1974). Anal. Chem. 92. 1491. Publisher USA edition Harper and Rwo.

Hilary T (2002). Sex hormones' link to stress. UBC reports 48(5).

Hill AG, Jacobson L, Gonzalez J, Rounds J, Majzoub JA, Wilmore DW (1996). Chronic central nervous system exposure to interleukin - 1 β causes catabolism in the rat. Am. J. Physiol. 271: 1142-1148.

Ling PR, Schwatz JH, Bistrain BR (1997). Mechanism of host wasting induced by administration of cytokines in rats. Am. J. Physiol. 272: 333-339.

Lunn JS, Chin W, Contacos PG, Coatney GR (1966). Changes in antibody titres and serum protein fractions guring the course of 107 prolonged infection with vavax or falciparum malaria. Am. J. Trop. Med. Hyg. 15: 1-13.

Mac SRNM, Whaley K (2001). Muirs text book of pathology .Thirteen edition Arnold-Hodderline Group. London. Co-published in USA by Oxford University press, Inc, New York.

Matsuda Y, Ogushi F, Ogaw K, Katunuma N (1986). Structure and properties of albumin. J. Biochem, 100, 375-379.

Meikle AW, Gardoso Dacosta N, Samlowski WE (1992). Direct and indirect effects of murine IL-2, -INF, and TNF on testosterone synthesis. J. Androl. 13: 437-443.

Michie HR, Manogue KR, Springs DR, Revhaug A, Dwyer SO, Dinarello CA, Cerami A, Wolff SM, Wilmore DW (1988). Detection of circulating tumor necrosis factor after endotoxin administration. N. Engl. J. Med. 318: 1481-1486.

Mitch WE, Goldberg AL (1996). Mechanism of muscle wasting. N. Engl. J. Med. 335: 1897-1905.

Orava M (1989). Comparison of the inhibitory effects of interferons on testosterone production in porcine leydig cell. Culture. J. interferon Res.9:135-141.

Prasad JA (1979). Method of hormone radioimmunoassay. Department of isotope, China Institute of Atomic Energy. Beijing 102413.

Resumo DT (1998). Evaluation of both cortisol and dehidroepiondro-sterone levels in patients with non-complicated malaria due to plasmodium falciparum Soc. Bra. Med. Trop. 13 (2): 243-244.

Robert W, SnowCarlos A, Guerra AM, Noor HY, Myint SI, Hay (2004). The global distribution of clinical episodes of plasmodium falciparum malaria. Nature.434: 214-217.

Roitt K (1993). Immunoglobulins structure and function. Immunol. 5th Ed. pp. 6-7, 109

Soini E, Kojala H (1983). Time-resolved flourometer for lanthanide chelates - Clin. Chem. 29: 65-68.

Stephen WJ (1999). Current st5atus of lestosterone replacement therapy in men. Arch Fam. Med. 8: 257-263.

Watters JM, Bessey PQ, Dinarello CA, Wolf SM, Wilmore DW (1986). Both inflammatory and endocrine mediators stimulate host responses to sepsis. Arch. Surg. 121: 79-190.

World Health Organisation (WHO) Report (1980). Identification of the four species of malaria parasites in blood films in: WHO ed. Manual of basic techniques for a health laboratory, Geneva Chap. 21:196-203.

Zhi Z, Chen L, Saito S, Kanagawa O, Sendo F (2000). Possile modulation by male sex hormone of Th1/Th2 function in protection against plasmodium chabaudi. As infection in mice. Exp. Parasitol. 96: 121-129.

29

The prevalence of malaria infection in pregnant women living in a suburb of Lagos, Nigeria

Raimi, O. G.[1,2]* and Kanu, C. P.[2]

[1]Division of Biological Chemistry and Drug Discovery, College of Life Sciences, University of Dundee, Scotland, DD1 5EH, United Kingdom.
[2]Tropical Disease Research Laboratory, Department of Biochemistry, Lagos State University, PMB 1087, Apapa, Lagos, Nigeria.

Malaria during pregnancy continues to be a major health problem in endemic countries with clinical consequences including death of both mother and child. In Nigeria, statistics shows that as many as 300,000 lives especially those of children and pregnant women are lost annually to malaria. This study was aimed at assessing the prevalence of malaria among pregnant women living in Ojo Local Council, a suburb of Lagos Nigeria, which is characterized by unstable transmission of malaria. 50 pregnant women attending the antenatal clinic of the popular saint Anthony Hospital and Maternity Home in Ojo Local Government Area of Lagos were recruited for this study. This study was carried out from April 2009 to September 2009 (period of high malaria transmission). The result showed that malaria infection was prevalent during pregnancy. A total of 26 (52%) of the pregnant women were positive to and showed symptoms of malaria while 24 (48%) were negative and showed no symptoms of malaria. Primigravidae were more susceptible to the parasite especially *Plasmodium falciparum* with mean parasite density of 2112.50 ± 420.90 (parasite/μl) than the multigravidas with parasite density 446.70 ± 296.90 (parasite/μl). The results showed the prevalence of malaria infection especially *P. falciparum* infection in pregnant women living in the area and that the younger women were more at risk. Malaria should therefore be recognized as a global priority in health care more so in pregnancy.

Key words: Malaria, pregnancy, prevalence, Lagos.

INTRODUCTION

Malaria remains the single most important infection causing morbidity and mortality in the world and is second only to Mycobacterium tuberculosis as the single most important infection agent (Greenwood, 1997a). It is one of the biggest impediments to progress in Africa and is the biggest killer in Africa, with 90% of the global malaria deaths occurring in this continent (Bulter, 1997). It is responsible for one in four deaths below the age of 5 years and could most times lead to miscarriage at the early stage of pregnancy (Bulter, 1997). In the endemic countries of Africa, children under the age of five and pregnant women bear the brunt of the burden of malaria disease, this is because they have lower immunity to the disease compared to other people in the same environment

(Raimi et al., 2004; Molyneux et al., 1989). In certain locations, the malaria situation is deteriorating as a result of environmental changes, including global warming, civil disturbances, increasing travel and increasing drug resistance (Greenwood, 1997a). According to World Health Organization report, malaria ranks 7th in the rank of leading selected causes of mortality with fatality rate put at 1.5 to 2.7 million annually while it comes second among the leading selected causes of morbidity with about 300 to 500 million people reporting to hospital due to the infection (WHO, 1997b).

Maternal mortality is twice in pregnant malaria women than among non-pregnant patients with severe malaria (Brain, 1998). Anaemia is the most common symptom of malaria in pregnancy and usually develops during the second trimester (Brain, 1998). Cerebral malaria is rare in adults except during pregnancy and is responsible for many maternal malaria deaths (Macleod, 1998).

---

*Corresponding author. E-mail: wale.raimi@gmail.com.

Severe *falciparum* malaria may cause deformities in the genital tract to make conception impossible or if conception does occur it may prevent normal implantation and development of the placenta (Burrow and Ferris, 1988).

Although so much work have been published on the prevalence of malaria in major cities of Nigeria but little information is available about the prevalence of this disease in the suburbs or outskirts of major cities where although transmission is unstable but high as a result of topography, attitude, rainfall, poor drainage system and high human-vector contact to mention a few. This work was therefore aimed at assessing the prevalence of malaria in pregnant women living in a suburb of Lagos, Lagos Nigeria.

MATERIALS AND METHODS

Data collection

A total of 50 pregnant women attending the antenatal clinic of the Saint Anthony Hospital in Ojo, were recruited after their consent has been sought. Questionnaires were then administered requesting information on age, parity and gestational age. Gestational age was calculated from the last menstrual period and confirmed by ultrasound scan and physical examination was conducted for symptoms of malaria.

Collection of blood and blood smear

2.0 ml of blood was collection from an ante-cubital vein of each woman. A minute quantity of blood obtained with the syringe was placed on a clean grease free slide that has been labeled. Thick films were prepared and the films were left to dry.

Staining / microscopy procedure

The dried slides were stained using 12% field stain A and B. The slides were examined under an x100 oil immersion objective of the light microscope. The thick film was used for detection and counting of the malaria parasite density.

Parasite density determination

Parasiteamia was measured by counting the number of the asexual parasite against the number of the leucocytes in the thick blood film. The numbers of asexual parasites were counted against 200 leukocytes. The parasite density was calculated by dividing the number of parasite by the number of leukocytes and multiplied by 6000, assuming each woman has 6000 leucocytes/μl.

RESULTS AND DISCUSSION

In this study, out of the 50 pregnant women, 52% were positive to and show symptoms of malaria while 48% were negative (Table 1). 84.6% of the infection was found to be due to *Plasmodium falciparum* while *Plasmodium malariae* was 15.4% of the women. Pregnant women

Table 1. Age distribution of pregnant women positive to malaria and those without malaria.

Age group (year)	Positive (%)	Negative (%)
20 - 30	16 (32)	14 (28)
31 - 40	10 (20)	9 (18)
Above 40	-	1 (2)
Total	26 (52)	24 (48)

Data as number (%).

within the age group of 20 to 30 years had the highest number of parasite density while those above 40 years had the least (Table 2). Primigravidae were found to be susceptible more to malaria with parasite density of 2112.50 ± 420.90 parasites/μl (Table 3). Women in their second trimester of pregnancy were found to have the highest parasite load (Table 4).

This study revealed a malaria prevalence of 52% in pregnant women living in this part of Lagos. Pregnant women within the age bracket of 20 to 30 years recorded the highest number of positive result while those of the age group of above 40 years recorded the lowest or no result at all (Table 2). This result supports the existing knowledge that high prevalence at lower ages and low prevalence at higher ages is due to the existence of natural immunity to infectious disease including malaria (Oduola et al., 1992; Rogerson et al., 2000; Bouyou-Akotet et al., 2003) which the pregnant women acquires as the age increases. Lander et al. (2002) however reported no significant association between malaria infection and maternal age (Lander et al., 2002).

An analysis of malaria in pregnancy in Africa revealed that parasiteamia is significantly common and heavier in primigravidae than multigravidae (McGregor, 1984). This study showed high level of infection in primigravidae (Table 3). This is because in an area where transmission is high and the level of acquired pregnancy immunity against malaria is expected to be significant, primigravidae is more affected (Brain, 1998; McGregor, 1984). Women in the second trimester had the highest level of parasiteamia (Table 4), which is line with other studies where the highest level of parasiteamia was recorded at the second and early third trimester (Menendez, 1995; Nosten et al., 1991).

Conclusions

The main epidemiological factor to *P. falciparum* infection in pregnancy should be considered in relation to the endemic malaria conditions under which women are living. Pregnancy is also one of the factors affecting the rate of malarial parasite infection in women living in malaria endemic communities. Malaria should therefore be recognized as a global priority in health care more so

Table 2. Mean ± SEM malaria parasite density in relation to age of pregnant women.

Age group (year)	Positive sample	Mean parasite density ± SEM (parasites/µl)
20 - 30	16	937 ± 331.50
31 - 40	10	166.80 ± 539.40
Above 40	-	-

SEM: Standard error of mean.

Table 3. Mean ± SEM malaria parasite density in relation to gravidity of pregnancy.

Gravidity	Number	Positive sample	Mean parasite density ± SEM (parasites/µl)
Primigravidae	18	12 (66.67%)	2112.50 ± 420.90
Multigravidae	32	14 (43.75%)	446.70 ± 296.90

SEM; Standard error of mean. Data as number or number (%) as appropriate.

Table 4. Mean ± SEM malaria parasite density in relation to trimesters of pregnancy.

Trimester	Number	Positive sample	Mean parasite density ± SEM (parasites/µl)
First	14	10 (71.43 %)	885.60 ± 364.10
Second	20	11 (55 %)	1913 ± 554.70
Third	16	5 (31.25 %)	577.40 ± 320.60

SEM; Standard error of mean. Data as number or number (%) as appropriate.

in pregnancy.

ACKNOWLEDGEMENTS

The authors would like to thank Dr. Osakwe, the Medical Director of Saint Anthony Medical Center and Maternity, Ojo, Lagos for allowing us the use of his hospital for the study. We also want to thank all the patients for their kind cooperation.

REFERENCES

Bouyou-Akotet MK, Ionete-Collard DE, Mabika-Manfoumbi M, Kendjo E, Matsiegui PB, Mavoungou E, Kombila M (2003). Prevalence of *Plasmodium falciparum* infection in pregnant women in Gabon. Malar. J., 2: 18.

Brain BJ (1998). An analysis of Malaria in Pregnancy in Africa. Bull. World Health Org., 61: 1005-1016.

Bulter D (1997). Time to Put Malaria Control on the Global Agenda. Nature, 386: 535-541.

Burrow NG, Ferris FT (1988). Medical complications during pregnancy, 3rd Edition, WB. Saunders Company, pp. 34-37, 425-427, 320-321.

Greenwood BM (1997a). The Epidemiology of Malaria. Ann. Trop. Med. Parasitol., 91: 763-769.

Lander J, Leroy V, Simonon A, Karita E, Bogaerats J, Clercq AD, Van de Perre P, Dabis F (2002). HIV infection, malaria and pregnancy: A prospective cohort study in Kigali, Rwanda. Am. J. Trop. Med. Hyg., 66: 56-60.

Macleod C (1988). Parasitic Infections in pregnancy and New born. Oxford Medical Publishers, pp. 10-25.

McGregor IA (1984). Epidemiology, Malaria and Pregnancy. Am. J. Trop. Med. Hyg., 33: 517-525.

Menendez C (1995). Malaria during pregnancy: A priority area of malaria research and control. Parasitol. Today, 11: 178-183.

Molyneux ME, Taylor TE, Wirima JJ, Borgstein A (1989). Clinical Features and Prognostic indicators in pediatric cerebral malaria: A study of 131 comatose Malawian children. QJM, 71: 369-371.

Nosten F, ter Kuile FO, Maelankirri L, Decludt B, White NJ (1991). Malaria during pregnancy in an area of unstable endemicity. Trans. Res. Soc. Trop. Med. Hyg., 85: 424-429.

Oduola AM, Sowunmi WR, Kyle DE, Martin RK, Walker O, Salako LA (1992). Innate resistance to new anti-malaria drugs in *Plasmodium falciparum* from Nigeria. Trans. Royal Soc. Trop. Med. Hyg., 86: 123-126.

Raimi OG, Elemo BO, Raheem L (2004). Malaria in Pregnancy: Serum Enzyme Level in Pregnant Malarial Patients in Lagos Nigeria. J. Sci. Technol. Res., 3(3): 60-63.

Rogerson SJ, Van den Broek NR, Chaluluka E, Qongwane C, Mhango CG, Molyneux ME (2000). Malaria and anemia in antenatal women in Blantyre, Malawi: A twelve-months survey. Am. J. Trop. Med. Hyg., 62: 335-340.

World Health Organization (WHO) (1997b): Malaria in tropics disease research. 13th programme report. WHO, Geneva, pp. 40-61.

30

Direct electron transfer of hemoglobin on nickel oxide nanoparticles modified graphite electrode

Rezaei-Zarchi S.[1], Imani S.[2]*, Javid A.[3], Zand A. M.[4], Saadati M.[5] and Zagari Z.[6]

[1]Department of Biology, Payam-e-Noor University, Yazd, Iran.
[2]Department of Biology, Basic Science Faculty, IHU, Tehran, Iran.
[3]Institute of Biochemistry and Biophysics, University of Tehran, Tehran, Iran.
[4]Departments of Biology, Basic Science Faculty, IHU, Tehran, Iran.
[5]Applied Biotechnology and Environmental Research Center, Baqiyatallah Medical Science University, and IHU, Tehran, Iran.
[6]Department of Biology, Payam-e-Noor University, Tehran, Iran.

Direct electron transfer of hemoglobin, immobilized on a nickel oxide nanoparticles modified graphite electrode, was studied. Nickel oxide nanoparticles synthesized by electrochemical methods. The prepared nanoparticles were characterized by scanning electron microscope (SEM) and transmission electron microscope (TEM). The resulting electrode displayed an excellent redox behavior for the hemoglobin. The hemoglobin showed a quasi-reversible electrochemical redox behavior with a formal potential of -48±5 mV (versus Ag/AgCl) in 50 mM potassium phosphate buffer solution at pH 7.0 and temperature 25°C. The cathodic transfer coefficient was 0.45 and electron transfer rate constant was evaluated to be 1.95 s^{-1}. Furthermore, the modified electrode was used as a biosensor and exhibited a satisfactory stability and sensitivity to H_2O_2. The linear range of this biosensor for H_2O_2 determination was from 15 to 650 µM while standard deviation in 40 µM H_2O_2 concentration was 2.8% for 4 repetitions.

Key words: Electron transfer, hemoglobin, nickel oxide, nanoparticles, biosensor.

INTRODUCTION

The alteration of electrode surface with the use of nanostructure materials as a mediator, in a word nanofabrication, is advantageous for the achievement of direct electron transfer (ET) between the biomolecule and the electrode (Sanz et al., 2005). The direct ET can be difficult to attain, because the prosthetic groups of the biomolecules are burried deeply in the biomolecules (Campuzano et al., 2003). Nanomaterials are commensurate in size to proteins and the multivalent functionalization on their surfaces holds great promise for controlling the biomolecular recognition (Salimi et al., 2006).

Direct ET has extended beyond the field of bioelectrochemistry. It is a new interdisciplinary area,

which combines biotechnology with electrochemical science and focuses on the structural organization and electron transfer functions of biointerfaces on electrode surfaces (Giovanelli et al., 2003). Bioelectrochemistry has proven to be both a useful way to understand the electrochemical properties and principles of biomolecules as well as a powerful method for the exploitation of these biomolecules at the biointerfaces of biosensors and bioelectronics (Dempsey et al., 2004; Gopel and Heiduschka, 1995).

Since that time, electrochemical devices have opened up new possibilities for studying the redox process of heme proteins (Gopel and Heiduschka, 1995; Turner et al., 1987). Many reports have described the electro-chemistry of heme proteins in terms of modifier-electrode and modifier-protein interactions (Campuzano et al., 2003; Turner et al., 1987). Many promoters, like some small organic compounds, amino acids together with some derived molecules, small peptides and conductive

---

*Corresponding author. E-mail: imani.saber@yahoo.com.

polymers, have been found to promote the direct electrochemistry of heme proteins on the electrode surface. Recently, the matrices used for the heme proteins immobilization are a series of inorganic porous and nanomaterials (Dempsey et al., 2004; Salimi et al., 2006). The study of direct electron transfer between protein and electrode represents not only a basic feature for the application of biocatalysts in chemical sensors and other electrochemical devices (Dempsey et al., 2004; Campuzano et al., 2003), but may also provide a model for the investigation of electron transfer processes in biological systems (Salimi et al., 2006; Turner et al., 1987).

In our previous studies, we had demonstrated that the nanosilver-modified graphite electrode could be used in the electrochemical ligand binding investigations of the hemoglobin. We had also revealed that the nanosilver facilitates electron transfer between hemoglobin and the surface of graphite electrode (Rezaei-Zarchi et al., 2007).

Here, we investigate the electrochemical behavior of hemoglobin in the presence of nickel oxide nanoparticle-modified graphite electrode and the design of a biosensor for the determination of hydrogen peroxide by Hb-immobilization onto the surface of modified graphite electrode.

EXPERIMENTAL

Reagents

Hemoglobin (Hb) and finer alumina powder was purchased from Sigma, St. Louis, USA. The phosphate buffer solution (PBS) consisted of a potassium phosphate solution (KH_2PO_4 and K_2HPO_4 from Merck, 0.05 mol L^{-1} total phosphate) at pH 7.0. An acetate buffer solution (CH_3COONa and CH_3COOH from Merck, 0.10 mol L^{-1}) was freshly prepared. $Ni(NO_3)_2.H_2O$ and the other reagents were reagent grade materials from Merck. Deionized water was used to prepare all solutions and to rinse the electrodes double distilled water was used to prepare all solutions and to rinse the electrodes. Stock solutions were stored at 4°C.

Apparatus and measurements

Electrochemical measurements were carried out with a potentiostat/ galvanostat (Model 263A, EG&G, USA) using a single compartment voltammetric cell, equipped with a platinum rod auxiliary electrode, an Ag/AgCl reference electrode (Metrohm & Co., Leinfelden-Echterdingen, Germany) and a nickel oxide nanoparticle-modified graphite electrode with a disk diameter of 2.0 mm as the working electrode (Azar Electrode Co., Orumiah, Iran). A three-electrode cell was also used, employing a graphite electrode or modified-graphite electrodes, acting as the working electrodes. A platinum wire was applied as the counter electrode. All potentials were reported with respect to this reference. All experiments were performed at 25±1°C. The experimental solutions were de-aerated using high purity nitrogen for 30 min and a nitrogen atmosphere was kept over the solutions during the measurements. All the electrochemical measurements were carried out in 0.05 M PBS, pH 7.0, at 25 ± 1°C.

Scanning electron microscopic images were recorded using a ZEISS DSM 960, while atomic force microscopic studies were performed with the help of a DME (controller, Dual Scope C-21) and a scanner (DS 95-50).

Preparation of Hb/nickel oxide NP/graphite electrode

Prior to the Hb immobilization, the surfaces of graphite electrode (1 mm in diameter) and alumina slurry (1.0, 0.3 and 0.05 μm) were polished, formerly to a mirror-finish with fine emery papers and then with a polishing cloth, thoroughly rinsed with de-ionized water. The electrode was then successively sonicated in ethanol and double distilled water to remove the adsorbed particles. Then, cyclic scans were carried out in PBS (0.05 mol L^{-1}, pH 7.0) in the potential range from -0.50 to 1.0 V, until the repetitive cyclic voltammograms (CVs) were obtained. The solution, in which the nickel deposition was conducted, typically consisted of 15 ml acetate buffer (pH 4.0). The nickel was initially electrodeposited (-0.80 V, 5 min. deposition time) on a graphite electrode from a 1 m ml^{-1} nickel nitrate pH 4.0 acetate buffer solution. Afterwards, the Ni– graphite electrode was placed into a fresh PBS (pH 7.0) and electrochemically passivated with the potential cycling method (protocol for the NiO NPs/ graphite electrode) (Gopel and Heiduschka, 1995; Laviron, 1979).

The NiO NPs/ graphite electrode was placed into a fresh PBS including 5 mg ml^{-1} Hb (pH 7.0, 3 to 5°C) for 8 h. At the end, the modified electrode was washed in deionized water and placed in PBS (PH 7.0) at a refrigerator (3 to 5°C), before being employed in the electrochemical measurements as the working electrode.

RESULTS AND DISCUSSION

Figure 1a shows the SEM image of graphite electrode surface before the construction of nickel oxide nanoparticles. Figure 1b illustrates the SEM image of nanometer-scale nickel oxide particles, generated onto the graphite electrode surface in various sizes. Figures 1c and d shows the TEM images of nickel oxide nanoparticles after being scraped from the electrode surface. Because the surface-to-volume ratio increases with the decreasing size, the smaller NPs are able to play a very important role during the immobilization process.

The electron transfer of the proteins, at the bare electrodes, is very slow so that the redox peak of proteins can usually not be observed (Dempsey et al., 2004; Gopel and Heiduschka, 1995). Figure 2a shows a typical cyclic voltammogram (CV) of the NiO-NPs/ graphite electrode. Figure 2b shows a cyclic voltammogram of a Hb/NiO-NPs/graphite electrode in 50 mM phosphate buffer at pH 7.0. The Hb showed quasi-reversible electro-chemical behavior with a formal potential of -48 ± 5 mV (vs Ag/AgCl), cathodic and anodic peaks were not observed using the bare graphite electrode. This shows that NiO-NPs acts as a facilitator of electron transfer from the redox species of Hb to the electrode surface and vice versa. These results are in line with the previous work that explains the behavior of nanoparticles as the facilitators of electron transfer (Rezaei-Zarchi et al., 2007).

To further investigate the Hb characteristics at the Hb/NiO-NPs/graphite electrode, the effect of scan rates on the Hb voltammetric behavior was studied in detail. The baseline subtraction procedure, for the cyclic

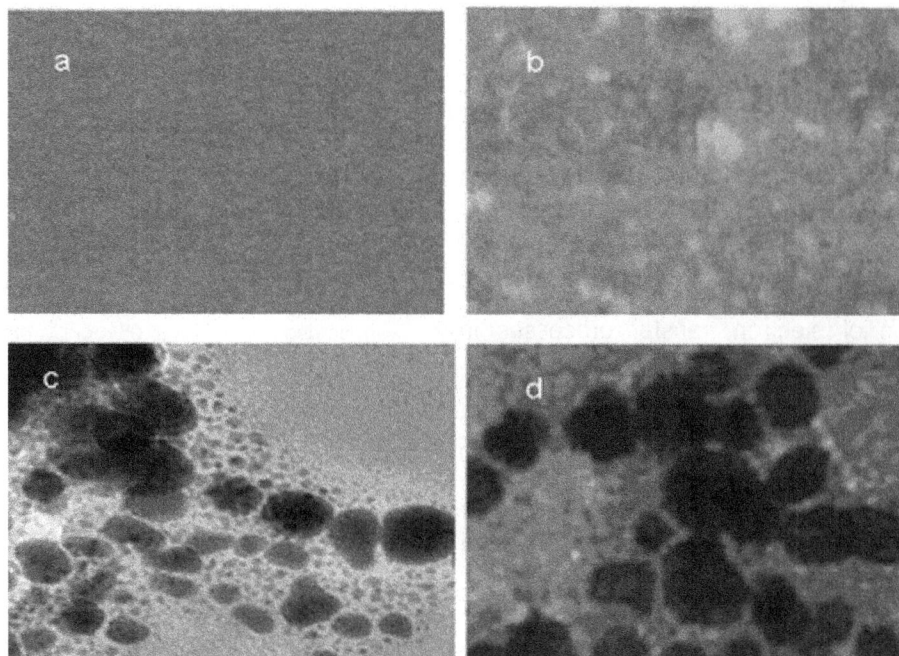

Figure 1. (a) SEM image of the graphite electrode surface before the construction of the NiO NPs. (b) The SEM image of electrodeposited NiO NPs on the graphite electrode surface, (c and d) TEM images of NiO NPs after being scraped from the electrode surface.

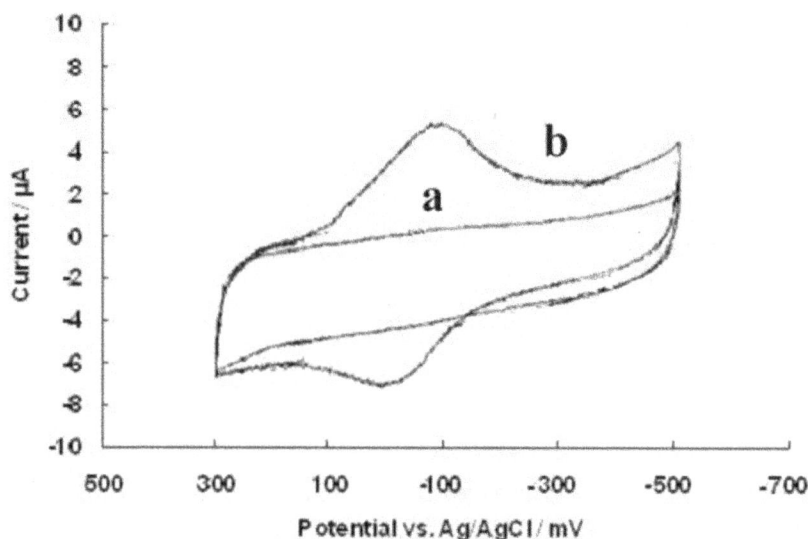

Figure 2. Cyclic voltammograms, using (a) the NiO NPs graphite electrode in 50 mM phosphate buffer and (b) Hb/ NiO -NPs/graphite electrode in 50 mM phosphate buffer (scan rate: 100 mV) s^{-1}).

voltammograms, was obtained in accordance with the method reported by Bard and Faulkner (2001). A linear dependence of anodic and cathodic peak currents on the scan rates is shown in Figures 3a and b. It can be seen that the redox peak currents increased linearly with the scan rate, the correlation coefficients were 0.993 and 0.984, respectively and high value of slopes were

obtained. This phenomenon suggested that the redox process was an adsorption-controlled one and the immobilized Hb was stable and was highly efficient. Also no decrease in the peak current was observed after repeated cycles of this experiment.

These findings indicate that Hb is strongly adsorbed onto the surface of modified electrode. As could be seen,

A

B

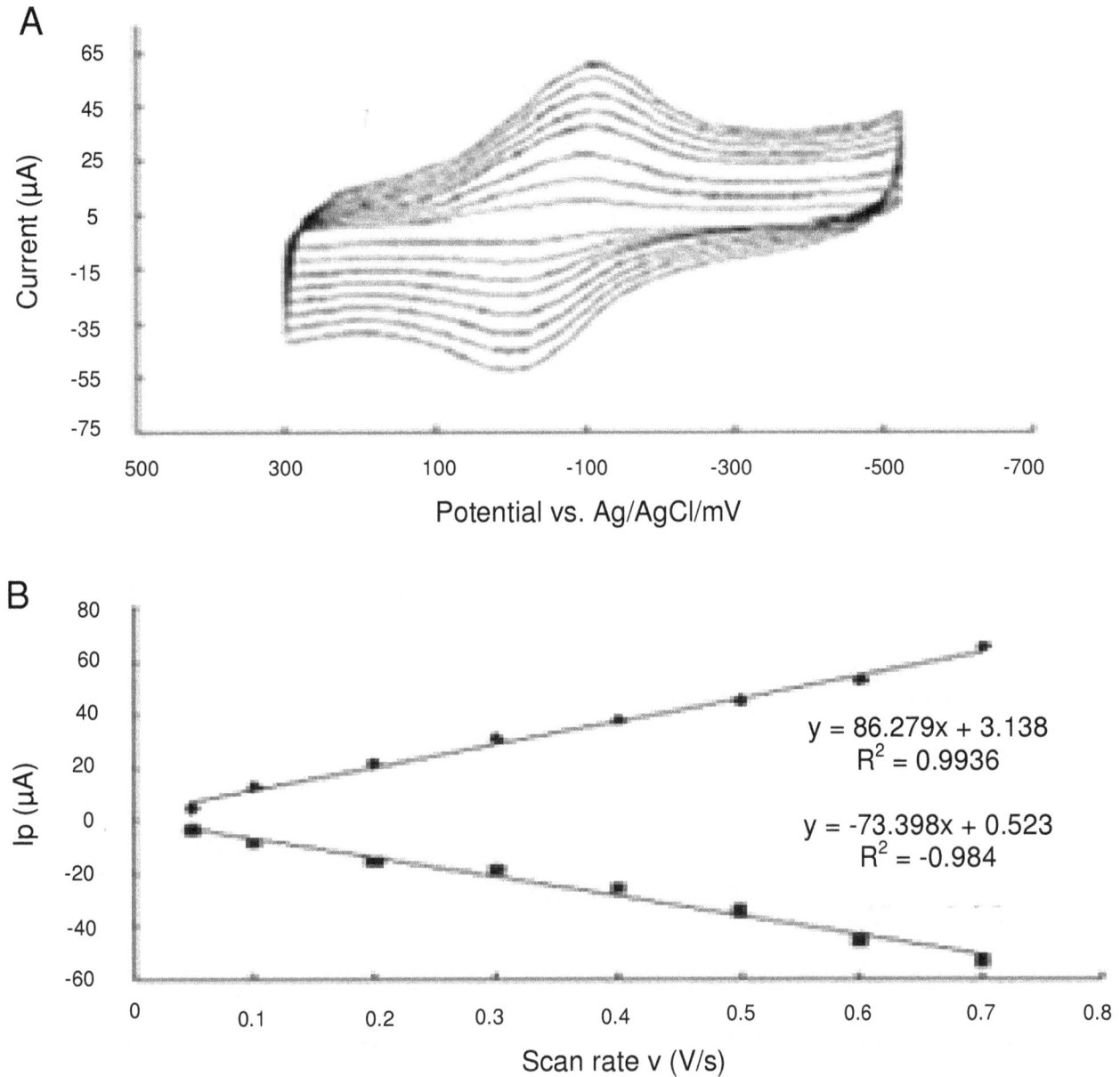

Figure 3. (a) Typical cyclic voltammograms of Hb/NiO-NPs/ graphite electrode at different scan rates. The voltammograms (from inner to outer) designate scan rates of 50, 100, 200, 300, 400, 500, 600 and 700 mV s^{-1}, respectively. (b) Dependence of the anodic and cathodic peak currents on the scan rates. All the data were obtained at pH 7.0 and in 50 mM phosphate buffer solution.

in the range from 400 to 1000 mV s^{-1}, the cathodic peak potential (E_{pc}) changed linearly versus ln v with a linear regression equation of y Z= 0.0424x + 0.3063 , r = 0.957. According to the following equation (Laviron, 1979):

$$E_p = E^{0'} + \frac{RT}{\alpha nF} - \frac{RT}{\alpha nF} \ln v'$$

Where α is the cathodic electron transfer coefficient, n the number of electrons, R, T and F are gas, temperature and Faraday constant,   respectively (R = 8.314 J mol^{-1}

K^{-1}, F = 96493 C/mol, T = 298 K), and αn is calculated to be 0.45. Given $0.3 < \alpha < 0.7$ in general (Ma et al., 2000), it could be concluded that $n = 1$ and $\alpha = 0.45$. From the width of the peak at mid height and low scan rate, we can also obtain $n = 1$ (Ma et al., 2000). So, the redox reaction between Hb and modified graphite electrode is a single electron transfer process. In order to calculate the value of apparent heterogeneous electron transfer rate constant (k_s), the following equation was used (Laviron, 1979):

$$\log k_s = \alpha \log(1-\alpha) + (1-\alpha)\log \alpha - \log(\frac{RT}{nFv}) - \alpha(1-\alpha)\frac{nF\Delta E_p}{2.3RT}$$

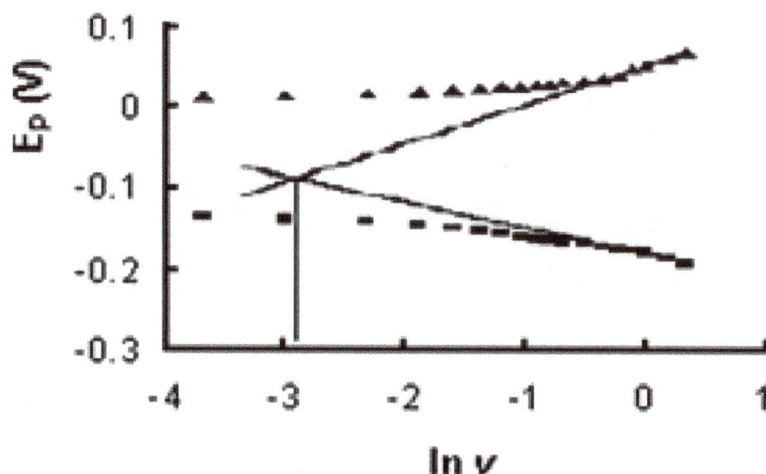

Figure 4. Relationship between the peak potential (Ep) and the natural logarithm of scan rate (ln v) for Hb/NiO-NPs/ graphite electrode in 50mM PBS (pH 7.0).

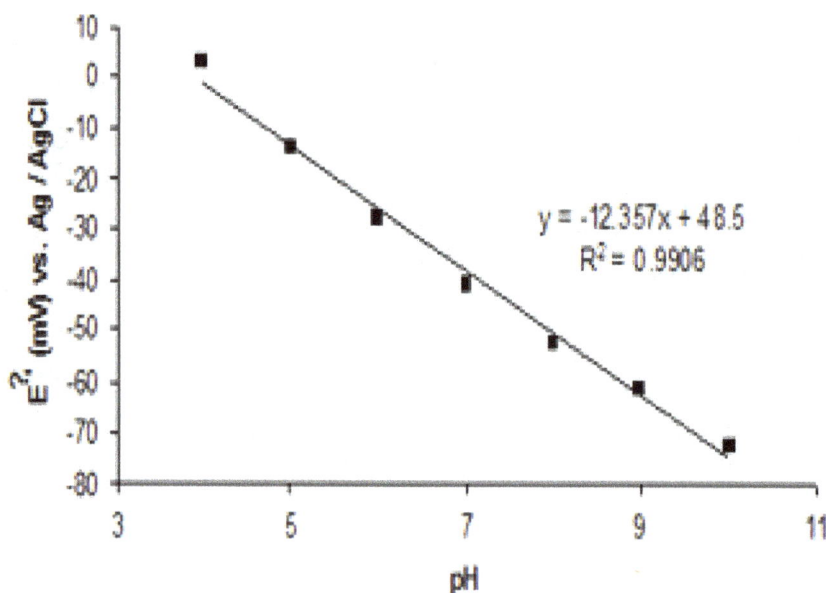

Figure 5. Effect of pH on the formal potential of Hb/NiO-NPs/graphite electrode.

According to Figure 4, in a range from 400 to 1000mV/s, the anodic peak potential (Epa) is also linear to ln v with a linear regression equation of $y = 0.0728x + 0.2411$, $r = 0.973$. Figure 4 also indicates that $nEp > 200$ mV so at cross point A, ln vc = lnva = −4.21, and ks was calculated to be 1.95 s^{-1}. This value is higher than the ks value (1.13 s^{-1}) reported for redox proteins in DNA film at graphite electrode (Cheng et al., 2000).

Figure 5 shows the formal potential of Hb, immobilized onto the NiO-NPs/ graphite electrode, in PBS has a strong dependence on the pH of solution. All the changes in the peak potentials and currents with solution pH were reversible in the pH range from 4 to 10. An increase in the solution pH caused a negative shift in both cathodic and anodic peak potentials. Plot of the formal potential versus pH (from 4 to 10) showed a line with the slope of -44 mV pH-1, which was close to the expected value of -57.8 mV pH-1 for a reversible proton-coupled single electron transfer at 291.15 K, indicating that one proton participated in the electron transfer process (Shang et al., 2003).

The cyclic voltammograms of the Hb/ NiO-NPs/ graphite electrode, in PBS, at pH 7.0, containing different concentrations of H_2O_2 are shown in Figure 6. Upon the

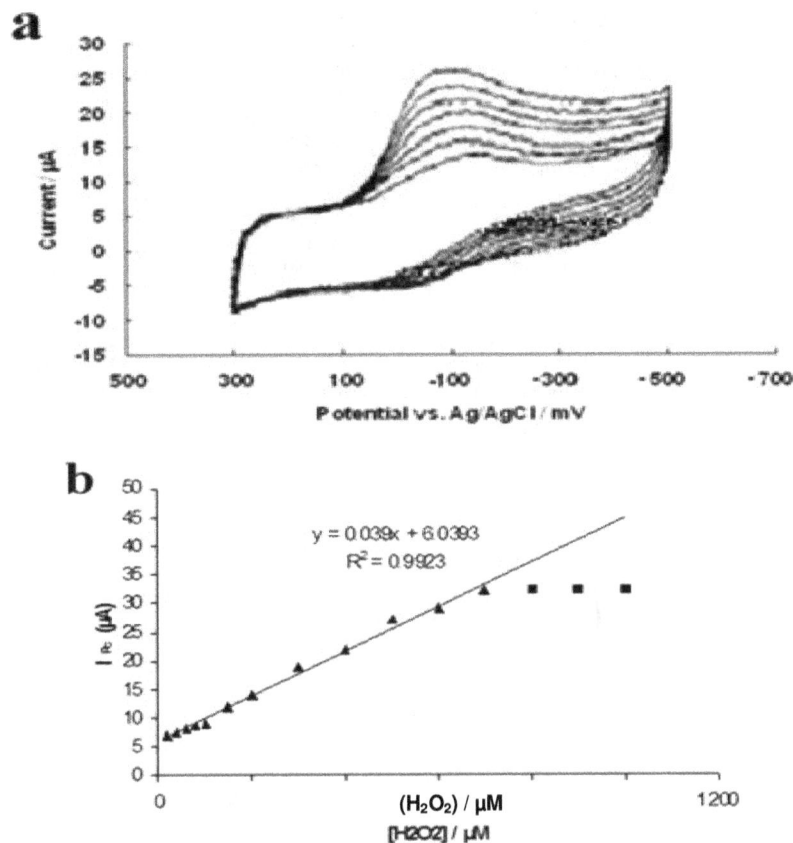

Figure 6. (a) Cyclic voltammograms obtained at an Hb/NiO-NPs/graphite electrode in 50 mM phosphate buffer solution (pH 7.0) for different concentrations of and (b) the relationship between cathodic peak current of Hb and different concentrations of H_2O_2 (scan rate: 100 mVs-)

addition of H_2O_2 to the electrochemical cell, the reduction peak current of the immobilized Hb increased, indicating a typical electro-catalytic behavior to the reduction of H_2O_2. The electro-catalytic process could be expressed as follows: (Turner, 2007) the calibration curve (Figure 6b) shows the linear dependence of the cathodic peak current on the H_2O_2 concentration in the range of 15 to 650 µM. In Figure 6 at higher concentration of H_2O_2, the cathodic peak current decreased and remains constant. This implies electrocatalytic property of electrode (Chaplin and Bucke, 1990). Thus, this experiment has introduced a new biosensor for the sensitive determination of H_2O_2 in solution.

The direct electron transfer of Hb at a bare graphite electrode is so slow that the redox peak could not be observed. The NiO-NPs significantly promotes the electron-transfer rate of Hb. It was suggested that when NiO-NPs acts as a promoter between the Hb and graphite electrode, a biocompatible microenvironment is offered. The proportionality of peak currents to sweep rate and the constancy of integration of reduction peaks at different scan rates represents the characteristics of diffusion less electro chemical behavior. When the prepared enzyme-electrode was used as a third

generation biosensor, it showed a satisfactory sensitivity and reproducibility for H_2O_2.

ACKNOWLEDGEMENTS

The financial supports of Payame Noor University, Yazd, Iran and IHU University for the present project are gratefully acknowledged.

REFERENCES

Bard AJ, Faulkner LR (2001). Electrochemical Methods, Fundamentals and Applications, 2nd ed.; John Wiley & Sons: New York.
Campuzano S, Serra B, Pedrero M, Villena F, Pingarron JM (2003). Amperometric flow-injection determination of phenolic compounds at self-assembled monolayer-based tyrosinase biosensors. J. Anal. Chim. Acta. 494: 187-197.
Chaplin MF, Bucke C (1990). Enzyme Technology, Cambridge University Press, 407 Cambridge, UK.
Cheng L, Liu JY, Dong S (2000). Layer-by-layer assembly of multilayer films consisting of silicotungstate and a cationic redox polymer on 4-aminobenzoic acid modified glassy carbon electrode and their electrocatalytic effects. J. Anal. Chim. Acta, 417: 133-142.
Dempsey E, Diamond D, Collier A (2004). Development of a biosensor for endocrine disrupting compounds based on tyrosinase entrapped

within a poly(thionine) film. Biosens. Bioelectron., 20: 367-377.

Giovanelli D, Lawrence NS, Jiang L, Jones TGJ, Compton RG (2003). Amperometric determination of sulfide at a pre-oxidised nickel electrode in acidic media. Analyst, 128: 173-177.

Giovanelli D, Lawrence NS, Wilkins SJ, Jiang L, Jones TGJ, Compton RG (2003). Anodic stripping voltammetry of sulphide at a nickel film: towards the development of a reagentless sensor. Talanta, 61: 211-220.

Gopel W, Heiduschka P (1995). Interface analysis in biosensor design. Biosens. Bioelectron. 10: 853-883.

Laviron E (1979). General expression of the linear potential sweep voltammogram in the case of diffusionless electrochemical systems. J. Electroanal. Chem. 101, 19-28.

Laviron E (1979). The use of linear potential sweep voltammetry and of a.c. voltammetry for the study of the surface electrochemical reaction of strongly adsorbed systems and of redox modified electrodes. J. Electroanal. Chem. 100: 263-270.

Ma HY, Hu NF, Rusling JF (2000). Electroactive Myoglobin Films Grown Layer-by-Layer with Poly(styrenesulfonate) on Pyrolytic Graphite Electrodes. Langmuir J. Electroanal. Chem., 16: 4969-4975.

Rezaei-Zarchi S, Saboury AA, Norouzi P, Hong J, Ahmadian S, Ganjali MR, Moosavi-Movahedi AA, Moghaddam AB, Javed A (2007). Use of silver nanoparticles as an electron transfer facilitator in electrochemical ligand-binding of haemoglobin. J. Appl. Electrochem., 37: 1021–1026.

Salimi A, Sharifi E, Noorbakhsh A, Soltanian S (2006). Direct voltammetry and electrocatalytic properties of hemoglobin immobilized on a glassy carbon electrode modified with nickel oxide nanoparticles. Electrochem. Commun., 8: 1499–1508.

Sanz VC, Mena ML, González-Cortés A, Yáñez-Sedeño P, Pingarrón JM (2005). Development of a tyrosinase biosensor based on gold nanoparticles-modified glassy carbon electrodes: Application to the measurement of a bioelectrochemical polyphenols index in wines. J. M. Anal. Chim. Acta., 528, 1-8.

Shang L, Liu X, Zhong J, Fan C, Suzuki I, Li G (2003). Fabrication of ultrathin, protein-containing films by layer-by-layer assembly and electrochemical characterization of hemoglobin entrapped in the film. Chem. Lett., 32, 296-297.

Turner APF, Karube I, Wilson GS (1987). Biosensors: Fundamentals and Applications. Oxford University Press, Oxford, 406.

Turner AFP (2007). Biochemistry - Biosensors sense and sensitivity. Am. Assoc. Advance. Sci., 290, 1315-1317.

Comparative methaemoglobin concentrations of three erythrocyte genotypes (HbAA, HbAS and HbSS) of male participants administered with five antimalarial drugs

P.C. Chikezie

Department of Biochemistry, Imo State University, Owerri, Imo State, Nigeria. E-mail: p_chikezie@yahoo.com.

In the present *in-vivo* study, the capacities of five antimalarial drugs (Fansidar, Halfan, Quinine, Coartem and Chloroquine phosphate) to alter/distort methaemoglobin concentrations of three human erythrocyte genotypes (HbAA, HbAS and HbSS) was investigated. Spectrophotometric method was used to ascertain this erythrocyte parameter. The male participants enrolled for this study were grouped according to their genotypes, pathologic status, (that is, non-malarious and malarious individuals). Determination of erythrocyte methaemoglobin concentration was carried out before (control; t = 0 h) and after (tests; that is, at t = 3, 6 and 18 h) the five (5) antimalarial drugs were administered to various corresponding groups of participants. The results showed that methaemoglobin concentrations of these individuals ranged between 1.45+/-0.13 and 2.50+/-0.43%; 8.27+/-2.41 and 14.78+/-2.45%, for non-malarious and malarious male individuals respectively. There was no significant difference (p > 0.05) between methaemoglobin concentrations of HbAA and HbAS erythrocyte of non-malarious participants. The doses of the five antimalarial drugs administered to non-malarious individuals did not cause toxic methaemoglobinemia. Under the same experimental conditions, erythrocytes obtained from persons of HbSS genotype exhibited significant (p < 0.05) elevation of methaemoglobin concentration. Relatively high levels of methaemoglobin concentration of parasitized red blood cells decreased in a time dependent manner after administration of the five antimalarial drugs. Therefore, erythrocyte methaemoglobin evaluation is a reliable biochemical marker and rational for diagnostic and therapeutic potential in malaria. Furthermore, moderate increases of erythrocyte methaemoglobin in HbSS individuals served as point of caution when administering these drugs to this category of human subjects.

Key words: Antimalarials, erythrocyte, malaria, genotype, methaemoglobin.

INTRODUCTION

Concisely, methaemoglobin is formed when the iron of deoxyhaemoglobin is oxidized from its ferrous (Fe^{2+}) to the ferric state (Fe^{3+}) (Murray et al., 2006). The oxidation of haemoglobin to methaemoglobin can arise from auto oxidation engendered by the activities of pro-oxidants and oxidizing substances (Callister, 2003), deficiency or impaired activity of methaemoglobin reductase enzymes (Yubisui and Takashita, 1980; Board, 1981,) and hereditary abnormal haemoglobin, haemoglobin M (Prchal and Jenkin, 2001). Methaemoglobin will not bind reversibly with oxygen. Under normal physiologic condition, methaemoglobin is continually formed in red blood cells (Tietz, 1976; Callister, 2003).

Malaria parasites are particularly vulnerable to oxidative stress during the erythrocytic life stages (Muller et al., 2003; Becker et al., 2004). This is not surprising as the parasites live in a pro-oxidant environment that contains oxygen and iron, the key prerequisite for the formation of reactive oxygen species (ROS) via the Fenton reaction.

The present study seeks to ascertain whether erythrocyte methaemoglobin concentration could serve as a reliable biochemical parameter for diagnosis and monitoring of therapeutic events in malarial disease. Furthermore, these parameters may give insight into the role, influence of these antimalarials on some functional status of the red blood cells, and unfold hitherto, unknown toxic/benefi-

Table 1. The participants administered with a single dose of each of five antimalarial drugs.

Male participants	Drugs/doses administered
1).Non-malarious	Fansidar: (pyrimethamine) = 14.9 mg/kg; (sulphadoxine)= 2.9mg/kg
	Halfan: (Halofantrine base) = 13.9 mg/kg
	Quinine = 5.9 mg/kg)
	Coartem:(artemether) = 1.2 mg/kg (lumefantrine)= 7.2mg/kg
	Chloroquine Phosphate = 14.9 mg/kg
2). Malarious	Fansidar: (pyrimethamine) = 14.5 mg/kg; (sulphadoxine)= 2.9 mg/kg
	Halfan: (Halofantrine base) =13.5 mg/kg
	Quinine = 5.8 mg/kg)
	Coartem: (artemether) = 1.2 mg/kg (lumefantrine) = 7.0 mg/kg
	Chloroquine Phosphate = 14.5 mg/kg

cial aspects of these drugs to humans who express the three red cell genotypes, for better-informed prescription and use.

MATERIALS AND METHODS

Anti-malarial drugs

Five (5) antimalarial drugs were used in this study: Fansidar[TM] {Swiss (Swipha) Pharmaceuticals Nigeria Ltd}, Coartem[TM], (Beijing Norvatis Pharmaceutical Company, Beijing, China) Chloroquine phosphate (May and Baker, Pharmaceutical Company, Nigeria Plc), Halfan[TM] (Smithkline Beecham Laboratories Pharmaceutical Company, France) and Quinine (BDH, UK).

Selection of volunteers/experimental design

Forty-three (43) non-malarious male (61-73 kg) participants of confirmed HbAA (15), HbAS (15) and HbSS (13) genotypes between the ages of 20-28years enrolled for this study. The malarious group consisted of forty-five (45) male (59 - 79 kg) participants – HbAA (15), HbAS (15) and HbSS (15). They aged between 21 - 34 years.

The participants administered with a single dose of each of five antimalarial drugs, were grouped according to their individual genotype and malarial status. The doses were administered in the following specifications: **Table** 1:

Blood samples were withdrawn from these participants at time intervals of 3, 6 and 18 h after dosage and analyses were carried out to ascertain for erythrocyte methaemoglobin concentration. The determinations of the red blood cell parameter prior to the administration of the five antimalarial drugs to participants constituted the control analysis.

Ethics

The institutional review board of the Department of Biochemistry, University of Port Harcourt, Port Harcourt, Nigeria, granted approval for this study and all participants involved signed an informed consent form. This conducted study was in accordance with the ethical principles that have their origins in the Declaration of Helsinki. Individuals drawn were from Imo State University, Owerri, Nigeria and environs. The research protocols were in collaboration with registered and specialized clinics and medical laboratories.

Collection of blood samples/preparation of erythrocyte haemolysate

Five milliliters (5.0 ml) of human venous blood of HbAA, HbAS, and HbSS genotypes obtained from participants by venipuncture was stored in EDTA anticoagulant tubes. Blood of HbSS genotype and malarious blood samples were from patients attending clinics at the Federal Medical Center (FMC), Imo State University Teaching Hospital (IMSUTH), Orlu, St. John Clinic/Medical Diagnostic Laboratories, Avigram Medical Diagnostic Laboratories, and Qualitech Medical Diagnostic Laboratories. These centers are located in Owerri, Imo State, Nigeria.

The erythrocytes were washed by methods as described by Tsakiris et al. (2005). Within 2 h of collection of blood samples, portions of 1.0 ml of the samples were introduced into centrifuge test tubes containing 3.0 ml of buffer solution pH = 7.4: 250 mM tris-HCl (Tris-HCl)/140 mMNaCl/1.0 mMMgCl$_2$/10 mMg lucose). The erythrocytes were separated from plasma by centrifugation at 1200 g for 10 min, washed three times by three similar centrifugations with the buffer solution. The erythrocytes re-suspended in 1.0 ml of this buffer were stored at 4 °C. The washed erythrocytes were lysed by freezing/thawing as described by Galbraith and Watts, (1980) and Kamber et al., (1984). The erythrocyte haemolysate was used for the determination of methaemoglobin concentration.

Determination of methaemoglobin concentration of erythrocyte lysate

Determination of methaemoglobin content of red cell lysate was by modification of the method of Evelyn and Malloy, (1938), as described by Akomopong et al. (2000). A total of 400 µl of 0.5 M Phosphate buffer (pH 6.5) was added to 600 µl of the cell lysate, and the mixture was centrifuged at 16,000 g for 5 min to sediment debris. A total of 700 µl of the supernatant fraction was used to measure the absorbance at λmax = 630 nm (the absorbance maximum for methaemoglobin), and the reading was recorded as SI. A total of 50 µl of 10gpercentageKCN was added, and after 5 min at room temperature (24 °C), a second reading (S2) was recorded. KCN converts methaemoglobin to cyanomethaemoglobin, which does not absorb at 630 nm; hence, the difference between absorbance readings S1 and S2 represents the absorbance due to methaemoglobin.

To measure total hemoglobin levels, all of the hemoglobin was converted to methaemoglobin, the absorbance of the sample at λmax = 630 nm was recorded, and then KCN was added to form cyanomethaemoglobin. Specifically, 70 µl of the supernatant fraction was diluted 10-fold into 600 µl of 0.1 M phosphate buffer (pH 6.5). Next, 30 µl of freshly prepared 20g% K$_3$Fe (CN)$_6$ was added and incubated for 5minutes at room temperature (24 °C) and an initial reading (T1) was recorded. A total of 50 µl of 10% KCN was subsequently added, and a second reading (T2) was recorded. The percent methaemoglobin in the sample was calculated as [100(S1-S2)] / [10(T1-T2)].

Statistical analyses

The experiments were designed in a completely randomized method and data collected were analyzed by the analysis of variance

Table 2. Methaemoglobin Concentration (Met. Hb %) of Male Erythrocyte Haemolysate.

Genotype	(Met. Hb %) (X±S.D)	
	Non-malarious	Malarious
1). HbAA (n=15[NM]; 15[M])	1.48 ± 0.14^a	14.07 ± 2.56^b
2). HbAS (n=15[NM]; 15[M])	1.45 ± 0.13^a	8.27 ± 2.41^a
3). HbSS (n=13[NM]; 15[M])	2.50 ± 0.43^b	14.78 ± 2.45^b

M and NM = number of malarious and non-malarious blood samples respectively. Means in the column with the same letter are not significantly different at $p < 0.05$ according to LSD.

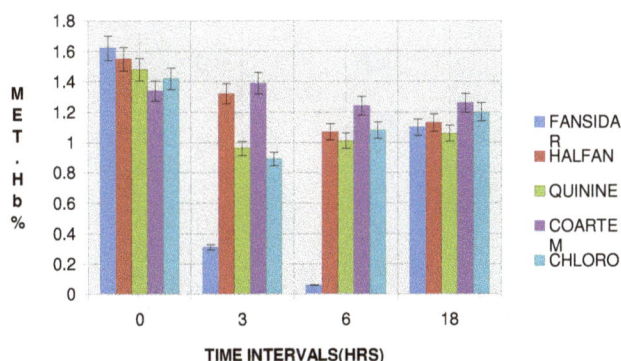

Figure 1. Comparative *in vivo* methaemoglobin concentrations of HBAA erythrocyte haemolysate of non-malarious male participants administered with five antimalarial drugs

procedure while treatment means were separated by the Least Significance Difference (LSD) incorporated in the Statistical Analysis System (SAS) package of 9.1 versions (2006).

RESULTS

The mean (+/-.S.D) methaemoglobin concentration, expressed as percentage (Met.Hb%) of total haemoglobin concentration of three erythrocyte genotypes (HbAA, HbAS and HbSS) of blood samples obtained from non-malarious and malarious male participants, before being administered with the corresponding five antimalarial drugs (control values) is presented in Table 2.

A cursory look at Table 2 showed erythrocyte obtained from blood sample of malarious male participant exhibited significantly ($p < 0.05$) higher levels of methaemoglobin concentrations than those of non-malarious individuals, irrespective of their genotype. There was no significant difference ($p > 0.05$) between methaemoglobin concentrations of HbAA and HbAS erythrocyte of non-malarious participants. An overview of the results showed that methaemoglobin concentrations of these individuals ranged between 1.45 ± 0.13 and $2.50 \pm 0.43\%$; 8.27 ± 2.41 and $14.78 \pm 2.45\%$, for non-malarious and malarious male

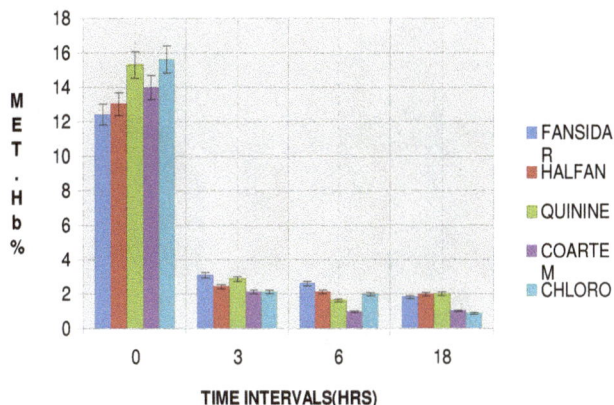

Figure 2. Comparative *in vivo* methaemoglobin concentrations of HBAA erythrocyte haemolysate of malarious male participants administered with five antimalarial drugs.

individuals respectively.

Non-malarious participants (Figure 1)

An overview of the results showed the five antimalarial drugs caused a time dependent decrease in erythrocyte methaemoglobin concentrations within approximately 6 h after administration. This followed a paradoxical increase in methaemoglobin concentration as the experimental time progressed.

Administered Fansidar elicited the lowest methaemoglobin concentration at the 6th h (Met. Hbpercentage = $0.06 \pm 0.02\%$). However, erythrocyte methaemoglobin concentration reached $1.18 \pm 0.47\%$ at the 18th h. The fall in methaemoglobin concentration in the presence of the drug between the 3rd and 6th hours was not significantly different ($p > 0.05$).

Although decreased level of erythrocyte methaemoglobin concentration occurred 3 h after Halfan was administered, it was not significantly different ($p > 0.05$) compared to the control/reference values. Furthermore, there was no significant difference ($p > 0.05$) in subsequent rise and fall in methaemoglobin concentration within the experimental period.

Coartem caused no significant difference ($p > 0.05$) in the levels of erythrocyte methaemoglobin throughout the period of the experiment.

Malarious participants (Figure 2)

The relatively high levels of methaemoglobin concentration of parasitized red blood cells decreased in a time dependent manner after administration of the five antimalarial drugs. It is worthwhile to note that the rate of decrease in erythrocyte methaemoglobin concentration of these individuals was by far more rapid in the first 3 h. In addition, methaemoglobin content of the red blood cells showed no significant difference ($p > 0.05$) after the 3rd,

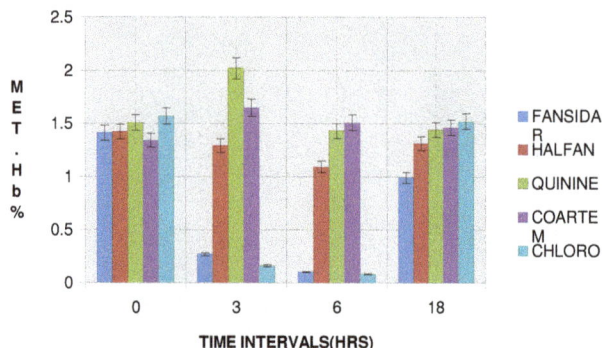

Figure 3. Comparative *in vivo* methaemoglobin concentrations of HbAS erythrocyte haemolysate of non-malarious male participants administered with five antimalarial drugs.

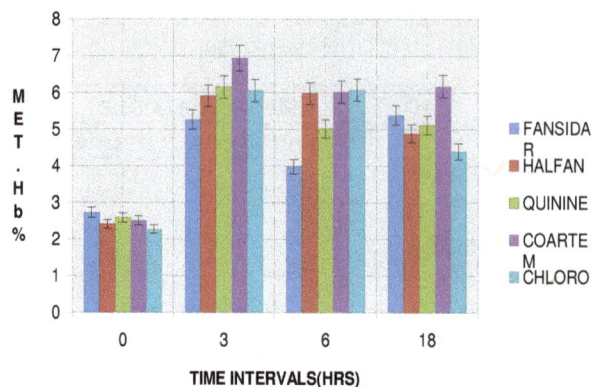

Figure 5. Comparative *in vivo* methaemoglobin concentrations of HbSS erythrocyte haemolysate of non-malarious male participants administered with five antimalarial drugs.

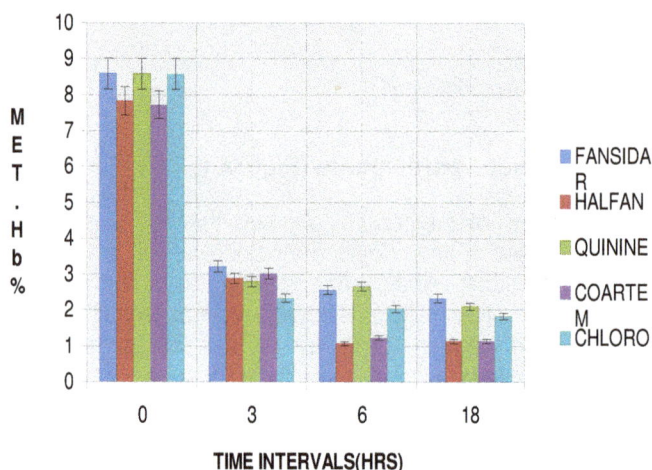

Figure 4. Comparative *in vivo* methaemoglobin concentrations of HbAS erythrocyte haemolysate of malarious male participants administered with five antimalarial drugs.

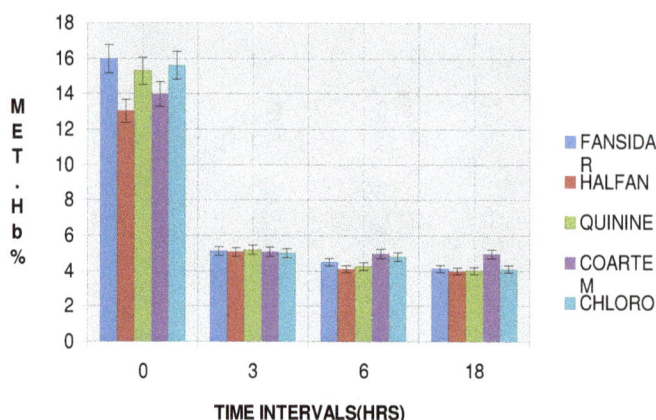

Figure 6. Comparative *in vivo* methaemoglobin concentrations of HbSS erythrocyte haemolysate of malarious male participants administered with five antimalarial drugs.

6th and 18th h, after the drugs were administered.

Non-malarious participants (Figure 3)

Three antimalarial drugs, Halfan, Chloroquine Phosphate and Fansidar promoted declining red blood cell methaemoglobin concentration within 6 h. However, no significant difference ($p < 0.05$) in methaemoglobin concentration was exhibited after 6 h of administration of Chloroquine Phosphate, Quinine, Coartem and Halfan to these individuals.

Conversely, Quinine and Coartem caused the elevation of methaemoglobin concentration. Specifically, administered Quinine engendered erythrocyte methaemoglobin concentration to increase from 1.51±0.09% to a peak value of 2.02±0.15% at zero and 3 h respectively.

Malarious participants (Figure 4)

Generally, there was a time dependent decline in meth-

aemoglobin concentration in the same manner as earlier described in malarious male participants of HbAA genotype. An overview of the results showed no significant ($p > 0.05$) corresponding capacities of the five antimalarial drugs to elicit declining methaemoglobin concentration with time (that is, at $t > 3$ h).

Non-malarious participants (Figure 5)

Elevation of erythrocyte methaemoglobin concentration occurred in these individuals approximately 3hours after dosing these participants with the five antimalarial drugs. Specifically, Coartem caused attainment of a peak value of 6.94±0.61% methaemoglobin concentration at the third hour. Further declining concentrations of methaemoglobin concentration was not significantly different ($p > 0.05$) after 3, 6 and 18 h after dosage.

Malarious participants (Figure 6)

In this experimental group, all corresponding erythrocyte

methaemoglobin concentrations, 3 – 18 h after the five antimalarial drugs were administered, were significantly different (p < 0.05) compared to the control/reference values (that is, at t = 0 h). Similar to non-malarious male individuals, further declining concentrations of methaemoglobin concentration was not significantly different (p > 0.05) after 3, 6 and 18 h of dosage.

DISCUSSION

The determination of erythrocyte methaemoglobin concentration as a toxic endpoint in chemical poisoning and pathologic conditions showed reliability and reproducibility in clinical diagnostic procedures (Hopkins, 2000; Bradberry, 2003; Uko et al., 2007). This present study reported significant (p < 0.05) elevation of methaemoglobin concentrations of human participants infected with malaria, irrespective of their genotype (Table 1). This observation was in agreement with the reports of Uko et al., (2007) who demonstrated high level of methaemoglobin in subjects with severe malaria parasitaemia and suggested routine estimation of methaemoglobin in malaria for clinical evaluation of patients. Furthermore, Anstey et al. (1996) documented elevated methaemoglobin concentration of Tanzanian children with severe and uncomplicated malaria. In the same vein, the present report was in concordance with observations of Friedman et al. (1979). These authors showed that cultures with high levels of parasitaemia contained between 3 to 10 times more methaemoglobin than those with low levels of parasitemia. Akompong et al. (1999) found isolated malarial parasites contained between 20 - 42% of methaemoglobin concentration. In contrast, uninfected red blood cells presented between 0.5 - 1.0% of methaemoglobin. Therefore, these reports suggest malaria parasite induced raised levels of erythrocyte methaemoglobin. In addition, these authors noted that increased methaemoglobin in malarial infection was a reflection of rapid oxidation of haemoglobin ingested by the parasites. It worthwhile to recall that raised oxidant levels compounded by compromised activity of erythrocyte redox enzymes further exacerbate the tendency towards spontaneous oxidation of haemoglobin molecule in parasitized red blood cells. De Rosa et al. (2003), used methaemoglobin levels as a marker molecule for the presence of nitrogen oxide generated by parasite as consequence of their metabolism (Clark et al., 1991). All these were pointers to the fact that parasitized erythrocytes exhibited raised levels of methaemoglobin. Similar to the reports of Neupane et al. (2008), the decreasing methaemoglobin concentrations with time after administration of the five antimalarial drugs suggest decreasing levels of ROS due to the killing of the parasites. Relationship exists between levels of parasitaemia and levels of ROS production (Friedman et al., 1979; Uko et al., 2007; Neupane et al., 2008).

The pattern of variability of basal erythrocyte methaemoglobin concentrations amongst non-malarious human participants showed that the dysfunctional red cell of HbSS genotype was significantly (p < 0.05) higher than the normal one, HbAA (Table 1). This result conformed to earlier reports by Tamer et al. (2000). They noted that the primary reason for the relatively raised concentration of oxidized haemoglobin (methaemoglobin) in HbSS erythrocytes was higher production of superoxide ion by these erythrocytes compared to those of HbAA and HbAS erythrocytes. Furthermore, erythrocyte endogenous oxidant (haemin) showed higher levels in HbSS than HbAA erythrocytes. Haemin has a profound capacity to activate certain erythrocyte redox enzymes e.g. NADH methaemoglobin reductase (Uwakwe, 1991) and its presence at high concentration was attributable to the high level of haemolytic phenomenon peculiar to this haemoglobin variant cell (Orjih et al., 1985). There is also the case of certain methaemoglobinopathies found in association with HbSS erythrocytes. These are $HbM_{Boston,}$ $HbM_{Iwate,}$ $HbM_{Hydepark}$ and $HbM_{Hammersmith,}$ noted for tendency towards spontaneous oxidation in vivo and resistant to enzymatic reduction.

Early studies have noted that certain xenobiotics were capable of eliciting elevation of erythrocyte methaemoglobin concentration, thereby distorting normal plasma haemoglobin (Fe^{2+})/methaemoglobin (Fe^{3+}) ratio. Callister, (2003) reported the nitrates and anilines as the most common cause of methaemoglobin toxicity in man. This physiologic dysfunctional state (methaemoglobinemia), presented in form of clinical cyanosis occur when plasma methaemoglobin concentration exceeds 15% (Hopkins, 1998). Methaemoglobin concentrations of HbAA and HbAS erythrocytes obtained from non-malarious human participants administered with the five antimalarial drugs fell within the normal physiologic range proposed by Tietz, (1976). Therefore, from methaemoglobin toxicity standpoint, these drugs did not induce methaemoglobinemia at the experimental doses while the investigation lasted. Similarly, Chikezie and Enemor, (2005), reported that guinea pigs administered with quinine did not present plasma methaemoglobin concentration that was diagnostic of toxic methaemoglobinemia. These authors averred that the capacity of the drug to oxidize ferrous (Fe^{2+}) haemoglobin did not overwhelm the physiologic systems responsible to maintain erythrocyte methaemoglobin concentration below the critical value of 15% that is diagnostic of toxic methaemoglobinemia. Furthermore, it is probable the redox potentials of these drugs and its metabolites were not high enough to engender the oxidation of considerable population of ferrous (Fe^{2+}) state haemoglobin. However, moderate increases of erythrocyte methaemoglobin in HbSS individuals (Figures 5 and 6) served as point of caution when administering these drugs to this category of human subjects. Decreasing erythrocyte methaemoglobin concentrations of malarious human participants of HbAA and HbAS genotypes administered with corresponding antimalarials confirmed the suitability of this parameter for evaluation of level of parasitaemia and drug efficacy (Alloueche et al., 2003; Uko et al., 2007).

REFERENCE

Akomopong T, Ghori N, Haldar K (2000). *In vitro* activity of riboflavin against the human malaria parasite *Plasmodium falciparum*. Antimicrob. Agents Chemother. (44). (1).88-96.

Alloueche A, Bailey W, Barton S, Bwika J (2003). Comparison of chlorproguanil-dapson with sulphadoxine-pyrimethamine for the treatment of uncomplicated falciparum malaria in young African children: double-blind randomized control trials. Lancet 63(9424): 1843-1848.

Anstey NM, Hassanali MY, Mlalasi J, Manyenga D, Mwaikambo ED (1996). Elevated levels of methaemoglobin in Tanzanian children with severe and uncomplicated malaria. Trans. R. Soc. Trop. Med. Hyg. 90(2): 147-151.

Becker K, Tilley L, Vennerstrom JL, Roberts D, Rogerson S, Ginsburg H (2004).Oxidative stress in malaria parasite infected erythrocytes: host – parasite interactions. Int. J. Parasitol. 34: 163 –189.

Board PG (1981). NaDH-ferricyanide reductase, a convenient approach to the evaluation of NADH-methaemoglobin reductase in human erythrocytes. Clin. Chem. Acta 18:233-237.

Bradberry SM (2003). Occupational methaemoglobinemia: Mechanisms of production, features, diagnosis and management including the use of methylene blue. Toxicol. Rev. 22(1): 13-27.

Callister R (2003). Methaemoglobin: Its causes and effects on pulmonary function and SPO_2 reading. http://www.mc.vanderbilt.edu/peds/pidl/hemeone/metheme.htm.

Chikezie PC, Enemor VH (2005). Effect of experimental quinine administration on plasma levels of haemoglobin and methaemoglobin in guinea pigs. Int. J. Nat. Appl. Sci. 1(2): 150-153.

Clark IA, Rockett KA, Cowden WB (1991). Proposed link between cytokines, nitric acid and human cerebral malaria. Parasitol. Today 7:205-207.

De Rosa M, Zarrilli S, Paesano L, Carbone U, Boggia B, Petretta M, Maisto A, Cimmino F, Puca G, Colao A, Lombardi G (2003).Traffic pollutants affect fertility in men. Hum Reprod. 18: 1055 -1061.

Evelyn K, Malloy H (1938). Micro determination of oxyhaemoglobin, methaemoglobin and sulfhaemoglobin in a single sample of blood. J. Biol. Chem. 126:655-662.

Friedman MJ, Roth EF, Nagel RL, Trager W (1979*). Plasmodium falciparum:* physiologic interaction with human sickle cell. Exp. Parasitol. 47:73-80.

Galbraith DA, Watts DC (1980).Changes in some cytoplasmic enzymes from Red blood cells fractionated into age groups by centrifugation in Ficoll™/Triosil™ gradients. Comparison of normal human and patients with Duchenne muscular dystrophy. Biochem. J. 191:63-70.

Hopkins U (2000): Methaemoglobinemia J. Tox. Clin. Tox. 36:6-12.

Kamber K, Poyiagi A, Delikonstantinos G (1984).Modifications in the activities of membrane-bound enzymes during *in vivo* ageing of human and rabbit erythrocytes. Comp. Biochem. Physiol. B. 77B: 95-99.

Murray RK, Granner DK, Rodwell VW (2006). Harper's Illustrated Biochemistry 27th Edition. McGraw Hill Companies. Asia.

Muller S, Lieban E, Walter RD, Krauth – Siegel RL (2003). Thiol-based radox mechanism of protozoan parasites Trends Parasitol. 19: 320 – 328.

Neupane DP, Majhi S, Chandra L, Rijal S, Baral N (2008). Erythrocyte glutathione status in human visceral Leishmaniasis. Indian. J. Clin. Biochem. 23(1): 95-97.

Orjih AU, Cheuli R, Fitch CD (1985). Toxic heme in sickle cells: An explanation for death of malaria parasites. Am. J. Trop. Med. Hyg. 34(2):223-227.

Prchal JF, Gregg XT (2005). Red cell enzymes. Haematology. American Society of Haematology Education Program Book.

Statistical Analytical System (SAS), (2006): Package 9.1 Version.

Tamer L, Gurbuz P, Guzide Y, Birol G, Fikri B (2000). Erythrocyte membrane Na^+K^+/Mg^{++} and Ca^{++}/Mg^{++} Adenosine 5' Tri-phosphate in patients with sickle cell anaemia. Turk. J. Haematol. 17(1):23-26.

Tietz N (1976). Fundamentals of Clinical Chemistry W. B Saunders Co., Philadephia.

Tsakiris S, Giannoulia-Karantana A, Simintzi I, Schulpis KH (2005).The effect of aspartame metabolites on human erythrocyte membrane acetylcholinesterase activity. Pharmacol. Res. 53: 1-5.

Uko EK, Udoka AE, Etukudoh MH (2007). Methaemoglobin profile in malaria infected children in Calabar. Niger. J. Med. 12(2): 94-97.

Uwakwe AA (1991). Activities of some redox enzymes from different human red blood cell genotype. Ph.D Thesis, University of Port Harcourt, Port Harcourt, Nigeria.

Yubisui T, Takeshita M (1980). Reduction of methaemoglobin through flavin at the physiological concentration by NADPH-flavin reductase of human erythrocytes. J. Biochem. 87(6): 1715-1720.

Permissions

The contributors of this book come from diverse backgrounds, making this book a truly international effort. This book will bring forth new frontiers with its revolutionizing research information and detailed analysis of the nascent developments around the world.

We would like to thank all the contributing authors for lending their expertise to make the book truly unique. They have played a crucial role in the development of this book. Without their invaluable contributions this book wouldn't have been possible. They have made vital efforts to compile up to date information on the varied aspects of this subject to make this book a valuable addition to the collection of many professionals and students.

This book was conceptualized with the vision of imparting up-to-date information and advanced data in this field. To ensure the same, a matchless editorial board was set up. Every individual on the board went through rigorous rounds of assessment to prove their worth. After which they invested a large part of their time researching and compiling the most relevant data for our readers.

The editorial board has been involved in producing this book since its inception. They have spent rigorous hours researching and exploring the diverse topics which have resulted in the successful publishing of this book. They have passed on their knowledge of decades through this book. To expedite this challenging task, the publisher supported the team at every step. A small team of assistant editors was also appointed to further simplify the editing procedure and attain best results for the readers.

Apart from the editorial board, the designing team has also invested a significant amount of their time in understanding the subject and creating the most relevant covers. They scrutinized every image to scout for the most suitable representation of the subject and create an appropriate cover for the book.

The publishing team has been an ardent support to the editorial, designing and production team. Their endless efforts to recruit the best for this project, has resulted in the accomplishment of this book. They are a veteran in the field of academics and their pool of knowledge is as vast as their experience in printing. Their expertise and guidance has proved useful at every step. Their uncompromising quality standards have made this book an exceptional effort. Their encouragement from time to time has been an inspiration for everyone.

The publisher and the editorial board hope that this book will prove to be a valuable piece of knowledge for researchers, students, practitioners and scholars across the globe.

List of Contributors

Omar Akil
Laboratoire de Biochimie et Biologie Moléculaire, Université Hassan II - Aïn Chock, Faculté des Sciences, Casablanca, Morocco

Zakaria El Kebbaj
INSERM U866; Université de Bourgogne, Laboratoire de Biochimie Métabolique et Nutritionnelle, Faculté des Sciences, 6 Bd Gabriel, 21000 Dijon cedex, France
Laboratoire de Biochimie et Biologie Moléculaire, Université Hassan II - Aïn Chock, Faculté des Sciences, Casablanca, Morocco

Norbert Latruffe
INSERM U866; Université de Bourgogne, Laboratoire de Biochimie Métabolique et Nutritionnelle, Faculté des Sciences, 6 Bd Gabriel, 21000 Dijon cedex, France

M'Hammed Saïd El Kebbaj
Laboratoire de Biochimie et Biologie Moléculaire, Université Hassan II - Aïn Chock, Faculté des Sciences, Casablanca, Morocco

C. E. Ugwu
Department of Biochemistry, Kogi State University, Anyigba, Nigeria

J. E. Olajide
Department of Biochemistry, Kogi State University, Anyigba, Nigeria

E. O. Alumana
Department of Biochemistry, University of Nigeria, Nsukka, Enugu State, Nigeria

L. U. S. Ezeanyika
Department of Biochemistry, University of Nigeria, Nsukka, Enugu State, Nigeria

Gabriel Oze
Institute of Neuroscience and Biomedical Research, College of Medicine, Imo State University, Owerri, Nigeria

Iheanyi Okoro
Institute of Neuroscience and Biomedical Research, College of Medicine, Imo State University, Owerri, Nigeria

Austin Obi
Institute of Neuroscience and Biomedical Research, College of Medicine, Imo State University, Owerri, Nigeria

Polycarp Nwoha
Institute of Neuroscience and Biomedical Research, College of Medicine, Imo State University, Owerri, Nigeria

Masoud Alirezaei
Department of Biochemistry, School of Veterinary Medicine, Shiraz University, Shiraz, 71345, Iran

Mehdi Saeb
Department of Biochemistry, School of Veterinary Medicine, Shiraz University, Shiraz, 71345, Iran

Katayoun Javidnia
Medicinal and Natural Products Chemistry Research Center, Shiraz University of Medical Sciences, Shiraz, Iran

Saeed Nazifi
Department of Clinical Pathology, School of Veterinary Medicine, Shiraz University, Shiraz, 71345, Iran

Najmeh Khalighyan
Medicinal and Natural Products Chemistry Research Center, Shiraz University of Medical Sciences, Shiraz, Iran

Saeedeh Saeb
Department of Biochemistry, School of Veterinary Medicine, Shiraz University, Shiraz, 71345, Iran

Nasar Yousuf Alwahaibi
Department of Pathology, College of Medicine and Health Sciences, Sultan Qaboos University, Muscat – Oman

Siti Belkis Budin
Faculty of Allied Health Sciences, Department of Biomedical Sciences, University of Kebangsaan, Kuala Lumpur Malaysia

Jamaludin Mohamed
Faculty of Allied Health Sciences, Department of Biomedical Sciences, University of Kebangsaan, Kuala Lumpur Malaysia

C. E Ugwu
Department of Biochemistry, Kogi State University, Anyigba, Nigeria

L. U. S. Ezeanyika
Department of Biochemistry, University of Nigeria, Nsukka, Nigeria

M. A. Daikwo
Department of Biochemistry, Kogi State University, Anyigba, Nigeria

R. Amana
Department of Biochemistry, Kogi State University, Anyigba, Nigeria

Francis M. Awah
Department of Biochemistry, University of Nigeria, Nsukka, Nigeria
Department of Biochemistry, Madonna University, Elele Campus, Rivers State, Nigeria

Onyinye Agughasi
Department of Biochemistry, Madonna University, Elele Campus, Rivers State, Nigeria

K. Yaqin
Department of Fisheries, Faculty of Marine Science and Fisheries, Hasanuddin University, Makassar, Indonesia

P. D. Hansen
Technische Universitaet Berlin, FB 7 -Intitute for Ecological Research and Technology, Department of Ecotoxicology, Berlin, Germany

U. A. Okon
Department of Physiology, College of Health Sciences, University of Uyo, Nigeria

S. O. Ita
Department of Physiology, College of Health Sciences, University of Uyo, Nigeria

C. E. Ekpenyong
Department of Physiology, College of Health Sciences, University of Uyo, Nigeria

K. G. Davies
Department of Physiology, College of Health Sciences, University of Uyo, Nigeria

O. I. Inyang
Department of Physiology, College of Health Sciences, University of Uyo, Nigeria

Mohamed H. Mahfouz
Biochemistry and National Institute of Diabetes and Endocrinology (NIDE), Cairo, Egypt

Ibrahim A. Emara
Biochemistry and National Institute of Diabetes and Endocrinology (NIDE), Cairo, Egypt

Mohamed S. Shouman
Internal Medicine Departments, National Institute of Diabetes and Endocrinology (NIDE), Cairo, Egypt

Magda K. Ezz
Biochemistry Department, Faculty of Science. Ain Shams University

Osama M. Abdel-Fatah
Department of Microbial Chemistry, National Research Centre, Dokki, Cairo, Egypt

Maysa A. Elsayed
Department of Microbial Chemistry, National Research Centre, Dokki, Cairo, Egypt

Ali M. Elshafei
Department of Microbial Chemistry, National Research Centre, Dokki, Cairo, Egypt

V. K. Dwivedi
Pre Clinical Division, Venus Medicine Research Centre, Baddi, H.P. India 173205, India

A. Arya
Pre Clinical Division, Venus Medicine Research Centre, Baddi, H.P. India 173205, India

H. Gupta
Pre Clinical Division, Venus Medicine Research Centre, Baddi, H.P. India 173205, India

A. Bhatnagar
Analytical Division, Venus Medicine Research Centre, Baddi, H.P. India 173205, India

P. Kumar
Pre Clinical Division, Venus Medicine Research Centre, Baddi, H.P. India 173205, India

M. Chaudhary
Pre Clinical Division, Venus Medicine Research Centre, Baddi, H.P. India 173205, India

I. A. Ibrahim
Department of Pharmacology, Faculty of Medicine, National University of Malaysia, Kuala Lumpur, Malaysia

F. S. Al-Joudi
Faculty of Allied Health Sciences, National University of Malaysia, Kuala Lumpur, Malaysia

R. Waleed Sulaiman
Department of Clinical Laboratory Science, Faculty of Pharmacy, University of Baghdad, Baghdad, Iraq

B. Hilal AL-Saffar
Department of Medicine, Hospital Dr. Abdul-Majeed, Karradah, Baghdad, Iraq

P. Bansal
Clinical Biochemistry Laboratory, National Institute of Ayurvedic Pharmaceutical Research, Moti Bagh Road, Patiala, Punjab, India

R. Sannd
National Institute of Ayurvedic Pharmaceutical Research, Moti Bagh Road, Patiala, Punjab, India

N. Srikanth
Central Council for Research in Ayurveda and Siddha, Ministry of Health and Family Welfare, Government of India, New Delhi, India

G. S. Lavekar
Central Council for Research in Ayurveda and Siddha, Department of AYUSH, Ministry of Health and Family Welfare, Government of India, New Delhi, India

Mukesh Sharma
Microbiology Laboratory, Department of Botany, University of Rajasthan, Jaipur-302004, India

Meenakshi Sharma
Microbiology Laboratory, Department of Botany, University of Rajasthan, Jaipur-302004, India

Vijay Mohan Rao
Microbiology Laboratory, Department of Botany, University of Rajasthan, Jaipur-302004, India

Saeed Nazifi
Department of Clinical Studies, School of Veterinary Medicine, Shiraz University, Shiraz, Iran

Mahdi Saeb
Department of Basic Sciences, School of Veterinary Medicine, Shiraz University, Shiraz, Iran

Hasan Baghshani
Department of Basic Sciences, School of Veterinary Medicine, Shiraz University, Shiraz, Iran

Saeedeh Saeb
Department of Basic Sciences, School of Veterinary Medicine, Shiraz University, Shiraz, Iran

N. W. K. Mungatana
Institute of Primate Research, P. O. Box 24481, Karen, Nairobi, Kenya
Department of Biochemistry, Egerton University, P. O. Box 536, Njoro, Kenya

R. M. Ngure
Department of Biochemistry, Egerton University, P. O. Box 536, Njoro, Kenya

A. A. Shitandi
Department of Biochemistry, Egerton University, P. O. Box 536, Njoro, Kenya

C. K. Mungatana
Department of Biochemistry, Egerton University, P. O. Box 536, Njoro, Kenya

D. S. Yole
Institute of Primate Research, P. O. Box 24481, Karen, Nairobi, Kenya

Ali Khaleghian
Department of Biochemistry and Hematology, Semnan University of Medical Sciences, Semnan, Iran
Department of Biochemistry and Biophysics, University of Tehran, 13145-1384, Iran

Gholam Hossein Riazi
Department of Biochemistry and Biophysics, University of Tehran, 13145-1384, Iran

Shahin Ahmadian
Department of Biochemistry and Biophysics, University of Tehran, 13145-1384, Iran

Mahmoud Ghafari
Department of Biochemistry and Biophysics, University of Tehran, 13145-1384, Iran

Marzieh Rezaie
Department of Biochemistry and Hematology, Semnan University of Medical Sciences, Semnan, Iran

Akira Takahashi
Department of Nutrition and Metabolism, University of Tokushima, Tokushima, Japan

Yutaka Nakaya
Department of Nutrition and Metabolism, University of Tokushima, Tokushima, Japan

Hossein Nazari
Department of Biochemistry and Hematology, Semnan University of Medical Sciences, Semnan, Iran
Department of Nutrition and Metabolism, University of Tokushima, Tokushima, Japan

Munish Garg
Department of Pharmaceutical Sciences, Maharshi Dayanand University, Rohtak-124001, Haryana, India

Chanchal Garg
Department of Pharmaceutical Chemistry, Faculty of Pharmacy, Jamia Hamdard, New Delhi-110062, India

V. J. Dhar
Department of Pharmacognosy, I. S. F. College of Pharmacy, Moga-142001, Punjab, India

A. N. Kalia
Department of Pharmacognosy, I. S. F. College of Pharmacy, Moga-142001, Punjab, India

Kelechukwu Clarence Obimba
Department of Biochemistry, College of Natural and Applied Sciences, Michael Okpara University of Agriculture Umudike. Abia State. Nigeria

E. O. Ibe
Department of Chem. Path, University of Nigeria Teaching Hospital, Enugu, Nigeria

A. C. J. Ezeoke
Department of Chem. Path, College of Medicine, University of Nigeria, Enugu Campus, Enugu, Nigeria

I. Emeodi
Department of Paediatrics, University of Nigeria, Enugu Campus, Enugu, Nigeria

E. I. Akubugwo
Department of Biochemistry, Abia State University, Uturu, Nigeria

E. Elekwa
Department of Biochemistry, Abia State University, Uturu, Nigeria

M. C. Ugonabo
Department of Chemical Pathology, College of Medicine, UNEC, Enugu, Nigeria

W. C. Ugbajah
Temple University Hospital, Philadelphia, PA, 19140, USA

Imafidon E. Kate
Department of Biochemistry, Faculty of Life Sciences, University of Benin, Benin City, Nigeria

Okunrobo O. Lucky
Department of Pharmaceutical Chemistry, Faculty of Pharmacy, University of Benin, Benin City, Nigeria

Haïtham Sghaier
Research Unit UR04CNSTN01 "Medical and Agricultural Applications of Nuclear Techniques", National Center for Nuclear Sciences and Technology (CNSTN), Sidi Thabet Technopark, 2020 Sidi Thabet, Tunisia

Katsuya Satoh
DNA Repair Protein Group, Research Unit for Quantum Beam Life Science Initiative, Quantum Beam Science Directorate, Japan Atomic Energy Agency, 1233 Watanuki, Takasaki, Gunma 370-1292, Japan
Gene Resource Research Group, Radiation-Applied Biology Division, Quantum Beam Science Directorate, Japan
Atomic Energy Agency, 1233 Watanuki, Takasaki, Gunma 370-1292, Japan

Hirofumi Ohba
DNA Repair Protein Group, Research Unit for Quantum Beam Life Science Initiative, Quantum Beam Science Directorate, Japan Atomic Energy Agency, 1233 Watanuki, Takasaki, Gunma 370-1292, Japan
Gene Resource Research Group, Radiation-Applied Biology Division, Quantum Beam Science Directorate, Japan
Atomic Energy Agency, 1233 Watanuki, Takasaki, Gunma 370-1292, Japan

Issay Narumi
DNA Repair Protein Group, Research Unit for Quantum Beam Life Science Initiative, Quantum Beam Science Directorate, Japan Atomic Energy Agency, 1233 Watanuki, Takasaki, Gunma 370-1292, Japan
Gene Resource Research Group, Radiation-Applied Biology Division, Quantum Beam Science Directorate, Japan
Atomic Energy Agency, 1233 Watanuki, Takasaki, Gunma 370-1292, Japan

C. O. Ibegbulem
Department of Biochemistry, Federal University of Technology, Owerri, Nigeria

E. U. Eyong
Department of Biochemistry, University of Calabar, Calabar, Nigeria

E. U. Essien
Department of Biochemistry, University of Calabar, Calabar, Nigeria

P. O. Ajah
Institute of Oceanography, University of Calabar, Calabar, Cross Rivers, Nigeria

M. U. Eteng
Department of Biochemistry, University of Calabar, Calabar, Cross Rivers, Nigeria

Amit Parashar
Department of Chemistry, Eshan College of Engineering, Agra-Mathura Highway Agra-282001, India
Department of Chemistry, Institute of Engineering and Management, Mathura, India, 281406 India

S. K. Gupta
Department of Chemistry, Bipin Bihari College, Jhansi, India

Ashok Kumar
Department of Chemistry, St. John's College, Agra, India

Okey A. Ojiako
Department of Biochemistry, Federal University of Technology, Owerri, Nigeria

G. O. C. Onyeze
Department of Biochemistry, Federal University of Technology, Owerri, Nigeria

Muawia A. Abdagalil
Department of Biochemistry, Faculty of science, University of West Kordufan, Sudan

Nabiela M. ElBagir
Department of Biochemistry, Faculty of Veterinary Medicine, University of Khartoum, Sudan

O. G. Raimi
Division of Biological Chemistry and Drug Discovery, College of Life Sciences, University of Dundee, Scotland, DD1 5EH, United Kingdom
Tropical Disease Research Laboratory, Department of Biochemistry, Lagos State University, PMB 1087, Apapa, Lagos, Nigeria

C. P. Kanu
Tropical Disease Research Laboratory, Department of Biochemistry, Lagos State University, PMB 1087, Apapa, Lagos, Nigeria

S. Rezaei-Zarchi
Department of Biology, Payam-e-Noor University, Yazd, Iran

S. Imani
Department of Biology, Basic Science Faculty, IHU, Tehran, Iran

A. Javid
Institute of Biochemistry and Biophysics, University of Tehran, Tehran, Iran

A. M. Zand
Departments of Biology, Basic Science Faculty, IHU, Tehran, Iran

M. Saadati
Applied Biotechnology and Environmental Research Center, Baqiyatallah Medical Science University, and IHU, Tehran, Iran

Z. Zagari
Department of Biology, Payam-e-Noor University, Tehran, Iran

P. C. Chikezie
Department of Biochemistry, Imo State University, Owerri, Imo State, Nigeria

www.ingramcontent.com/pod-product-compliance
Lightning Source LLC
Chambersburg PA
CBHW080258230326
41458CB00097B/5105

9 781682 860380